Jin-Hua Li • Lixing Sun • Peter M. Kappeler
Editors

The Behavioral Ecology of the Tibetan Macaque

Editors
Jin-Hua Li
School of Resources
and Environmental Engineering
Anhui University
Hefei, Anhui, China

International Collaborative Research
Center for Huangshan Biodiversity
and Tibetan Macaque Behavioral Ecology
Anhui, China

School of Life Sciences
Hefei Normal University
Hefei, Anhui, China

Peter M. Kappeler
Behavioral Ecology and Sociobiology
Unit, German Primate Center
Leibniz Institute for Primate Research
Göttingen, Germany

Department of Anthropology/Sociobiology
University of Göttingen
Göttingen, Germany

Lixing Sun
Department of Biological Sciences, Primate
Behavior and Ecology Program
Central Washington University
Ellensburg, WA, USA

https://doi.org/10.1007/978-3-030-27920-2

This book is an open access publication.

Their [Tibetan macaques'] flat, broad, bearded faces provide perhaps the most humanlike countenance I have ever seen in a monkey. . . . Apart from chimpanzees, I had never seen primate males so intensely involved with each other. In chimpanzees, too, males are at the same time rivals and friends, and I would argue the same for human males.

—Frans de Waal, *The Ape and the Sushi Master* (2001)

Foreword

Mystery surrounded Tibetan macaques for a long time, even for experts. The species was not identified until the last third of the nineteenth century, and nothing more than its geographic distribution and external characters were known for the next hundred years. It was long referred to as Père David's macaque, a rather odd name in reference to the French missionary and naturalist Father Armand David, who first collected the species. Moreover, the current name, Tibetan macaque, is misleading since the species is typically found in east-central China and not within the boundaries of Tibet. This is due to the fact that Père David initially located the species at a place close to the Sino-Tibetan border of his time.

We had to wait until the 1980s to see Mount Emei and Mount Huangshan come to light on the primatology map. This is where Qikun Zhao and Ziyun Deng from the Kunming Institute of Zoology and Qishan Wang and Jinhua Li from Anhui University began to study the behavior and life history of Tibetan macaques, definitively adding a new dimension to the macaque landscape. I still have the reprints of their publications in my bibliography, some written in Chinese. The works of Hideshi Ogawa, Carol Berman, and a new generation of primatologists soon followed. Now appears this multi-authored volume entirely devoted to the Tibetan macaque. This combined effort of two dozen scientists to review 40 years of research and present new findings about a single species should be viewed as a celebration of the species. It frees Tibetan macaques from the purgatory of scientific papers scattered across various journals and collections to join the small club of primate species that are honored with this attention. Many people would consider that brown monkeys like Tibetan macaques all look similar. Although they do not have the immediate visual appeal of more brightly colored primates, brown monkeys have different but equally attractive assets. With their fiery gaze and prominent beards, Tibetan macaques are no exception, and their adaptations and behaviors attract a great deal of research interest.

As scientists we are expected to test hypotheses and theories. Some of the contributors to this book do so, addressing broad issues such as cooperative strategies, social dynamics, collective decisions, feeding ecology, and pathogen transmission. They use Tibetan macaques as a model to investigate mainstream research questions in the field of behavioral ecology and evolution. Every animal species has its singularities, however, and deserves to be studied for itself. This is why other contributors seek to identify what makes Tibetan macaques special, investigating patterns such as social play, call types, or the fascinating "bridging" interactions in which infants play a role as buffers to reduce tension between adults. Science generally values the testing of general theories more than the humble seeking of what gives a species its own touch. In the end, however, both of these approaches are necessary. Years ago, I was struggling to rank the different species of macaque according to their levels of social tolerance. Quantitative measures were available in a limited number of species and I had to rely on qualitative data for others. I remember asking Qikun Zhao about the behaviors particular to Tibetan macaques at a conference held in Japan in 1996. I was trying to guess their social style, i.e., their own touch. This resulted in a tentative scaling of macaque species which would later be amended when quantitative data became available in Tibetan macaques. It should be emphasized that the story is far from over. As discussed in the book, why and to what extent the different behavioral traits constituting social styles may covary during the evolutionary process still remains to be elucidated. I am delighted to see how the study of the particular meets the general by yielding new perspectives and hypotheses to be tested.

As the editors point out, this book should not be considered an end, but rather a beginning. This highlighting of research into Tibetan macaques has the potential to strengthen Chinese primatology and favor its development at the national and international level. It may in turn help the Tibetan macaques. Like other species of non-human primates, their populations are threatened by the loss and fragmentation of their habitat. Admiring, knowledge, and conservation should go hand in hand to save the future of this unique species.

University of Strasbourg, Strasbourg
France

Bernard Thierry

The original version of the book frontmatter was revised: For detailed information please see Correction. The correction to the book frontmatter is available at https://doi.org/10.1007/978-3-030-27920-2_15

Preface

This book is mainly based on research papers presented in a spirited international primatology symposium held in the scenic area of Mt. Huangshan, China, in the summer of 2017. The chapters were grouped into five logical parts. Part I consists of a single chapter, which offers a brief introduction to recent developments in Chinese primatology and a short history of research on the primates of China in general and the Tibetan macaque in particular.

Part II contains seven chapters (Chaps. 2–8) focusing on social behavior and social dynamics in Tibetan macaques. In Chap. 2, Jin-Hua Li and Peter M. Kappeler provide a comprehensive review of three decades of field research in Tibetan macaques at the Valley of Monkeys, highlighting the significance of this species as a model for understanding broader questions in primate behavior and evolution. Lixing Sun, Dong-Po Xia, and Jin-Hua Li follow up in Chap. 3 by introducing a new way to analyze the dynamics of macaque social hierarchy from a social mobility perspective with new insights unveiled through comparing Tibetan macaques with Japanese macaques. In Chap. 4, Dong-Po Xia, Paul A. Garber, Cédric Sueur, and Jin-Hua Li look into the internal behavioral mechanisms promoting group stability in Tibetan macaque from a behavioral exchange and biological market point of view. In Chap. 5, Xi Wang, Claudia Fichtel, Lixing Sun, and Jin-Hua Li investigate how Tibetan macaques make collective decisions during group movements. In Chap. 6, Jessica A. Mayhew, Jake A. Funkhouser, and Kaitlin R. Wright explore the significance of play behavior in the development of social cognition in juvenile Tibetan macaques. This chapter is followed by Sofia K. Blue's analysis of vocal communication in Chap. 7, which generates insights from comparing Tibetan macaques with other macaque species. In Chap. 8, Krishna N. Balasubramaniam, Hideshi Ogawa, Jin-Hua Li, Consuel Ionica, and Carol M. Berman offer a comprehensive review of Tibetan macaque's social structure, with insights from their previous work on social styles, comparative studies with other macaque species, and male–male social tolerance.

Part III contains two highly focused studies about ritualized behavior of Tibetan macaques with the implication about how culture evolves. In Chap. 9, Grant J. Clifton, Lori K. Sheeran, R. Steven Wagner, Jake A. Funkhouser, and Jin-Hua

Li examine how infants are used for the regularly observed behavior of bridging between adult females. This is further pursued in Chap. 10, where Hideshi Ogawa compares bridging behavior in two populations of *Macaca assamensis*, which are then compared with Tibetan macaques to explore the evolutionary origins of this highly ritualized behavior.

Part IV is composed of four chapters, focusing on how Tibetan macaques live with microbes, parasites, and diseases. In Chap. 11, Binghua Sun, Michael A. Huffman, and Jin-Hua Li take us into the microbial world inside the gut of Tibetan macaques and show how microbes adapt to the social behavior of the species. Then, Michael A. Huffman, Binghua Sun, and Jin-Hua Li present data in Chap. 12 to test the hypothesis that the diet of Tibetan macaques may incorporate self-medicative aspects to better survive in their environment, a proposition that has never been examined in this species before. Broadening the scope in Chap. 13, Krishna N. Balasubramaniam, Cédric Sueur, Michael A. Huffman, and Andrew J. J. MacIntosh review previous work on infectious agents at human–macaque interfaces and offer several key future directions for research in this area.

Many recent discoveries in primatology involve technological advancements in research, which is the content of Part V. In a single chapter (Chap. 14), Yong Zhu and Paul A. Garber explore the great potential of the high field MRI technology in the study of primate behavior and cognition. While promising, this new imaging technology has several obvious limitations at present.

All in all, the contributors of this volume examine a broad range of topics about the behavioral ecology of the Tibetan macaque. Although data are still far from adequate and some conclusions are tentative, we hope this volume will help remove the Tibetan macaque from the list of little known primate species. We expect that the information presented here can stimulate further comparative study of behavioral, ecological, and evolutionary questions about macaques and other primates and hope that this contribution will facilitate the integration of Chinese primatology into the mainstream field.

Hefei, Anhui, China	Jin-Hua Li
Ellensburg, WA, USA	Lixing Sun
Göttingen, Germany	Peter M. Kappeler

Acknowledgments

This volume is based mainly on the research papers presented during the 2017 International Primatological Symposium at Mt. Huangshan, China. The National Natural Science Foundation of China sponsored the meeting and also provided funding to support open access publication of this volume, as well as the publishing fund of Hefei Normal University. Hefei Normal University also provided a fund to defray the cost of publication including book purchase. We are grateful to all the contributors for sharing their work, without which the timely publication of this volume would have been impossible.

All the chapters in the volume were peer reviewed and benefited from the sharp comments and constructive suggestions by colleagues who generously donated their time and offered professional help. We are particularly thankful for the following external reviewers: Filippo Aureli, Louise Barrett, Fred Bercovitch, Marco Gamba, Andrew King, Daoying Lan, Bonaventura Majolo, Nadine Müller, Charles Nunn, Paula Pebsworth, Odile Petit, Gabriele Schino, Masaki Shimada, Wencheng Song, Bernhard Thierry, Kazuo Wada, Qi Wu, and Hongyi Yang. We also thank Rose Amrhein who offered language help for most of the chapters. We are lucky to have Srinivasan Manavalan as our in-house editor who oversaw the book project from the beginning to the end and answered all of our questions and inquiries throughout the process.

April 2019

Jin-Hua Li
Lixing Sun
Peter M. Kappeler

Contents

List of Contributors

Krishna Balasubramaniam Department of Population Health and Reproduction, School of Veterinary Medicine, University of California at Davis, Davis, CA, USA

Carol M. Berman Department of Anthropology and Graduate Program in Evolution Ecology and Behavior, State University of New York at Buffalo, Buffalo, NY, USA

Sofia K. Blue Primate Behavior and Ecology Program, Department of Anthropology and Museum Studies, Central Washington University, Ellensburg, WA, USA

Grant J. Clifton Primate Behavior and Ecology Program, Central Washington University, Ellensburg, WA, USA

Claudia Fichtel Behavioral Ecology and Sociobiology Unit, German Primate Center, Leibniz Institute for Primate Research, Göttingen, Germany

Jake A. Funkhouser Primate Behavior and Ecology Program, Central Washington University, Ellensburg, WA, USA
Department of Anthropology, Washington University in St. Louis, St. Louis, MO, USA

Paul A. Garber Department of Anthropology, Program in Ecology, Evolution, and Conservation Biology, University of Illinois, Urbana, IL, USA

Michael A. Huffman Primate Research Institute, Kyoto University, Kyoto, Japan

Consuel Ionica Biomedical Department, F. I. Rainer Anthropology Institute, Romanian Academy, Bucureşti, Romania

Peter M. Kappeler Behavioral Ecology and Sociobiology Unit, German Primate Center, Leibniz Institute for Primate Research, Göttingen, Germany
Department of Anthropology/Sociobiology, University of Göttingen, Göttingen, Germany

Jin-Hua Li School of Resources and Environmental Engineering, Anhui University, Hefei, Anhui, China
International Collaborative Research Center for Huangshan Biodiversity and Tibetan Macaque Behavioral Ecology, Anhui, China
School of Life Sciences, Hefei Normal University, Hefei, Anhui, China

Andrew J. J. MacIntosh Kyoto University Primate Research Institute, Kyoto, Japan

Jessica A. Mayhew Primate Behavior and Ecology Program, Department of Anthropology and Museum Studies, Central Washington University, Ellensburg, WA, USA

Hideshi Ogawa School of International Liberal Studies, Chukyo University, Toyota, Japan

Lori K. Sheeran Primate Behavior and Ecology Program and Department of Anthropology and Museum Studies, Central Washington University, Ellensburg, WA, USA

Cédric Sueur Université de Strasbourg, CNRS, IPHC, UMR 7178, Strasbourg, France

Binghua Sun School of Resources and Environmental Engineering, Anhui University, Hefei, China

Lixing Sun Department of Biological Sciences, Primate Behavior and Ecology Program, Central Washington University, Ellensburg, WA, USA

R. Steven Wagner Department of Biological Sciences, Central Washington University, Ellensburg, WA, USA

Xi Wang School of Resources and Environmental Engineering, Anhui University, Hefei, China
International Collaborative Research Center for Huangshan Biodiversity and Tibetan Macaque Behavioral Ecology, Anhui, China

Kaitlin R. Wright Primate Behavior and Ecology Program, Central Washington University, Ellensburg, WA, USA

Dong-Po Xia School of Life Sciences, Anhui University, Hefei, China
International Collaborative Research Center for Huangshan Biodiversity and Tibetan Macaque Behavioral Ecology, Anhui, China

Yong Zhu High Magnetic Field Laboratory, Chinese Academy of Sciences, Hefei, China
School of Life Sciences, Hefei Normal University, Hefei, Anhui, China

Part I
Introduction

Chapter 1
Recent Developments in Primatology and Their Relevance to the Study of Tibetan Macaques

Lixing Sun, Jin-Hua Li, Cédric Sueur, Paul A. Garber, Claudia Fichtel, and Peter M. Kappeler

L. Sun (✉)
Department of Biological Sciences, Primate Behavior and Ecology Program, Central Washington University, Ellensburg, WA, USA
e-mail: Lixing@cwu.edu

J.-H. Li
School of Resources and Environmental Engineering, Anhui University, Hefei, Anhui, China

International Collaborative Research Center for Huangshan Biodiversity and Tibetan Macaque Behavioral Ecology, Anhui, China

School of Life Sciences, Hefei Normal University, Hefei, Anhui, China
e-mail: jhli@ahu.edu.cn

C. Sueur
CNRS, IPHC, UMR, Université de Strasbourg, Strasbourg, France
e-mail: cedric.sueur@iphc.cnrs.fr

P. A. Garber
Department of Anthropology, Program in Ecology, Evolution, and Conservation Biology, University of Illinois, Urbana, IL, USA
e-mail: p-garber@illinois.edu

C. Fichtel
Behavioral Ecology and Sociobiology Unit, German Primate Center, Leibniz Institute for Primate Research, Göttingen, Germany
e-mail: Claudia.Fichtel@gwdg.de

P. M. Kappeler
Behavioral Ecology and Sociobiology Unit, German Primate Center, Leibniz Institute for Primate Research, Göttingen, Germany

Department of Anthropology/Sociobiology, University of Göttingen, Göttingen, Germany
e-mail: pkappel@gwdg.de

J.-H. Li et al. (eds.), *The Behavioral Ecology of the Tibetan Macaque*, Fascinating Life Sciences, https://doi.org/10.1007/978-3-030-27920-2_1

1.1 Recent Trends and Developments in Primatology

Given their shared evolutionary history with humans, nonhuman primates play an exceptional role in the study of animal behavior, ecology, and evolution. This close phylogenetic relationship has led scholars from a diverse set of disciplines (e.g., biological and social sciences, notably psychology and anthropology) and theoretical perspectives (e.g., kinship theory, multilevel selection, social interactions, cultural traditions, competition, cooperation, innovation) to examine a broad range of research topics and methodologies in primatology. It is hardly an exaggeration to say that primatology is an intellectual "melting pot" in the study of animals.

The integration of different disciplines into the science of primatology has led to a major paradigm shift in the philosophy of science. Traditionally, scientists tended to assume animals had limited agency and behavioral flexibility or were incapable of engaging in complex forms of decision-making. They would be accused of committing a major scientific sin, namely, anthropomorphism, if they empathized with their research subjects or thought animals share emotions, social strategies, or cognitive abilities with humans (e.g., Masson and McCarthy 1996). Such a philosophical standpoint has become increasingly tenuous, as recent evidence in many mammals, especially nonhuman primates, have identified that they do indeed exhibit emotions, empathy, behavioral strategies, social bonding, cognitive abilities, and, in some instances, a moral sense of fairness (e.g., de Waal 1988, 2010; Kappeler and van Schaik 2006; Kappeler and Silk 2009; van Schaik 2016). In light of these findings, anthropomorphism and highlighting the behavioral and cognitive continuity between humans and other mammals may provide a more parsimonious null hypothesis than the alternative view, namely, that mammals in general and primates in particular have limited ability to respond to changes in their social and ecological environments (de Waal et al. 2006).

Primatologists have a keen and profound understanding that the sensations and emotions of researchers, as well as their subjective experiences as primates, are an essential variable in pursuing scientific questions. As in the case of quantum mechanics, where measurements and the behavior of quanta are entwined no matter what controls researchers place on the experiment, scientific objectivity in primatology is accomplished by including their own sensations and emotions as critical variables in research. We argue there exists no clear demarcation between being overtly subjective and being prudently self-reflective, and therefore in the case of primate research, we attempt to balance anthropomorphism without losing scientific objectivity.

Accompanying this philosophical shift, broad comparative studies, grounded in evolutionary, ecological, and behavioral perspectives, have led, in recent years, to new and exciting discoveries focused on social strategies, problem-solving, and cognitive abilities of nonhuman primates including complex social networks, cooperation between kin and nonkin, collective behavior, biological markets, behavioral economics, and culture (Byrne and Whiten 1988; van Schaik et al. 1999; Byrne and Bates 2007; Dufour et al. 2008; Sueur et al. 2010; Balasubramaniam et al.

2011; Sussman and Garber 2011; Pasquaretta et al. 2014; Garber 2019). Using social cognition as an example, major developments have been made in understanding primate intelligence (Machiavelli intelligence), social development, personality, empathy, theory of mind, trading, intuitive moral sense (particularly, fairness), and others (Dufour et al. 2008; Devaine et al. 2017). These discoveries would have been impossible without careful self-examination of our own behaviors, societies, and cognitive abilities. In fact, the study of our focal species in this volume, the Tibetan macaque (*Macaca thibetana*), clearly reflects such introspective, anthropomorphic thinking across a wide range of topics from social interactions and feeding ecology to gut microbe communities and self-medication. Despite the many insights anthropomorphic approaches provide, we must refrain from assuming that all similarities in social behaviors, social organizations, and cognitive abilities are derived from shared ancestry between humans and nonhuman primates because these similarities could also have arisen independently through convergent or parallel evolution. As such, distinguishing between homology and homoplasy will continue to pose a major challenge to primatologists.

1.2 Why Macaques, Especially Tibetan Macaques?

New developments in primatology in recent years have reinforced the view that nonhuman primates offer a window for us to look into human behavior, biology, and sociality from an evolutionary perspective (e.g., Kappeler and van Schaik 2006; Kappeler and Silk 2009; van Schaik 2016). Primates not only provide an instructive model to better understand human evolution, they may also provide information and insights into a range of practical issues in our society from promoting cooperation to preventing disease transmission (Romano et al. 2016). In fact, it is hard for us to underestimate how much we can learn from behavioral, ecological, and evolutionary knowledge gained from nonhuman primates.

Although nonhuman great apes represent our closest living relatives, all primate radiations can provide important model systems that unveil evolutionary patterns and processes. Currently there are over 500 species of living primates (Estrada et al. 2017). Many of these species are more abundant and evolutionarily successful (in numbers, populations, and species) than are great apes. As such they can provide a broader range of demographic and socioecological scenarios for testing evolutionary hypotheses, especially using comparative approaches (e.g., Harvey and Pagel 1991). In this sense, macaques (genus *Macaca*) may be unmatched as a model group for evolutionary studies of social behavior, social organization, and social cognition in primates. With 20–23 species, macaques are among the most successful primate radiations, with the largest geographical distribution of any taxa (Thierry et al. 2004; Fleagle 2013). The genus forms a monophyletic clade (Morales and Melnick 1998), and the evolutionary relationships among species have been mapped out with a reasonable level of certainty (Purvis 1995; Li and Zhang 2005; Jiang et al. 2016).

The availability of this critical information has paved the way for pursuing questions about the evolution of their behaviors using comparative methods.

One major challenge now for macaque behavioral ecologists is to obtain quality long-term data on behavioral variability and trade-offs between affiliative and agonistic alliances, social cohesion, and reproductive success. This level of information is missing for virtually all species, including the Tibetan macaque. In an attempt to find a general pattern of social style in macaques, for instance, Thierry (2004) classified all macaque species into four grades of social structure, from despotic (Grade 1) to egalitarian societies (Grade 4). These grades were associated with traits such as degree of social tolerance, symmetrical or asymmetrical conflict, a linear-like dominance hierarchy, and the strength of kin bonds. Clearly, there are significant challenges in attempting to quantify these variables. Based on limited information, Thierry (2004) identified Tibetan macaque as a Grade 3 species. Close examination, however, revealed that the aggressive behavior and dominance hierarchy of this species are more consistent with Grade 2 (Berman et al. 2004, 2006), which led to the revision of Thierry's classification scheme (Thierry 2011). This example illustrates the challenges of attempting to answer evolutionary questions using limited information from little known species. As such, this volume aspires to fill some glaring gaps in our knowledge of Tibetan macaques. Furthermore, several chapters attempt to directly address evolutionary questions from comparative perspectives between the Tibetan macaque and its sister species.

The Tibetan macaque is endemic to China and is listed in the IUCN Red List as near threatened (see Chap. 2). It is one of the most widespread primate species in China, distributed across 13 provinces. Its estimated population size is 20,000 individuals (Li et al. unpublished data). While rhesus macaques have been intensively studied in the field and laboratory, Tibetan macaques are far less known. As such, they provide a good comparison for understanding macaque ecology and behavior across a range of demographic, social, and ecological conditions. Also, Tibetan macaques have several unique features in the genus *Macaca*. They are the largest in body size (adult male body mass ~ 15 kg) and are found at elevations up to 2400 m (see Berman et al. 2006). They live in relatively small groups with a strict linear dominance hierarchy. Home range size is 1.62–3.62 km^2 with a pattern of habitat utilization (feeding in particular) particularly favorable for long-term observation and data collection. These features make the species well-suited for a wide range of studies for testing hypotheses related to behavior, ecology, evolution, conservation, management, human-animal interaction, and infectious disease transmission (see Chap. 14 Balasubramaniam, Sueur, and MacIntosh). Additionally, given that some populations are present in national parks or protected areas, such as those in Mt. Huangshan and Mt. Emei, they are highly accessible for educational and research purposes. (See the elaboration by Jin-Hua Li and Peter Kappeler in Chap. 2.)

1.3 A Short History of Tibetan Macaque Research

Primate research in China was largely absent until the nineteenth century, when European and American naturalists and missionaries came to China and reported their discoveries of exotic species. Following these initial descriptions, there remained no systematic studies of the behavior and ecology of Chinese primates until the 1970s, when a small number of pioneers of Chinese zoologists began to study and observe endemic species such as snub-nosed monkeys (*Rhinopithecus* spp.) and Tibetan macaques. During this period, the research was often sporadic and was published principally in Chinese journals.

The Tibetan macaque was first described to the scientific community as *M. thibetanus* by A. Milne-Edwards in 1870 based on a specimen collected by French missionary Abbé Armand David at Baoxing County in Sichuan Province (see Fooden 1983). However, in 1938, G. M. Allen believed it to be a subspecies of the stump-tailed macaque (formerly *M. speciosa* but currently *M. arctoides*) and named it *M. speciosus thibetanus*. It was not until 1983 that the Tibetan macaque regained its current species status as *M. thibetana*. This changed consensus was based on Jack Fooden's work on the anatomy of the reproductive system of the species. Recent studies using mtDNA have corroborated and validated this taxonomic distinction (Liu et al. 2006).

Despite the fact that Tibetan macaques are found in several small and isolated locations across China (a small population was recently found in eastern India) today, they were once widely distributed across a broad strip running from the foothills of southeastern Tibet to the coastal regions of East China including 13 provinces: Zhejiang, Anhui, Fujian, Jiangxi, Hubei, Hunan, Guangdong, Guangxi, Sichuan, Guizhou, Yunnan, Gansu, and Tibet (Jiang et al. 2015). Currently, four geographic subspecies are identified based on morphological characters and mtDNA (Jiang et al. 1996; Liu et al. 2006; Sun et al. 2010). Genomic studies show that the Tibetan macaque diverged from its congenic species some 0.5 Ma after a long bottleneck for the genus (Fan et al. 2014).

Field studies on Tibetan macaques have been carried out primarily at two sites, Mt. Emei in Sichuan Province and Mt. Huangshan in Anhui Province. Research on the Mt. Emei population has been led by Qikun Zhao and Ziyun Deng based at the Kunming Institute of Zoology of the Chinese Academy of Sciences. The bulk of the data from the population at this site was collected between 1987 and 1999. Research at the Mt. Huangshan site began in 1983 and led by Qishan Wang and Jin-Hua Li from Anhui University, in long-term collaboration (since 2003) with researchers from Central Washington University including Lixing Sun, Lori Sheeran, and colleagues. These research efforts have led to most of our current understanding of the species's social behavior, ecology, and population biology.

Research on ecology and population biology of Tibetan macaques has focused on morphological adaptation to ecological factors in its habitat (Xiong 1984), the current distribution of the species (Wada et al. 1987; Jiang et al. 1996), home range (Wang and Xiong 1989), adaptive relationships between body mass and elevation (Zhao and Deng 1988a, b; Zhao 1994a), effects of climate, vegetation, and slope on food resources (Zhao et al. 1989), selection of sleeping sites (Li and Wang 1994), ecological factors linked to group fission and reformation (Li et al. 1996a, b), diet (Li 1999; Zhao 1999), social organization (Deng and Zhao 1987; Li and Wang 1996), population dynamics (Wang et al. 1994), age structure and life expectancy and mortality (Li et al. 1995), and population growth in relation to population density, processes of group fission, and disease (Li et al. 1996a, b).

Several other research projects have focused on social bond formation, grooming relationships, social networks, and reproductive strategies. This has been made possible by long-term observation and monitoring of known individuals in several groups, aided by provisioning. This part of research has compared species differences in mating tactics between Tibetan and Japanese macaques (Xiong and Wang 1991; Zhao 1994c), birth timing in relation to socioecological factors such as dominance rank and altitude (Li et al. 1994; Zhao 1994c; Li et al. 2005), birth seasonality (Zhao and Deng 1988a, b; Li et al. 2005), and descriptions and evaluations of behaviors that are critical measures of social relationships such as bridging (defined as two individuals ritualistically lifting an infant accompanied by affiliative behaviors such as teeth chattering, see Zhao 1996) and grooming (Li et al. 1996a, b). These efforts have led to a complete ethogram of Tibetan macaques consisting of 32 distinct patterns of social behavior (Li 1999; Li et al. 2004). Using this ethogram, researchers can conduct in-depth analyses of behavior, social structure, and social dynamics of the species. A series of papers have been published on sociality and group stability analyzed at various levels from the individual, to dyads, social networks (cliques), and to the entire group. The topics addressed include collective decision-making and leadership in group movement (Wang et al. 2015, 2016; Fratellone et al. 2019), personality (Pritchard et al. 2014), benefit-cost analyses of grooming exchanges (Xia et al. 2012, 2013), and social networks among group members (Fratellone et al. 2019).

In recent years, new field and laboratory techniques from other disciplines have been applied to behavioral, ecological, and evolutionary studies of the species. They include the development and application of fecal DNA analysis (Zhao and Li 2008), extraction and analysis of fecal steroid hormones (Xia et al. 2015, 2018), and gut microbe analysis (Sun et al. 2016). Several of the chapters in this volume reflect these new and exciting research developments.

1.4 Tibetan Macaques at Mt. Huangshan Research Site

Many advancements in our understanding of primate behavior, biology, and evolution have been made from field studies. In this respect, long-term field studies are particularly valuable (Kappeler and Watts 2012) because they allow us to observe especially rare behaviors and biological events that may be missed based on short-term observations. For instance, long-term field studies have enabled researchers to document the cultural transmission of information within a group of Japanese macaques on Koshima Island and warfare in chimpanzees (*Pan troglodytes*). Likewise, much of our understanding of the behavior ecology and reproductive strategies of male and female Tibetan macaques reported in this volume would not be possible without long-term field studies and the dedication of primatologists working at Mt. Huangshan.

The Mt. Huangshan research site was first established in 1983, when the late Qishan Wang, then head of the Biology Department at Anhui University, founded a primate research program in collaboration with Kazuo Wada from the Primate Research Institute at Kyoto University in Japan. They jointly led a team to explore southern Anhui for a field site suitable for long-term research of the species. After some scouting, they settled on a location, now known as the Valley of Monkeys, at Yulingken, just a 15-min walk from the village of Fuxi, within the scenic area of Mt. Huangshan. This was the beginning of the first and in the meantime longest running primate research site in China, and long-term systematic observations of the Tibetan macaques have continued for more than 30 years. Today, the site is recognized as one of the eight long-term primatological study sites in the world (Kappeler and Watts 2012). Over the course of three decades, more than 150 primatologists from Japan, the United States, Germany, England, and Canada have come and conducted research in collaboration with Chinese colleagues at Mt. Huangshan.

Kazuo Wada's insights and contributions were essential, especially in getting this long-term research project started. One of the major inaugural events took place in October, 1986, when Wada presented a series of talks during the First Primatological Research Symposium of China hosted by Anhui University. The meeting was attended by only 12 participants (including Jin-Hua Li and Lixing Sun, then both beginning graduate students), representing probably fewer than ten qualified primatological researchers in the nation. At that time, some 27 primate species naturally occurred in China. In comparison, there were about 500 primatologists in Japan, working on a single endemic species, the Japanese macaque (*M. fuscata*). Wada was a dedicated field primatologist who had published extensively about the biology of macaques. Though skeptical about sociobiology, he was quite open in sharing his research ideas and experience, especially under the mild impact of Chinese whiskey. Among what he brought to Chinese primatology were two simple yet brilliant Japanese methodological perspectives that changed the face of Chinese primatological research: he recommended that primates be recognized individually and helped establish a naming system that reflected the sex, generation, and lineage information for all group members. At the time, these practices were still considered an inappropriate use of anthropomorphism by many in the West.

Since then, primatological research has thrived in China with increasingly well-trained primatologists working on virtually all of China's primate species across a range of topics from ecology, behavior, and conservation to molecular genetics and cognitive science with several highly productive and visible research groups. So, a few trickling streams of effort, initiated just a few decades ago, have grown to be a torrent of vibrant primatological research in China today, culminating in the formal launch of the Chinese Primatological Society in 2017. The inauguration conference of the society was attended by well over 200 researchers from all corners of the nation.

The chapters presented here partly illustrate the status quo of Chinese primatology from the research work conducted mainly in Tibetan macaques at Mt. Huangshan. They are also a demonstration for the fruitfulness of international collaboration by sharing research resources, ideas, and methodologies, reflecting the multidisciplinary nature of primatology. We expect that collaborations among researchers from a diverse range of academic and cultural backgrounds will continue to grow and deepen in the future.

References

Balasubramaniam KN, Berman CM, Ogawa H et al (2011) Using biological market principles to examine patterns of grooming exchange in *Macaca thibetana*. Am J Primatol 73:1269–1279

Berman CM, Ionica CS, Li J (2004) Dominance style among *Macaca thibetana* on Mt. Huangshan, China. Int J Primatol 25:1283–1312

Berman CM, Ionica CS, Dorner M et al (2006) Postconflict affiliation between former opponents in *Macaca thibetana* on Mt. Huangshan, China. Int J Primatol 27:827–854

Byrne RW, Bates LA (2007) Sociality, evolution, and cognition. Curr Biol 17:R714–R723

Byrne RW, Whiten A (1988) Machiavellian intelligence: social expertise and the evolution of intellect in monkeys, apes and humans. Clarendon Press, Oxford

De Waal FBM (1988) Chimpanzee politics: power and sex among apes. Johns Hopkins University Press, Baltimore, MD

De Waal FBM (2010) The age of empathy: nature's lessons for a kinder society. Broadway Books, New York

De Waal FBM, Wright R, Korsgaard CM et al (2006) Primates and philosophers: how morality evolved. Princeton University Press, Princeton, NJ

Deng Z, Zhao Q (1987) Social structure in a wild group of *Macaca thibetana* at Mount Emei, China. Folia Primatol 49:1–10

Devaine M, San-Galli A, Trapanese C et al (2017) Reading wild minds: a computational assay of theory of mind sophistication across seven primate species. PLoS Comput Biol 13(11): e1005833

Dufour V, Pelé M, Neumann M et al (2008) Calculated reciprocity after all: computation behind token transfers in orang-utans. Biol Lett 5(2):172–175

Estrada A, Garber PA, Rylands AB et al (2017) Impending extinction crisis of the world's primates: why primates matter. Sci Adv 3:e1600946

Fan Z, Zhao G, Li P et al (2014) Whole-genome sequencing of Tibetan macaque (*Macaca thibetana*) provides new insight into the macaque evolutionary history. Mol Biol Evol 31:1475–1489

Fleagle JG (2013) Primate adaptation and evolution, 3rd edn. Academic, San Diego, CA

Fooden J (1983) Taxonomy and evolution of the *sinica* group of macaques: 4. Species account of *Macaca thibetana*. Fieldiana Zool 11:1–20

Fratellone GP, Li J-H, Sheeran LK, Wagner RS, Wang X, Sun L (2019) Social connectivity facilitates collective decision making in wild Tibetan macaques (*Macaca thibetana*). Primates 60:183–189

Garber PA (2019) Primate cognitive ecology: challenges and solutions to locating and acquiring resources in social foragers. In: Lambert JE, Rothman JM (eds) Primate diet and nutrition: needing, finding, and using food. University of Chicago Press, Chicago, IL

Harvey PH, Pagel MD (1991) The comparative method in evolutionary biology. Oxford University Press, Oxford

Jiang X, Wang Y, Wang Q (1996) Taxonomy and distribution of Tibetan macaque (*Macaca thibetana*). Zool Res 17:361–369

Jiang Z, Ma Y, Wu Y et al (2015) China's mammal diversity and geographic distribution. Science Press, Beijing

Jiang J, Yu J, Li J et al (2016) Mitochondrial genome and nuclear markers provide new insight into the evolutionary history of macaques. PLoS One 11(5):e0154665

Kappeler PM, Silk JB (eds) (2009) Mind the gap: tracing the origins of human universals. Springer, Heidelberg

Kappeler PM, van Schaik CP (eds) (2006) Cooperation in primates and humans. Springer, Berlin

Kappeler PM, Watts DP (eds) (2012) Long-term field studies of primates. Springer, Berlin

Li J (1999) The Tibetan macaque society: a field study. Anhui University Press, Hefei

Li J, Wang Q (1994) Selection of sleeping site in Tibetan macaques in the summer. Chin J Zool 29:58

Li J, Wang Q (1996) Dominance hierarchy and its chronic changes in adult male Tibetan macaque (*Macaca thibetana*). Acta Zool Sin 42:330–334

Li Q, Zhang Y (2005) Phylogenetic relationships of the Macaques (Cercopithecidae: *Macaca*), inferred from mitochondrial DNA sequences. Biochem Genet 43:375–386

Li J, Wang Q, Li M (1994) Population ecology of Tibetan macaques II: patterns of reproduction. Acta Theriol Sin 14:255–259

Li J, Wang Q, Li M (1995) Population biology of Tibetan macaques III: age structure and life table. Acta Theriol Sin 15:31–35

Li J, Wang Q, Han D (1996a) Fission in a free-ranging Tibetan macaque troop at Huangshan Mountains, China. Chin Sci Bull 16:1377–1381

Li J, Wang Q, Li M (1996b) Migration of male Tibetan monkeys (*Macaca thibetana*) at Mt. Huangshan, Anhui Province, China. Acta Theriol Sin 16:1–6

Li J, Yin H, Zhou L et al (2004) Social behaviors and relationships among Tibetan macaques. Chin J Zool 39:40–44

Li J, Yin H, Wang Q (2005) Seasonality of reproduction and sexual activity in female Tibetan macaques *Macaca thibetana* at Huangshan, China. Acta Zool Sin 51:365–375

Liu Y, Li J, Zhao J (2006) Divergence and phylogeny of mitochondrial cytochrome B gene from Tibetan macaque and stump-tailed macaque. Ecol Sci 25:426–429

Masson JM, McCarthy S (1996) When elephants weep: emotional lives of animals. Vintage, New York

Morales JC, Melnick DJ (1998) Phylogenetic relationships of the macaques (Cercopithecidae: *Macaca*), as revealed by high resolution restriction site mapping of mitochondrial ribosomal genes. J Hum Evol 34:1–23

Pasquaretta C, Levé M, Claidiere N et al (2014) Social networks in primates: smart and tolerant species have more efficient networks. Sci Rep 4:7600

Pritchard AJ, Sheeran LK, Gabriel KI et al (2014) Behaviors that predict personality components in adult free-ranging Tibetan macaques *Macaca thibetana*. Curr Zool 60:362–372

Purvis A (1995) A composite estimate of primate phylogeny. Philos Trans R Soc B 348:405–421

Romano V, Duboscq J, Sarabian C et al (2016) Modeling infection transmission in primate networks to predict centrality-based risk. Am J Primatol 78:767–779

Sueur C, Deneubourg JL, Petit O (2010) Sequence of quorums during collective decision making in macaques. Behav Ecol Sociobiol 64:1875–1885

Sun B, Li J, Zhu Y et al (2010) Mitochondrial DNA variation in Tibetan macaque (*Macaca thibetana*). Folia Zool 59:301–307

Sun B, Wang X, Bernstein S et al (2016) Marked variation between winter and spring gut microbiota in free-ranging Tibetan macaques (*Macaca thibetana*). Sci Rep 6. https://doi.org/10.1038/srep26035

Sussman RW, Garber PA (2011) Cooperation, collective action, and competition in primate social interactions. In: Campbell CJ, Fuentes A, Mackinnon KC et al (eds) Primates in perspective, 2nd edn. Oxford University Press, New York, pp 587–599

Thierry B (2004) Social epigenesis. In: Thierry BR, Singh M, Kaumanns W (eds) Macaque societies: a model for the study of social organization. Cambridge University Press, New York, pp 267–290
Thierry B (2011) The macaques: a double-layered social organization. In: Campbell CJ, Fuentes A, MacKinnon KC et al (eds) Primates in perspective, 2nd edn. Oxford University Press, New York, pp 229–241
Thierry B, Singh M, Kaumanns W (eds) (2004) Macaque societies: a model for the study of social organization. Cambridge University Press, New York, pp 80–83
Van Schaik CP (2016) The primate origins of human nature. Wiley-Blackwell, Hoboken, NJ
Van Schaik CP, Deaner RO, Merrill MY (1999) The conditions for tool use in primates: implications for the evolution of material culture. J Hum Evol 36:719–741
Wada K, Xiong C, Wang Q (1987) On the distribution of Tibetan and rhesus monkeys in southern Anhui. Acta Theriol Sin 7:168–176
Wang Q, Xiong C (1989) Seasonal home range changes of Tibetan macaques at Yulingken. Acta Theriol Sin 9:239–246
Wang Q, Li J, Yang Z (1994) Tibetan macaques in China. Bull Biol 29:5–7
Wang X, Sun L, Li J et al (2015) Collective movement in the Tibetan macaques (*Macaca thibetana*): early joiners write the rule of the game. PLoS One 10(5):e0127459
Wang X, Sun L, Sheeran LK et al (2016) Social rank versus affiliation: which is more closely related to leadership of group movements in Tibetan macaques (*Macaca thibetana*)? Am J Primatol 78:816–824
Xia D, Li J, Garber PA et al (2012) Grooming reciprocity in female Tibetan macaques *Macaca thibetana*. Am J Primatol 74:569–579
Xia D, Li J, Garber PA et al (2013) Grooming reciprocity in male Tibetan macaques. Am J Primatol 75:1009–1020
Xia D, Li J, Sun B et al (2015) Evaluation of fecal testosterone, rank and copulatory behavior in wild male *Macaca thibetana* at Huangshan, China. Pak J Zool 47:1445–1454
Xia D, Wang X, Zhang Q et al (2018) Progesterone levels in seasonally breeding, free-ranging male *Macaca thibetana*. Mammal Res 63:99–106
Xiong C (1984) Ecological research of the Tibetan macaque. Acta Theriol Sin 4:1–9
Xiong C, Wang Q (1991) A comparative study on the male sexual behavior in Tibetan and Japanese monkeys. Acta Theriol Sin 11:13–22
Zhao Q (1994a) Birth timing shift with altitude and its ecological implication in Tibetan macaques at Mt. Emei. Oecol Mont 3:24–26
Zhao Q (1994b) Seasonal changes in body weight of *Macaca thibetana* at Mt. Emei, China. Am J Primatol 32:223–226
Zhao Q (1994c) Mating competition and intergroup transfer of males in Tibetan macaques (*Macaca thibetana*) at Mt. Emei, China. Primates 35:57–68
Zhao Q (1996) Male-infant-male interactions in Tibetan macaques. Primates 37:135–143
Zhao Q (1999) Responses to seasonal changes in nutrient quality and patchiness of food in a multigroup community of Tibetan macaques at Mt. Emei. Int J Primatol 20:511–524
Zhao Q, Deng Z (1988a) *Macaca thibetana* at Mt. Emei, China: II. birth seasonality. Am J Primatol 16:261–268
Zhao Q, Deng Z (1988b) *Macaca thibetana* at Mt. Emei, China: I. A cross-sectional study of growth and development. Am J Primatol 16:251–260
Zhao J, Li J (2008) Analysis of factors affecting DNA extracting from mammalian faecal samples. J Biol 25:5–8
Zhao Q, Xu JM, Deng Z (1989) Climate, vegetation and topography of the slope habitat of *Macaca thibetana* at Mt. Emei, China. Zool Res 10(supplement):91–99

Part II
Social Behavior and Dynamics in Tibetan Macaques

Chapter 2
Social and Life History Strategies of Tibetan Macaques at Mt. Huangshan

Jin-Hua Li and Peter M. Kappeler

2.1 Introduction

Among the more than 25 species of *Macaca*, the Tibetan macaque (*Macaca thibetana*) is relatively late to be known in primatology. There are several reasons for this. First, it had been considered as a subspecies of *Macaca speciosa* until Fooden (1983), based on his re-examination of the form and size of the glans penis and baculum and the structure of the female reproductive tract, elevated it to species status. Second, Tibetan macaques are endemic to east central China, where most of them inhabit high mountains with dense forests and cliff ledges, making field studies very challenging. Third, as argued above (Sun et al. 2019), there were few Chinese researchers interested in studying primates until the 1980s. As a result, the population status, life history, and social organization of *M. thibetana* have remained relatively poorly known.

Existing information, though still limited, indicates that the Tibetan macaque may be special in many ways. Morphologically, it resembles the stump-tailed macaque, *M. arctoides* (Delson 1980). Phylogenetically, it is close to the Assamese macaque, *M. assamensis*, as indicated by both morphological (Delson 1980) and genetic

J.-H. Li (✉)
School of Resources and Environmental Engineering, Anhui University, Hefei, Anhui, China

International Collaborative Research Center for Huangshan Biodiversity and Tibetan Macaque Behavioral Ecology, Anhui, China

School of Life Sciences, Hefei Normal University, Hefei, Anhui, China
e-mail: jhli@ahu.edu.cn

P. M. Kappeler
Behavioral Ecology and Sociobiology Unit, German Primate Center, Leibniz Institute for Primate Research, Göttingen, Germany

Department of Anthropology/Sociobiology, University of Göttingen, Göttingen, Germany
e-mail: pkappel@gwdg.de

J.-H. Li et al. (eds.), *The Behavioral Ecology of the Tibetan Macaque*, Fascinating Life Sciences, https://doi.org/10.1007/978-3-030-27920-2_2

analyses (Hoelzer et al. 1992). Ecologically and behaviorally, it is more similar to the Barbary macaque, *M. sylvanus*, because both live in montane habitats near the subtropical/temperate boundary, have a similar diet, and share intensive infant care with "bridging behavior," which involves two adults simultaneously lifting up an infant (Ogawa 2019). Yet, the four abovementioned species have been placed into different species groups within the genus *Macaca* by Fooden (1980) and Delson (1980), leading Wada et al. (1987) to suggest that the Tibetan macaque is a key species for understanding the evolution of Asian *Macaca*. These are the main reasons why *M. thibetana* has attracted much interest among primatologists lately.

Tibetan macaques have been listed as a Class II protected species in China since 1988, and they were classified as "Near Threatened" by the IUCN Red List of Threatened Species in 2017. Four subspecies of Tibetan macaque (*M. t. thibetana*, *M. t. guizhonensis*, *M. t. huangshanensis*, and *M. t. pullus*) have been identified based on morphological comparisons (Jiang et al. 1996) and mtDNA analyses (Sun et al. 2010). These subspecies represent distinct conservation units (Liu et al. 2006). Since *M. t. huangshanensis* is the most genetically distinctive and geographically isolated taxon (Sun et al. 2010), it deserves particular attention (Li et al. 2008) (Fig. 2.1).

2.2 Long-term Study of Tibetan Macaques at Mt. Huangshan

The Mt. Huangshan study site is located in southern Anhui province in East China (118.3E, 30.2 N, elevation 1841 m), about 1000 km south of Beijing. It covers an area of 154 km^2 with a south-north dimension of 40 km and an east-west extension of 30 km. This region is primarily made up of granite with many separate, sharp peaks and cliffs, which prevent people from gaining access to most parts. The weather in the mountain changes with altitude, from subtropical near the bottom, to temperate on the slopes, and to cold at the peak. Correspondingly, the vegetation changes from evergreen broad-leaved forest near the base, to deciduous and evergreen broad-leaved mixed forest on the slopes, and to montane grassland at the top. Mean temperature is 7.8 °C with the highest mean temperature in August (25.6 °C) and the lowest in January (−19.8 °C) (Li 1999). Mt. Huangshan is a well-known scenic spot and a popular tourist destination in China and is listed as a World Cultural and Natural Heritage Site, a World Geological Park, and is in the Man and the Biosphere Programme (MAB) (Fig. 2.2).

Two primate species inhabit these mountains: the Tibetan macaque and the rhesus macaque (*M. mulatta*). Tibetan macaques inhabit higher altitudes above 600 m asl with rocky cliffs, whereas the rhesus macaques live in lower areas with a relatively continuous distribution (Wada et al. 1987). Both have been strictly protected from hunting and trapping since the 1940s and have no known large predators.

Nine social groups of Tibetan macaques are currently living in Mt. Huangshan, where they maintain apparently nonoverlapping home ranges (Wada et al. 1987).

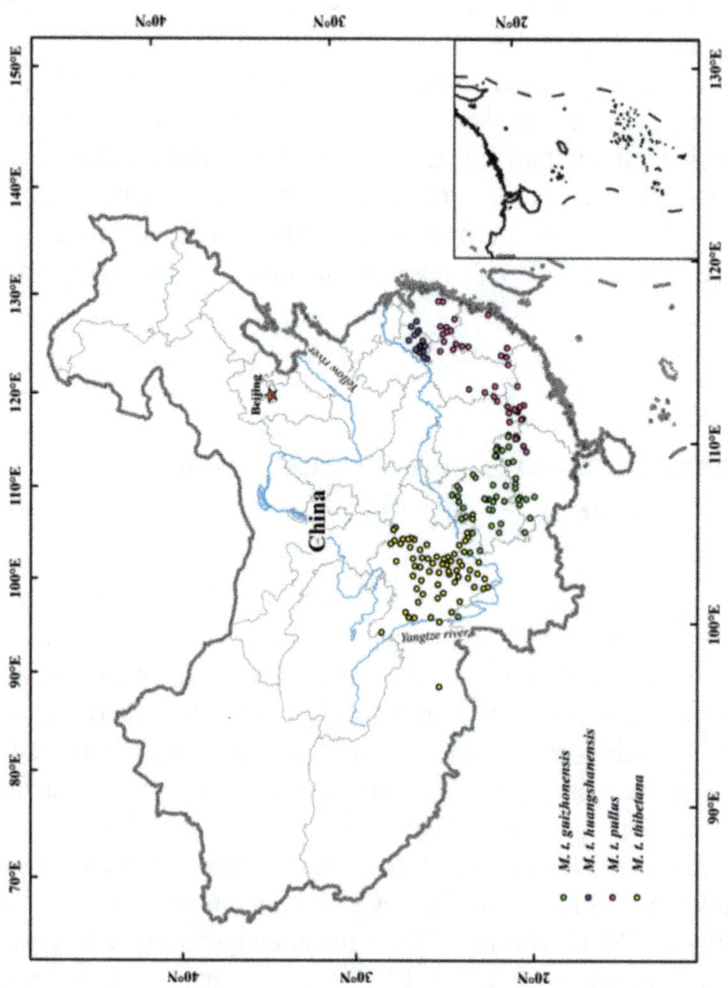

Fig. 2.1 Geographical distribution of the four subspecies of *Macaca thibetana* in China. The species sporadically inhabits in the mid-east region of China, along the Yangtze River

Fig. 2.2 Tibetan macaques inhabit high mountains with evergreen broad-leaved forest (left) and rocky cliffs (right)

They occur in two types of forest and mixed forest: the evergreen broad-forest between 500 and 1000 m asl and deciduous and evergreen mixed broad-forest between 800 and 1200 m asl. They are heavily reliant on structural plant parts for food with marked seasonality: bamboo and grass shoots in spring; fruits, nuts, acorns, and tubers in fall; and bark and mature leaves in winter, supplemented with invertebrates year-round (Xiong and Wang 1988). Geophagy has been observed in terms of ingesting yellow clay from well-established spots and licking rocks that are in contact with soil (Wang et al. 2007). In winter, Tibetan macaques huddle together and sleep on the terraces of rocky cliffs (Xiong 1984), but in summer, they spend the night in trees, which are cool and are safe from venomous snakes (Li and Wang 1994).

Our long-term study group is named Yulingken group (lately Yulingken A or YA1 following a recent group fission) at the village of Fuxi in the southern part of the mountain. The group has been monitored continuously since 1986. To facilitate observation, we have provisioned the group with dried maize to attract them to an open area by a stream. An amount of 5–6 kg of maize per day, which is about 1/3 of the daily food intake of the group, has been given by local wardens 3–4 times a day on a set schedule. When not being fed, the macaques spend most of their time in the forest near the provisioning area. Provisioning has made it possible for us to study the group in a steep terrain and has prolonged our observational time, though it may inevitably affect their natural activities to some extent (Matheson et al. 2006).

We set up a standard study protocol for long-term collection of consistent data within and among researchers at the beginning of the study in 1986. Since then, we have recognized individuals from birth on (natal individuals) or first discovery (immigrated individuals) and given them Chinese names. We have identified each group member based on individually specific physical characteristics such as body size, fur color, facial features, and body scars and taken pictures for confirmation and comparison over time. Furthermore, since Tibetan macaques have broad faces and long fur, both of which change with age, we can fairly reliably sort them into age-sex groups: infant (<1 year), juvenile (1–3 years), adolescent (♂3–7 years, ♀3–5 years), young adult (♂7–10 years, ♀5–10 years), middle-aged adult (10–15 years), and old adult (>15 years). Our records show that a male can live up to 28 years (Gaoshan) and a female up to 33 years old (Hua). They both died of old age. Because all females in the group have been identified after their births, we have been able to trace matrilineal relationships among all group members since 1983. Recently, we used microsatellite DNA markers to determine paternity for most group members (Appendix I). For behavioral data collection, we have followed Altmann's methods (Altmann 1974) and recognized and defined 33 distinct social behaviors (Li 1999, see Appendix II). In order to conduct genetic, physiological, and microbiological analyses, we have developed noninvasive techniques to collect cell or hormone samples from the feces and saliva of living monkeys (Zhao et al. 2005; Simons et al. 2012).

The long-term study of Tibetan macaques at Mt. Huangshan has been ongoing for over 30 years. It is the longest research project for wild primates in China. More than 150 researchers and students from the United States, Japan, Australia, England, and Germany as well as China have visited or conducted research at our field site, and 50 theses and 12 dissertations have been completed. This project is now recognized among the eight sites in the world for the long-term study of free-living monkeys (Kappeler and Watts 2012). Recently, as an integral part of the International Research Center for Huangshan Biodiversity and Tibetan Macaque Behavioral Ecology, the project has been approved by the Anhui provincial government.

2.3 Social Life History Strategies

2.3.1 The Largest Macaca

It is difficult to study Tibetan macaques in the wild because of their elusiveness and aggression, so catching and weighing them is nearly impossible. An unexpected opportunity to obtain morphometric data presented itself in July 1988, however, when 17 individuals of our study group died within 10 days for a reason that is still unknown. We obtained morphometric data from 14 of the 17 corpses. The average weight for adult males, adult females, and neonates are 16.4 ($n = 4$), 11.0 ($n = 3$), and 0.60 kg ($n = 2$), respectively. They were all heavier than members of the same age-sex classes in other *Macaca* species (Table 2.1). Apparently, the Tibetan macaque is the heaviest species in the genus *Macaca*. This may be the reason why local people call them "bear monkeys." This result has been verified by morphometric analysis of 72 linear dental and cranial variables of 11 macaque species, showing that the two stump-tailed species (*M. thibetana* and *M. arctoides*) are the largest of the macaques (Pan et al. 1998).

Table 2.1 Body weights (kg) of some *Macaca* species

Species	Adult male	Adult female	Neonate
M. thibetana	16.4	11.0	0.60
M. fuscatta	11.7	9.1	0.50
M. sylvanus	11.2	10.0	–
M. nemestrina	10.4	7.8	0.47
M. nigra	10.4	6.6	0.49
M. maura	9.5	5.1	–
M. arctoides	9.2	8.0	0.49
M. silenus	6.8	5.0	–
M. radiata	6.6	3.7	0.40
M. sinica	6.5	3.4	–
M. mulatta	6.2	3.0	0.48
M. fascicularis	5.9	4.1	0.35

Note: the data for species other than *M. thibetana* are from Smuts et al. (1987)

2.3.2 Medium-Sized Group with Even Adult Sex Ratios

We monitored the size and composition of Yulingken group annually and found a fluctuation of 21–51 individuals with an average of 35.3 individuals ($SE = 1.6$) over the 30 years from 1987 to 2017. Whenever group size reached about 50 individuals, it fissioned into two subgroups and the smaller one of the two subgroups left the study area. During this period, we recorded four fission events happening in 1993, 1996, 2001, and 2003, respectively (Li et al. 1996; Li 1999; Berman and Li 2002), and noticed that males and females who were relatives tended to separate on these occasions (Li et al. 1996). Thus, Tibetan macaques at Mt. Huangshan appear to live in medium-sized groups, in comparison with the 70–80 and sometimes more than 100 individuals found in groups of other macaque species (Macintosh et al. 2012; Waters et al. 2015). Group size is an important component of social organization that determines how many individuals a group member can potentially interact with, which in turn contributes to social complexity (Kappeler and Watts 2012).

Like other species of *Macaca*, Tibetan macaques are organized into multi-male, multi-female groups with female philopatry and male dispersal (Li et al. 1996; Li and Wang 1996). However, male Tibetan macaques do not leave their natal groups until adulthood, later than most *Macaca* in which male transfers typically occur at puberty (Zhao 1993; Li et al. 1996). Thus, Tibetan macaque groups tend to include large proportions of natal adult males and relatively even sex ratios (0.91 ± 0.05), with an average of 8.52 ($SE = 0.67$) adult males to 9.35 ($SE = 0.51$) adult females (Fig. 2.3).

2.3.3 A Rich Repertoire of Affiliative and Ritualized Behaviors

An important feature of Tibetan macaques is their rich repertoire of affiliative and ritualized behaviors. Of the 33 social behaviors we have identified, at least

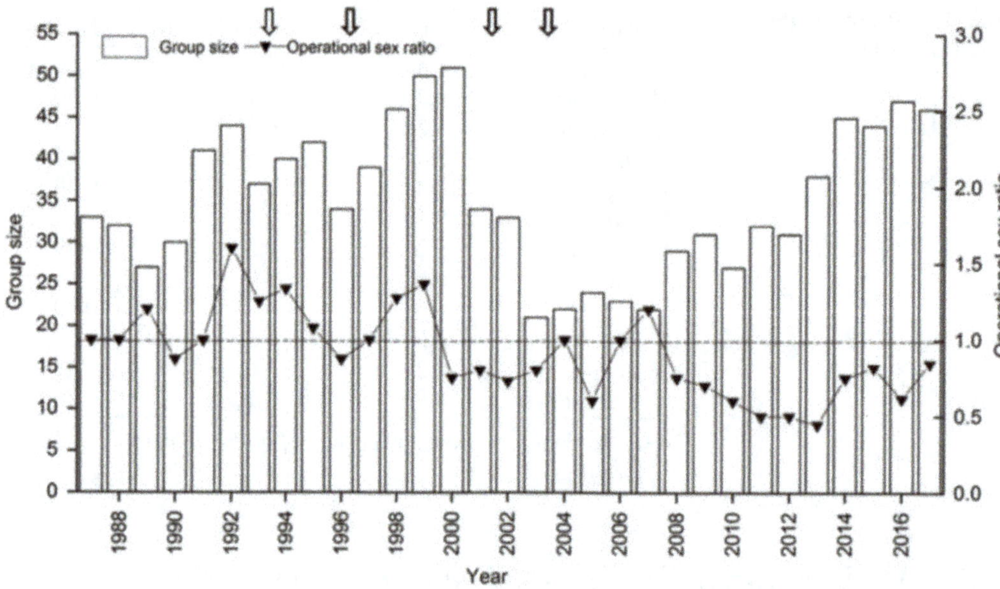

Fig. 2.3 Dynamic of group size and operational sex ratio of male to female in Yulingken A1 group during 1986 and 2017. Down arrow refers group fission events

17 (> 50%) include affiliation (Li 1999). Like other macaques, *allogrooming* (social grooming) is widespread among all group members and takes up about 20% of daily time expenditure (Wang et al. 2007). Females and juveniles are more active groomers than males in peaceful situations. Grooming is more frequent between sexual partners than between other male–female pairs. Specifically, higher-ranking males, including alpha males, groom females for a short time after copulation or during consortship. Grooming between adult males is also frequent (Xia et al. 2013). When two individuals meet, the lower-ranking individual usually *presents* to the high-ranging individual. If the two individuals are both males, they may *embrace* each other, sometimes with their hands stretching out to grasp the genitals of their partners. Additionally, the lower-ranking individual may approach a higher-ranking one, *showing its penis* to be sucked by the partner, or the higher-ranking individual *touches* a part of the body (head, back, or shoulder) of the lower-ranking individual or *mounts* the lower-ranking individual with *teeth chattering*. If the two individuals are of the opposite sex, the male may *grimace* at the female, and the female approaches the male for *genital inspection*, before copulation ensues. Also, during conflicts between two females, a third, higher-ranging female may often *approach* and *hold the bottom* of the attacking female to stop further aggression.

Tibetan macaques show an intense interest in infants. Group members, regardless of sex or age, often *hold infants*. Sometimes, an individual picks up an infant and carries it to another individual to perform the ritualized behavior of *bridging*. This behavior can occur between males, between females, and between a male and a female (Ogawa 1995a; Li 1999). The infants used in bridging are predominantly younger than 6 months, but occasionally 2–3-year-old juveniles may also be used

Fig. 2.4 Infant-holding behavior in Tibetan macaques is common. Left: three infants are attracted to two adult males, apparently waiting to be held or used in bridging. Right: a male Tibetan macaque living alone in Kowloon in Hong Kong holds an infant long-tailed macaque

(Zhang et al. 2018). Infants appear to be willing for bridging and mothers appear to be tolerant of infant handling by all group members (Fig. 2.4).

2.3.4 Despotic Dominance Style

The dominance style concept has proven useful for understanding covariation patterns in relationship qualities, particularly among macaques (Berman et al. 2004). As a member of the *sinica* lineage, Tibetan macaques are predicted to have a relaxed dominance style (Matsumura 1999; Thierry 2000). Previous studies did indicate relaxed dominance in this species. For example, males frequently engage in ritualized greetings in which they groom, mount, embrace, or touch each other's genitalia (Li 1999). In addition, both sexes engage in frequent bridging, and infants' mothers appear to be tolerant of infant handling by a wide range of group members (Ogawa 1995a).

However, a detailed study indicated that Tibetan macaques are more despotic than previously suspected. Bidirectional aggression, including counter-aggression (1.9%) and conciliatory tendency (6.4%), was consistently low in frequency across partner combinations, seasons, and locations (Berman et al. 2006). Females consistently displayed high levels of kin bias in affiliation and tolerance. Compared to other macaque species with better known dominance styles, data from Tibetan macaques generally fall within the range characteristic of despotic species and outside the range of relaxed species (Berman et al. 2004).

Despotic dominance of Tibetan macaques is consistent with our field observations. First, mating of low-ranking males was inhibited by high-ranking males. If a low-ranking male surreptitiously mated with a female in the forest but happened to be seen by a high-ranking male, the high-ranking male would immediately rush to punish him by scratching and biting. Second, when a series of conflicts happened,

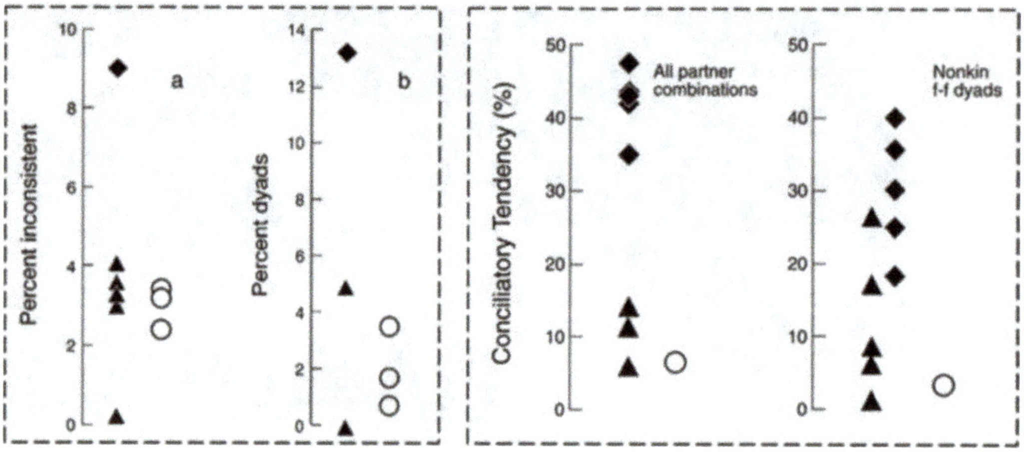

Fig. 2.5 Bidirectionality of aggression (left) and conciliatory tendencies of Tibetan macaques (right). Both a directional inconsistency index (a) and dyads-up index (percentage of dyads in which the primary direction of aggression is up the hierarchy of three data points representing three study periods for Tibetan macaques (circles)) fall within the range of despotic macaques (rhesus, long-tailed and Japanese macaques) (triangles) and are considerably lower than the one relatively relaxed species (stump-tailed macaques) (diamonds). Conciliatory tendencies of Tibetan macaque in (1) all partner combinations and degrees of relatedness combined, and (2) unrelated female–female partners (circles) are relatively low, even compared to values for despotic macaques (triangles). Data from Berman et al. (2004)

male Tibetan macaques often formed a "power coalition," in which several high-ranking males support one another to defeat low-ranking ones (Li 1999). Third, serious fresh wounds on Tibetan macaques were frequently seen immediately after intense aggressive interactions (Fig. 2.5).

2.3.5 Reproductive Pattern with Year-Round Mating but Seasonal Births

It took a long time for us to understand the reproductive pattern of Tibetan macaques. At first, based on intermittent field observations, they were thought to be nonseasonal breeders (Wada and Xiong 1996). After intensive year-round observations, we found that, although mating indeed takes place throughout the year, mating with high frequency and with ejaculation occurs only between July and December, with subsequent births mostly occurring between January and April (Li 1999; Li et al. 2005). Thus, Tibetan macaques are seasonal breeders (Fig. 2.6).

In order to explain this unusual reproductive pattern of Tibetan macaques, we investigated changes in the female sexual skin and nonreproductive matings in more detail. We found that adult females had a slight but detectable sexual skin in the perineal region, but the sexual skin lacked the regular changes characteristic of the physiological cycle of a female or the reproductive season. Females also had no typical behaviors to show their sexual motivation during estrus. This led to the

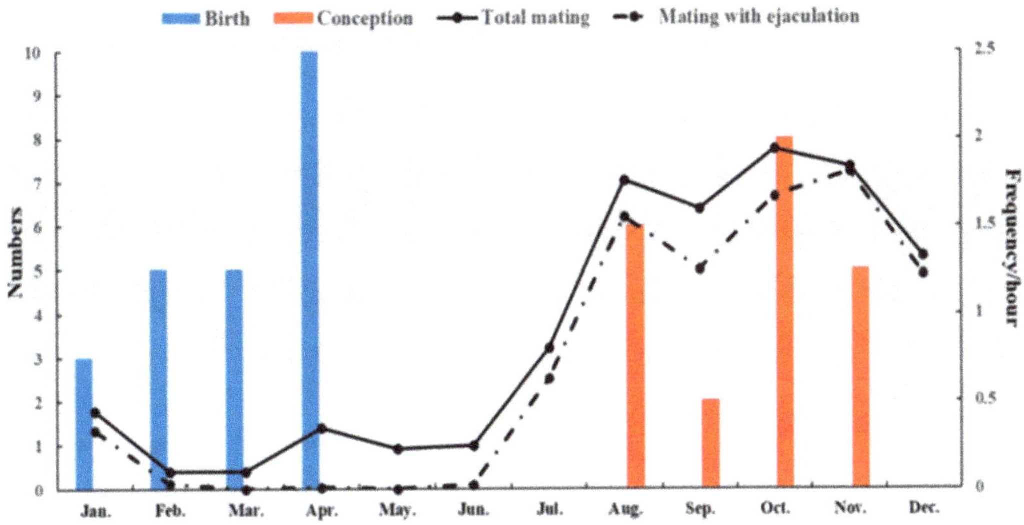

Fig. 2.6 Annual reproductive cycle of Tibetan macaques. Though Tibetan macaques mate year-round (solid line), mating with ejaculation (dotted line) is confined to a period from July to December. Therefore, the mating season is from July to December, and the birth season is from January to April

conclusion that Tibetan macaques have concealed ovulation (Li et al. 2005). Further study revealed that males could discriminate between potentially receptive females, but that they could not recognize the reproductive status of a given female (Li 1999; Zhang et al. 2010). Concealed ovulation may result in a higher mating frequency during a female's receptive period.

Compared with sexual behavior in the mating season, nonreproductive matings occurred at a lower frequency, with less frequent ejaculation, less harassment, shorter mount duration, and an absence of pauses with vocalization. It often took place in a situation in which non-lactating females were involved in social conflict or approached males for mating. Neither pregnant nor lactating females were observed to mate in the birth seasons. Copulation during the birth season did not affect a sexually receptive female's delivery the next year, nor was it associated with increased proximity, grooming, or agonistic aid within the mating pair. However, mating pairs spent more time co-feeding, presumably reflecting increased male tolerance (Li et al. 2007). Adolescent males, who rarely copulated in the mating season, engaged in mating activities during birth seasons as well. Therefore, even though birth season copulations have no reproductive function, they might fulfill social functions for females, such as post-aggression appeasement by males or improving access to resources. They may also offer good opportunities for adolescent males and females to develop their sexual skills (Li et al. 2007) (Fig. 2.7).

It is worth noting that an important aspect of the mating pattern of Tibetan macaques at Mt. Huangshan is a single mount ejaculation (SME), with average duration and thrust numbers of 23.2 s and 20 times, respectively (Xiong and Wang 1991). High-ranking males copulate more often with more females than do low-ranking males, and alpha males monopolize over 2/3 of all matings (Xiong

Fig. 2.7 Inconspicuous sexual skin in adult females (left) and large testes in adult males (right) in Tibetan macaque

and Wang 1991; Li et al. 2005). However, it has been reported that Tibetan macaques at Mt. Emei exhibit serial or multiple mount ejaculation with a mean duration and thrust number of mounts with 30.3 s and 43.3 times (MME), respectively (Zhao 1993). Since no data on the number of mounts in a MME have been provided, we do not know whether the difference is due to a difference in populations per se or observation methods.

Finally, females at the age of 7–12 years have the highest birth rate. Not only are birth rates of high-ranking females higher, but those of high-ranking females are also earlier than those of low-ranking ones in the birth season (Li et al. 1994).

2.4 Potential Contributions to Understanding Behavioral Mechanisms in Primate Societies

2.4.1 Bridge and Bond Role of Infant

Tibetan macaques exhibit some interesting and unique behavior patterns. Bridging behavior is one of them. Bridging involves two individuals and an infant and is a triadic interaction. Bridging behavior is common in Tibetan macaque society, and it can be performed between males, between females, between a male and a female, and even between an adult and a juvenile. Infants of both sexes are involved in the "bridge." Despite this variation in bridging partners, bridging lasts only a few seconds. Once bridging has ended, the two bridging individuals often groom each other or sit together, ignoring the infant. Obviously, the "bridge" between monkeys is not for infant care, but for social contact or communication. In other words, infants are a "social tool" used to facilitate older group members' associations and interactions (Ogawa 1995b). Zhao (1996) even compared infants in bridging behavior to a cigarette-like social facilitator in human social interactions.

Primatologists have always paid particular attention to the role of infants in nonhuman primate societies. However, most studies have focused on infant care or handling (Ogawa 1995a). For example, in capped langurs, *Presbytis pileata*, newborn infants of less than one month old spend nearly the same amount of time with alloparental females as with their own mothers (Stanford 1992). Likewise, young, female vervet monkeys, *Chlorocebus aethiops sabaeus*, without infants hold or carry newborn infants, which has been explained as alloparental investment (Fairbanks 1990). Bridging behavior has also been observed in *M. sylvanus* (Deag and Crook 1971). Clearly, infants in Tibetan and Barbary macaques play an active part in enhancing social relationships among group members. This is a relatively new and important research direction that is worthy of further exploration.

2.4.2 Male-Female Friendships in a Promiscuous Mating System

During the mating season from July to December, high-ranking Tibetan macaque males often follow sexually receptive females (usually those without unweaned infants). They move, feed, and rest together, often accompanied by mating; that is, they form a temporary consortship. The duration of such a consortship varies from a few days up to over 2 weeks (Li 1999). During a consortship, the male always keeps close with the female, and the male's activities are restricted by the female's movements. Sometimes, when a consorting male realizes that his female has disappeared, he shows signs of anxiety and searches for the female everywhere.

The role of consortship is unclear at present. On the one hand, the duration of consortship in Tibetan macaque is rather long, lasting as much as 18 continuous days (Li 1999). This does not only exceed the time needed to impregnate a female, but is also longer than the duration in other *Macaca* species (Li 1999). On the other hand, although a male follows one female at a time, usually he cannot monopolize all of her matings even though he is the alpha male. So why do high-ranking males use consortship as a mating tactic?

We already mentioned that female Tibetan macaques tend to conceal their ovulation by neither showing cyclical fluctuation of sexual skin nor estrus-related behaviors (Li et al. 2005). Although high-ranking males form consortship with females, they are able to opportunistically mate with other sexually receptive females, too. Thus, the co-occurrence of consort behavior and opportunistic mating may indicate a male mating tactic that is effective when mating is promiscuous and males cannot accurately detect ovulation in females from morphological and behavioral signs. That is, consortship may be a tactic related to the assurance of paternity. From an evolutionary perspective, consortship in Tibetan macaques appears to be a special type of sexual relationship, which has also been reported for some baboons, where they have been labeled as "friendships" (Smuts 1985).

2.4.3 Competitive and Cooperative Relationships Among Males

Male Tibetan macaques exhibit stable, linear dominance hierarchies typical of despotic species. On the one hand, aggression is frequent and can be accompanied by serious wounds in all sex/age classes (Berman et al. 2004; Li 1999). Additionally, the large body, large testes, and short tenures of alpha males (on average about 10 months) indicate intense competition among males (Li and Wang 1996). On the other hand, Tibetan macaque males frequently engage in friendly behaviors, such as grooming and sitting together, and engage in ritualized greetings involving mounting, embracing, and sucking penis. Many of these behaviors are rarely seen in other *Macaca* males. Tibetan macaque males also quickly reconcile after aggression. The conciliatory tendency is 19.7% in male–male dyads, compared to 4.2% in female–female dyads, and most male–male reconciliations occur within 1 min after aggression (Berman et al. 2004). It is evident that male Tibetan macaques show both competitive and cooperative relationships. This appears to agree with observations taken from other primate societies with relatively male-biased adult sex ratios.

2.4.4 Behavioral Mechanism Promoting Genetic Diversity in a Small Group

Like most nonhuman primates in China, Tibetan macaques live in small, isolated populations. Theoretically, a small population or group may suffer the detrimental effects of inbreeding and the loss of genetic diversity (Spielman et al. 2004). Our study group is a medium-sized one with 20–50 individuals and that ranges in a small habitat (Li et al. 1996). However, we have not found conspicuous changes in behavior during the 30-year period, except the sudden death of 17 individuals in the spring of 1988. Although we have noticed a decline in the frequency of affiliative interactions among group members and an increase in infant mortality in recent years, these appear to be a side effect of provisioning, which increases the rate of aggression among group members (Berman and Li 2002).

Additionally, we found that our study group has a relatively high haplotype diversity (Hd = 0.341) when compared to other subspecies (*M. t. pullus*, Hd = 0.222; *M. t. guizhouensis*, Hd = 0.478) (Sun et al. 2010). As such, loss of genetic diversity may not be a major concern in our study group at present. It is likely that some behavioral mechanisms in this group may play an important role in preventing or slowing down the loss of genetic diversity that is often observed in small groups. For instance, we found that the transfer of males into and out of the group occurs annually (Li et al. 1996). Also, kin recognition may prevent mating between matrilineal relatives. In fact, one of our studies demonstrates that no copulation involved mother-son dyads, and only 2.1% (7/329) of copulations were between maternal siblings. Females may be more averse to copulating with relatives than are males (Zhu et al. 2008).

2.5 Conclusions

Long-term studies of primate populations are essential for documenting important events or phases of individual life histories that are critical for our understanding of subtle and complex relationships of group members that determine the nature and dynamic of social organization and, as a result, variation in individual reproductive success. The medium-sized group of Tibetan macaques we study facilitates censuses and unambiguous identification of all group members in the field. As the largest macaque species with a relatively male-biased adult sex ratio and year-round mating but seasonal birth, this species offers a large spectrum of sociodemographic conditions conducive for testing hypotheses related to dominance style and group fission. Furthermore, a rich repertoire of affiliative and ritualized behaviors and a complex interplay of competitive and cooperative relationships, especially among males, can provide primatologists with a wealth of opportunities to pursue additional behavioral, ecological, and evolutionary research projects focused on Tibetan macaques.

Acknowledgments Our long-term field study has been supported by Anhui University and the Huangshan Garden Bureau. Our research would be impossible without their assistance in terms of permission, personnel, funding, and other logistic support. For more than three decades, many researchers and students, especially those from Anhui University, Kyoto University, State University of New York at Buffalo, University of Washington, and Central Washington University, have joined and contributed to the study. We cordially thank Qi-Shan Wang, Kazuo Wada, Cheng-Pei Xiong, Guo-Qiang Quan, He-Ling Shen, Ren-Mei Ren, Ming Li, Jiao Shao, Zu-Wang Wang, Zhi-Gang Jiang, Kunio Watanabe, Hideshi Ogawa, Frans de Wall, Noel Rowe, Carol Berman, Consuel Ionica, Lei Zhang, Toshisada Nishida, Randall Kyes, Lisa Jones-Engel, Lixing Sun, Lori Sheeran, Megan Matheson, Steven Wagner, Dong-Po Xia, Bing-Hua Sun, Yong Zhu, Xi Wang, Yang Liu, Jian-Yuan Zhao, Jessica Mayhew, and Paul Garber for their full-hearted supports and noted contributions. We also express our thanks to Fu-Wen Wei, Bao-Guo Li, Yan-Jie Su, Cheng-Ming Huang, Xue-Long Jiang, Ji-Qi Lu, and other members of China Primatology Society. This work was financially supported by the National Natural Science Foundation of China, the China Scholarship Council, the Key Teacher Program of the Ministry of Education of China, the Outstanding Youth Foundation of Anhui Province, the International Science and Technology Cooperation Plan of Anhui Province, Natural Science Foundation of Anhui Province, Inoue Scientific fund, Primate Conservation Inc., the L.S.B. Leakey Foundation, the Wenner-Gren Foundation, the ASP Conservation Small Grant, and the National Natural Science Foundation. Special thanks to Mr. Cheng's family and Fuxi villagers for their outstanding logistic support during our field observation for more than three decades.

Appendix I

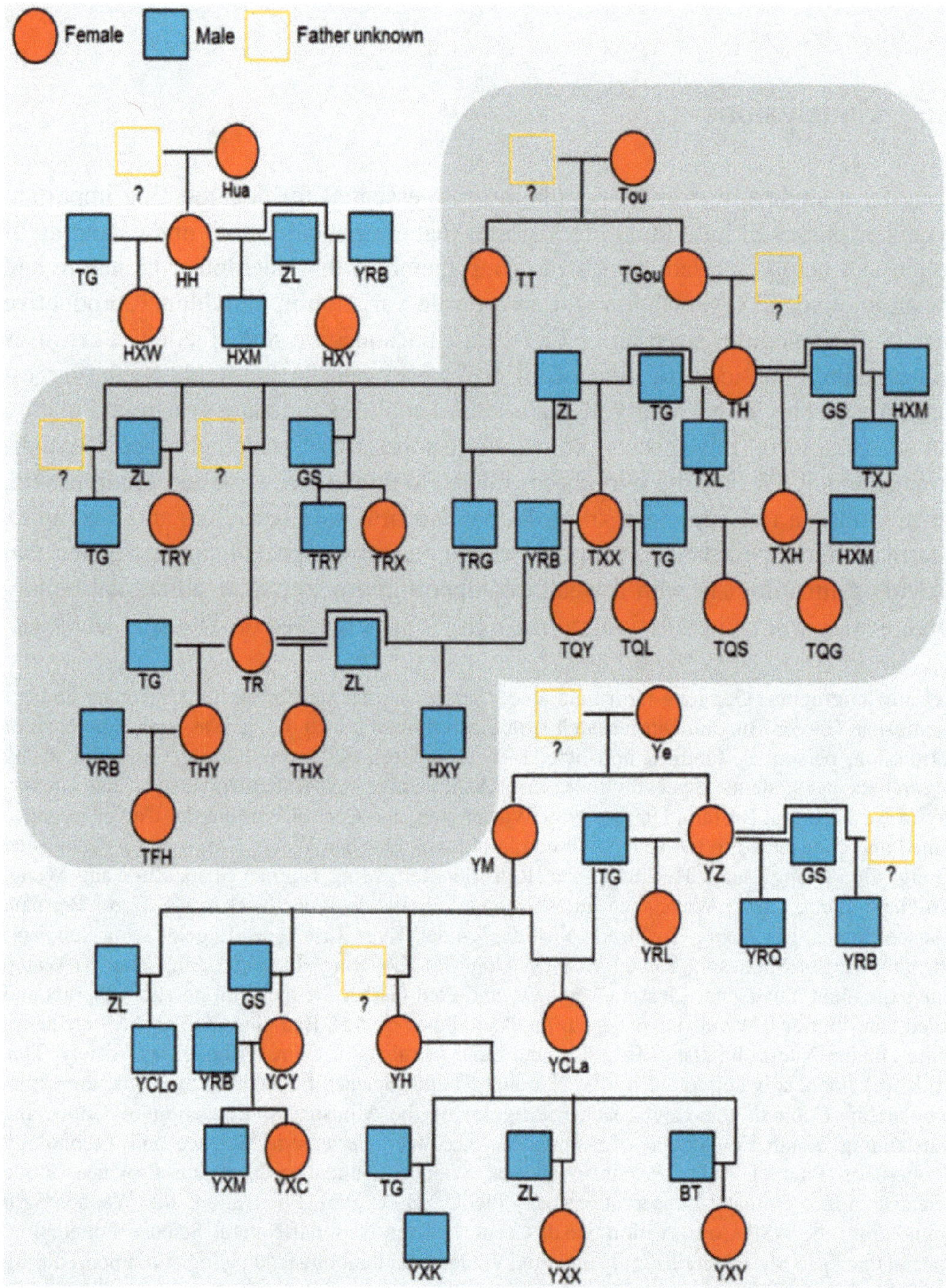

Pedigree of the YA1 social group of Tibetan macaques in Huangshan (September, 2017). Note: The pedigree was mapped based on maternal relationships of three matrilines. Males with known identities were also included. Letters under the blue squares (males) or red circles (females) represent individual names. Question marks indicate the identity of males are yet to be unknown

Appendix II

Ethogram of social behaviors in Tibetan macaque

No.	Repertoire	Definition	Picture
1	Stare	An individual looks directly at another individual with its eyes wide open and with its shoulders raised for about 3–5 s. The staring individual appears as if it is preparing to lunge or chase the recipient of the stare.	Picture 1
2	Ground slap	An individual places one hand on the ground and slaps the ground, a rock, or some grassy vegetation with the other while staring at the recipient. The ground slap may be repeated several times.	Picture 2
3	Chase	An individual stares at the recipient and rushes at him/her at great speed. The recipient typically flees.	
4	Seize	The performer grabs the body, face, neck, or ear of the recipient with one hand, shakes it, and then releases it. Sometimes the performer also comes very close (face to face) and stares at the recipient.	Picture 3
5	Bite	The performer grabs the receiver tightly, preventing him/her from fleeing, and bites the recipient vigorously.	Picture 4
6	Avoid	The performer turns its body away from the attacker as if preparing to flee while displaying a "horrified" facial expression toward the attacker.	Picture 5
7	Flee	The target of an attack will run in the opposite direction from the attacker.	
8	Scream	A vocal reaction to an attacker. The scream is high pitched and of long duration. This may be accompanied by fleeing.	Picture 6
9	Redirection	When A is attacked by B, A responds immediately by attacking C. C is an individual that is nearby and lower-ranking than both A and B. In some cases, B will join the attack against C.	Picture 7
10	Solicit support	When an individual is attacked by another individual, it intermittently scans the attacker and a higher-ranking individual nearby with screams. The higher-ranking bystander may respond by interrupting the attack on the scanner.	Picture 8
11	Hold bottom	The performer (usually the alpha female) approaches one of two other females that are engaged in a fight. She approaches from behind and holds the bottom of the female. Both females typically teeth-chatter to one another. This tends to interrupt the fight.	Picture 9
12	Approach	An individual moves directly toward another individual, coming within 1 m.	
13	Leave	An individual moves from within 1 m to more than 1 m of another individual.	Picture 10
14	Proximity	Two or more individuals are sitting or lying within 1 m or between 1 and 5 m (depending on studies).	Picture 11

(continued)

No.	Repertoire	Definition	Picture
15	Following	When one individual begins to travel from one location to another, another individual nearby immediately goes in the same direction.	Picture 12
16	Teeth chatter	The performer opens his/her mouth, places the tongue close to the teeth, and moves his/her jaw rapidly, thereby making clicking noises with his/her teeth. The eyelids are lowered, the chin is raised, and the tongue is moved rapidly back and forth across the teeth. This behavior accompanies same-sex mounting, embracing, and infant carrying. It appears to function as a friendly invitation to interact.	Picture 13
17	Play	Brief sequence of rapid, repetitive, and exaggerated movements, without clear objectives. Interacting play movements involving two or more individuals include chase, mock fight, and combinations of previous behavior.	Picture 14
18	Social grooming/ allogrooming	One individual uses his/her fingers and palms to groom the fur of another individual. The groomer may pick out small objects from the recipients fur and eat them.	Picture 15
19	Embrace	Two individuals, usually adult males, hold each other while face to face. Each partner will reach with one hand and attempt to touch the genitals of the other. Both partners typically teeth-chatter and vocalize excitedly.	Picture 16
20	Touch	A ritualistic behavior between males in which the lower-ranking individual approaches the higher-ranking individual in front. The higher-ranking male touches some part of the body of the lower-ranking individual (head, back, shoulder). Then the lower-ranking individual leaves. This may be used in a tense situation and appears to be a means by which the lower-ranking individual obtains "permission" to proceed on its pathway or with its apparent goal.	
21	Present	One individual approaches the front of the other and shows his bottom to the other. Usually the presenter is the lower-ranking of the two partners. Females also present to males.	Picture 17
22	Mount	One male (usually the lower-ranking) grabs the back hair of another male and mounts from behind, using the full ankle clasp posture. Both males teeth-chatter and scream excitedly. Then the mounter dismounts. The duration of the mount is about 3–5 s. This may be a simple friendly gesture or a post-conflict behavior.	Picture 18
23	Show penis	This is a ritualized behavior shown by juvenile males toward adult males. Usually, the lower-ranking male approaches the higher-ranking male, raises one leg, and displays his penis. The higher-ranking male puts his head on the belly of the lower-ranking male and licks or touches the penis with his hand. The juvenile may show his penis from a reclining position.	Picture 19
24	Suck penis	A young male approaches an adult male and jumps on his head. The adult male holds the younger male by the waist in such a way that his mouth can reach the young male's penis. The adult sucks the young male's penis and then the younger male leaves.	Picture 20

(continued)

No.	Repertoire	Definition	Picture
25	Hold infant	An adult male holds an infant and may carry it ventrally. Usually the infant is male. This apparently serves to invite other adult males to engage in social interaction with the adult male.	Picture 21
26	Bridge	Bridging involves three individuals, an infant or young juvenile and (a) two adult males, (b) one adult male and one subadult male, or (c) two adult females. The two older individuals hold the infant on its back, lower their heads, and lick the belly and/or genitals of the infant. They will often teeth-chatter and vocalize excitedly. The infant is usually male and his penis becomes erect immediately. Other infants, particularly females, will approach a bridging triad with excitement as if they wish to participate.	Picture 22
27	Genital inspection	One individual will touch the genitals—the vagina of a female or the anus of a male—and sniff it or lick it directly. When a male inspects a female in estrus, it is called a sexual inspection.	Picture 23
28	Sexual inspection	A genital inspection by an adult male of an estrus female.	Picture 24
29	Sexual chase	An adult male chases a female and attempts to mate with her. The sexual chase differs from the aggressive chase in that it is slower and is interrupted by pauses. In addition, although the female attempts to avoid the male, she is less likely to scream.	Picture 25
30	Sexual grimace	An individual directs a grimace toward an adult of the opposite sex that may be either nearby or far away. The male sways his body in a manner that makes him more noticeable to the female. The grimace differs from a fear grimace in that the corners of the mouth are not drawn back as far. This is a sexual solicitation.	Picture 26
31	Mating	Three types of mating are described: (1) intromission and thrusting only; (2) intromission, thrusting, and pausing; and (3) intromission, thrusting, pausing, and calling.	Picture 27
32	Harassment	Individuals approach a mating pair excitedly, touch the heads, lick, or stick their faces in the faces of the mating partners. They may also grimace or scream. This does not generally interrupt the copulation of the mating partners.	Picture 28
33	Consortship	An adult male and female form a temporary close affiliative and sexual relationship during the mating season. They will move, feed, mate, and groom together, but do not necessarily mate exclusively with one another. Their frequencies of following one another are much higher than during non-mating periods.	Picture 29

Picture 1 An individual stares another animal

Picture 2 The right adult male slaps toward his front individual

Picture 3 An adult male (middle) seizes another adult male (right)

Picture 4 An adult male bites another adult male (two monkeys in right)

Picture 5 An adult male (right) avoids adult male's (middle) aggression

Picture 6 An adult female (right) screams when she is attacked by adult male (left)

Picture 7 The middle adult male redirects aggression from right adult male to the left adult male

Picture 8 The middle adult male is getting the left male's support when he is in conflict with the right adult male

Picture 9 An adult male holds the button of a juvenile male (left)

Picture 10 The right adult male is leaving left adult male

Picture 11 An adult male (left) is in proximity with an adult female within 1 m

Picture 12 An adult female (left) is following an adult male

Picture 13 Teeth chatting

Picture 14 Three juveniles play together

Picture 15 Social grooming between two females

Picture 16 Embrace

Picture 17 An adult male presents to another male

Picture 18 Male–male mount

Picture 19 Show the penis

Picture 20 A juvenile (left) is sucking an infant's penis

Picture 21 An adult male holds an infant

Picture 22 Male–male bridging

Picture 23 Genital inspection

Picture 24 Sexual inspection

Picture 25 An adult male (left) chases an adult female (right) for mating

Picture 26 An adult male (right) grimace to an adult female (left)

Picture 27 An adult male (left) is mating with an adult female (right)

Picture 28 An adult male (left) is harassing a mating pair

Picture 29 An adult male (right) forms consortship with an adult female (left)

References

Altmann J (1974) Observational study of behavior: sampling methods. Behaviour 49:227–266

Berman C, Li JH (2002) Impact of translocation, provisioning and range restriction on a group of *Macaca thibetana*. Int J Primatol 23:283–397

Berman C, Ionica C, Li JH (2004) Dominance style among *Macaca thibetana* on Mt. Huangshan, China. Int J Primatol 25:1283–1312

Berman C, Ionica C, Dorner M, Li JH (2006) Postconflict affiliation between former opponents in *Macaca thibetana* on Mt. Huangshan, China. Int J Primatol 27:827–854

Deag J, Crook J (1971) Social behaviour and "agonistic buffering" in the wild Barbary macaque *Macaca sylvana* L. Folia Primatol 15:183–200

Delson E (1980) Fossil macaques, phyletic relationships and a scenario of development. In: Lindburg D (ed) The macaques: studies in ecology, behavior, and evolution. Van Nostrand Reinhold, New York, pp 31–51

Fairbanks L (1990) Reciprocal benefits of allomothering for female vervet monkeys. Anim Behav 9:425–441

Fooden J (1980) Classification and distribution of living macaques. In: Lindburg D (ed) The macaques: studies in ecology, behavior, and evolution. Van Nostrand Reinhold, New York, pp 1–9

Fooden J (1983) Taxonomy and evolution of the *sinica* group of macaques: 4. species account of *Macaca thibetana*. Fieldiana Zool 17:1–20

Hoelzer G, Hoelzer M, Melnick D (1992) The evolutionary history of the *sinica* group of macaque monkeys as revealed by mtDNA restriction site analysis. Mol Phylogenet Evol 1:215–222

Jiang XL, Wang YX, Wang QS (1996) Taxonomy and distribution of Tibetan macaque (*Macaca thibetana*). Zool Res 17:361–369

Kappeler P, Watts D (2012) Long-term field studies of primates. Springer, New York

Li JH (1999) The Tibetan macaque society: a field study. Anhui University Press, Hefei. (In Chinese)

Li JH, Wang QS (1994) Sleeping site of Tibetan macaques in summer. Chin J Zool 29(5): 58

Li JH, Wang QS (1996) Dominance hierarchy and its chronic change in adult male Tibetan macaques (*Macaca thibetana*). Acta Zool Sin 42:330–334

Li JH, Wang QS, Li M (1994) Studies on population ecology of Tibetan monkeys (*Macaca thibetana*). II. Reproductive pattern of Tibetan monkeys. Acta Therlol Sin 14(4):255–259

Li JH, Wang QS, Han DM (1996) Fission in a free-ranging Tibetan macaque group at Huangshan Mountain, China. Chin Sci Bull 41:1377–1381

Li JH, Yin HB, Wang QS (2005) Seasonality of reproduction and sexual activity in female Tibetan macaques (*Macaca thibetana*) at Huangshan, China. Acta Zool Sin 51:365–375

Li JH, Yin HB, Zhou LZ (2007) Non-reproductive copulation behavior among Tibetan macaques (*Macaca thibetana*) at Huangshan, China. Primates 48:64–72

Li DM, Fan LG, Ran JH, Yin HL, Wang HX, Wu SB, Yue BS (2008) Genetic diversity analysis of *Macaca thibetana* based on mitochondrial DNA control region sequences. DNA Seq 19 (5):446–452

Liu Y, Li JH, Zhao JY (2006) Sequence variation of mitochondrial DNA control region and population genetic diversity of Tibetan macaques *Macaca thibetana* in the Huangshan Mountain. Acta Zool Sin 52(4):724–730

Macintosh A, Huffman M, Nishiwaki K, Miyabe-Nishiwaki T (2012) Urological screening of a wild group of Japanese macaques (*Macaca fuscata yakui*): investigating trends in nutrition and health. Int J Primatol 33(2):460–478

Matheson M, Sheeran L, Li JH, Wagner R (2006) Tourist impact on Tibetan macaques. Anthrozoös 19:158–168

Matsumura S (1999) The evolution of "egalitarian" and "despotic" social systems among macaques. Primates 40:23–31

Ogawa H (1995a) Bridging behavior and other affiliative interactions among male Tibetan macaques (*Macaca thibetana*). Int J Primatol 16:707–729

Ogawa H (1995b) Recognition of social relationships in bridging behavior among Tibetan macaques (*Macaca thibetana*). Am J Primatol 35:305–310

Ogawa H (2019) Bridging behavior and male-infant interactions in *Macaca thibetana* and *M. assamensis*: insight into the evolution of social behavior in the sinica species-group of macaques. In: Li J-H, Sun L, Kappeler P (eds) The behavioral ecology of the Tibetan Macaque. Springer, Heidelberg

Pan RL, Jablonski N, Oxnard C, Freedman L (1998) Morphometric analysis of *Macaca arctoides* and *M. thibetana* in relation to Other Macaque species. Primates 39(4):519–537

Simons N, Lorenz J, Sheeran L, Li JH, Xia DP, Wagner R (2012) Noninvasive saliva collection for DNA analysis from free-ranging Tibetan macaques (*Macaca thibetana*). Am J Primatol 74:1064–1070

Smuts BB (1985) Sex and friendship in baboons. Aldine, Hawthorne, NY

Smuts B, Cheney D, Seyfarth R, Wrangham R, Struhsaker T (1987) Primate society. University of Chicago Press, Chicago, IL

Spielman D, Brook B, Frankham R (2004) Most species are not driven to extinction before genetic factors impact them. Proc Natl Acad Sci U S A 101:15261–15264

Stanford C (1992) The costs and benefits of allomothering in wild capped langurs (*Presbytis pileata*). Behav Ecol Sociobiol 30:29–34

Sun BH, Li JH, Zhu Y, Xia DP (2010) Mitochondrial DNA variation in Tibetan macaque (*Macaca thibetana*). Folia Zool 59:301–307

Sun L, Xia D-P, Li J-H (2019) Size matters in primate societies: how social mobility relates to social stability in Tibetan and Japanese macaques. In: Li J-H, Sun L, Kappeler P (eds) The behavioral ecology of the Tibetan macaque. Springer, Heidelberg

Thierry B (2000) Covariation of conflict management patterns in macaque societies. In: Aureli F, de Waal FBM (eds) Natural conflict resolution. University of California Press, Berkeley, CA, pp 106–128

Wada K, Xiong CP (1996) Population changes of Tibetan monkeys with special regard to birth interval. In: Shotake T, Wada K (eds) Variations in the Asian macaques. Tokai University Press, Tokyo, pp 133–145

Wada K, Xiong CP, Wang QS (1987) On the distribution of Tibetan and rhesus monkeys in southern Anhui Province, China. Acta Theiol Sin 7:148–176

Wang GL, Yin HB, Yu G, Wu M (2007) Time budget of adult Tibetan macaques in a day in spring. Chin J Wildlife 29:6–10

Waters S, Harrad A, Chetuan M, Amhaouch Z (2015) Barbary macaque group size and composition in Bouhachem forest, north Morocco. Afr Primates 10:53–56

Xia DP, Li JH, Garber P, Matheson M, Sun BH, Zhu Y (2013) Grooming reciprocity in male Tibetan macaques. Am J Primatol 75:1009–1020

Xiong CP (1984) Studies on the ecology of Tibetan monkeys. Acta Theriol Sin 4:1–9

Xiong CP, Wang QS (1988) Seasonal habitat used by Tibetan monkeys. Acta Theriol Sin 8:176–183

Xiong CP, Wang QS (1991) A comparative study on the male sexual behavior in Thibetan and Japanese monkeys. Acta Theriol Sin 11:13–22

Zhang M, Li JH, Zhu Y, Wang X, Wang S (2010) Male mate choice in Tibetan macaques *Macaca thibetana* at Mt. Huangshan, China. Curr Zool 56:213–221

Zhang D, Xia DP, Wang X, Zhang QX, Sun BH, Li JH (2018) Bridging may help young female Tibetan macaques *Macaca thibetana* learn to be a mother. Sci Rep 8:16102

Zhao QK (1993) Sexual behavior of Tibetan macaques at Mt. Emei, China. Primates 34(4):431–444

Zhao QK (1996) Male-infant-male interactions in Tibetan macaques. Primates 37(2):135–143

Zhao JY, Li JH, Liu Y, Yin HB (2005) Research on DNA extraction from old feces of *Macaca thibetana*. Acta Theriol Sin 25:410–413

Zhu Y, Li JH, Xia DP, Chen R, Sun BH (2008) Inbreeding avoidance by female Tibetan macaques *Macaca thibetana* at Huangshan, China. Acta Zool Sin 54:183–190

Chapter 3
Size Matters in Primate Societies: How Social Mobility Relates to Social Stability in Tibetan and Japanese Macaques

Lixing Sun, Dong-Po Xia, and Jin-Hua Li

3.1 Introduction

Social mobility refers to vertical movement of positional status of members in a society. In human societies, it refers to a wide range of upward or downward change in metrics such as income, social stature, education, and others that show some level of stratification (Lipset and Bendix 1992). As a focal issue in equality, social justice, economic development, and social stability in human societies (e.g., Wilkinson and Pickett 2009; Breen 2010; Cox 2012; Matthys 2012; Corak 2013; Clark 2014; Piketty 2014), social mobility has drawn a broad and sustained interest from social thinkers, social activists, and concerned citizens and is among the most important current research topics in social science disciplines including economics, sociology, and political science. Recently, interest in social mobility has also been on the rise in biological, psychological, and medical sciences. This is largely due to the discoveries that, along with other effects, social status can affect hormones (especially those

L. Sun (✉)
Department of Biological Sciences, Primate Behavior and Ecology Program, Central Washington University, Ellensburg, WA, USA
e-mail: Lixing@cwu.edu

D.-P. Xia
School of Life Sciences, Anhui University, Hefei, China

International Collaborative Research Center for Huangshan Biodiversity and Tibetan Macaque Behavioral Ecology, Anhui, China

J.-H. Li
School of Resources and Environmental Engineering, Anhui University, Hefei, Anhui, China

International Collaborative Research Center for Huangshan Biodiversity and Tibetan Macaque Behavioral Ecology, Anhui, China

School of Life Sciences, Hefei Normal University, Hefei, Anhui, China
e-mail: jhli@ahu.edu.cn

J.-H. Li et al. (eds.), *The Behavioral Ecology of the Tibetan Macaque*, Fascinating Life Sciences, https://doi.org/10.1007/978-3-030-27920-2_3

related to stress such as glucocorticoids), immunity, and health in humans and nonhuman primates (e.g., Marmot et al. 1991; Sapolsky 2005; Seabrook and Avison 2012; Snyder-Mackler et al. 2016). As such, social mobility is among the key factors determining the well-being of primates and other social animals.

The level of social mobility is critically important for a society, especially from the long-term perspective. Both low and high social mobility are believed to weaken and disrupt social stability. A persistently low mobility will necessarily deprive low-ranking members of opportunities for social advancement, which in turn may increase the probability of revolts and revolutions. An extremely high mobility, on the other hand, will make a society constantly in flux. Therefore, a stable society is a dynamic one with the level of mobility bounded within a certain range, beyond which society may become chaotic or even collapse. Such a system dynamics view thereorized for human societies has gained some popularity recently (Acemoglu et al. 2018). New conception aside, empirical tests are hard to carry out for the obvious reason that human societies are usually too complex in structure and too diverse in culture to pursue such studies. In this respect, testing mobility-related hypotheses is more feasible in nonhuman primates because their societies are usually smaller in size and simpler in structure than human societies, including most hunter-gatherer societies (Price and Brown 1985; Hamilton et al. 2007). Furthermore, understanding how social mobility influences primate societies is interesting and important in its own right because it can give us a new tool to investigate the dynamics and evolution of primate societies.

(Note that social primates can live in groups of varying complexity in social structure, ranging from simple linear hierarchy in some macaque species (*Macaca* spp.), multilevel societies such as those in geladas (*Theropithecus gelada*) and snub-nosed monkeys (*Rhinopithecus* spp.), to highly organized human communities governed by formal or informal norms and laws. Because social mobility is a feature of an operationally definable social unit or organization regardless of the level of its organizational complexity, we here use the word "society" synonymously with "social group" for both human and nonhuman primates for the purpose of finding general and comparable patterns in social mobility.)

There are two types of social mobilities: *intragenerational* and *intergenerational*. Intragenerational mobility refers to upward and downward movement of social stature for members in a society within their lifetimes. For instance, in a Tibetan macaque society, each member has a 3-to-1 ratio of upward versus downward mobility in a year (Sun et al. 2017). Because status change tends to accumulate over time, intragenerational mobility is often measured per time unit, such as a month, a year, or a decade, to be comparable across studies. For this, it is more accurately known as the rate of social mobility (Clark 2014). Intergenerational mobility, on the other hand, refers to the status change between two consecutive generations. In American society, for instance, income mobility declined from 92% for people born in 1940 to 50% for people born in 1985 (Chetty et al. 2017). In this chapter, we will focus on intragenerational mobility.

Social mobility can be *absolute* or *relative*. In social sciences, absolute mobility refers to the total number of individuals, whereas relative mobility refers to the probability of an individual (or percentage of all individuals) moving from one social stratum to another. A society can be high in absolute mobility but low in relative mobility or vice versa. The key behind this paradox lies in the size of a society. For example, in terms of absolute mobility, two members moving up one step in rank in a group of 30 are twice as mobile as one member for the same rank change in a group of 15 during the same period of time, but the two groups are equally mobile in relative mobility, both 6.67% for this period of time. As this hypothetical example illustrates, absolute mobility does not consider the size of a society. Thus, relative mobility can better reflect the overall condition for a society and is most commonly used in characterizing upward or downward mobility (Heckman and Mosso 2014; Simandan 2018). Accordingly, social mobility, unless specified, refers to relative mobility hereafter.

The difference between absolute and relative mobility indicates that society size can play an important role in social mobility. Unfortunately, few attempts have been made to address this issue due to the reason that most mobility research has been conducted on large human societies in industrial nations. As such, society size is usually not of great interest because it has no perceivable effect on social mobility. In nonhuman primate societies, however, group size can be magnitudes smaller, typically a few dozen (up to a few hundred in some species such as baboons and snub-nosed monkeys). A group of a few dozen, such as in Tibetan macaques, *M. thibetana* (Li 1999), can be vastly different in social organization from a group of a few hundred, such as in Sichuan snub-nosed monkeys, *R. roxellana* (Zhang et al. 2006) or Japanese macaques, *M. fuscata* (Mori et al. 1989). Therefore, to understand how group size affects mobility, we compared two macaque species, the Tibetan macaque and the Japanese macaque, in this study.

Before we present our results, we briefly review some key issues related to the study of social mobility.

3.2 Social Mobility and Opportunity

In the study of animal social behavior, proximate benefits and costs for a set of alternative behaviors (such as mating versus feeding or fighting) are often estimated in terms of time and energy. However, although researchers are keenly aware of the benefit and cost in terms of opportunity intrinsic to every behavioral decision (e.g., when an animal hides in a den, it loses the opportunity to feed or mate), quantitative measurement has not been attempted due to practical difficulties in empirical studies. Measuring social mobility may provide just such a way to gauge gain or loss in opportunity for social advancement for social animals such as primates. Because mobility is an aggregate measure for a society, it can provide a baseline (the average or the expected) for opportunity in social advancement for all members in a social group (Sun et al. 2017). With this baseline as a benchmark, the opportunity consequences of many behaviors, biological identities, and social relationships of individuals (such as dispersal, conflict, cooperation, kinship, age, and sex) can be quantitatively assessed in terms of gains or losses in social opportunity relative to this benchmark. In other words, these behaviors, identities, or relationships can alter an individual's expected upward and downward mobility; that is, the benchmark. For instance, a monkey deciding to transfer from Group A to Group B must accept a complete loss in opportunity for social advancement in Group A, in exchange for a gain in opportunity in Group B. Whether this dispersal event is adaptive depends on the difference between the two opportunity measurements.

3.3 Social Mobility and Social Stability

In his classic, *Democracy in America*, the French political scientist Alexis de Tocqueville first noticed that mobility has a profound effect on the stability of human societies (de Tocqueville 1835/2002). Social scientists have since developed several explanations for why mobility is essential for stability (e.g., Lipset and Bendix 1992; Erikson and Goldthorpe 1994; Chetty et al. 2017). Studies in developed nations show that high mobility is closely related to equality and stability, and low mobility engenders inequality, which may cause a spectrum of social problems detrimental to society in general and democracy in particular (e.g., Wilkinson and Pickett 2009). It is a consensus now that a certain level of mobility is essential for the persistence of a society in the long run. Such a broadly held belief, however, hinges on the veracity of the hypothesis that mobility leads to stability. Unfortunately, this hypothesis, though logically compelling, is still short of empirical evidence (Bai and Jia 2016).

In reality, the relationship between mobility and stability is profound. On the one hand, an absolutely stable society without mobility can deprive its low-ranking members of opportunities to advance their social status, resulting ultimately in a loss in their fitness relative to that of high-ranking members (Sun 2013; Sun et al. 2017). As such, low-ranking members in a stagnant society will have a strong incentive to resort to revolt against the existing order so as to gain critical resources (such as food and mates) that are otherwise inaccessible for them. Therefore, an extremely unequal society with little mobility is evolutionarily unstable (Sun 2013). It tends to periodically experience major disruptions in the form of rebellion and revolution in humans (Bai and Jia 2016). It is in this sense that a certain level of mobility is essential for long-term social stability (Sun et al. 2017). On the other hand, mobility is the opposite of stability. That is, a higher level of mobility will necessarily decrease stability. This implies that too much mobility may also destabilize a society, which, beyond a certain point, may cascade into social upheaval (e.g., Clark 2014). However, this hypothesis, despite a broad theoretical interest, remains untested as well.

3.4 Measuring Social Mobility in Primate Societies

As mentioned earlier (Sect. 3.1), intragenerational mobility is normally calculated as the rate of status change in society within a specified time period (e.g., Lammam et al. 2012; Clark 2014). Technically, it can be measured either as the mean value in the rate of rank change or using a linear regression analysis between the ranks of all individuals in two consecutive time periods, t and $t + 1$. For the latter method, the rate of social mobility is thus 1-b, where b is the slope of the regression line ranging from 0 to 1 (Clark 2014; Sun et al. 2017). If b is 0, there is no mobility and, in our case, no rank change between two consecutive years for all adults involved. If b is 1, there is no stability. In primates, social mobility has been measured only in Tibetan macaques, where the rate of rank mobility was calculated for all adults on a yearly basis (Sun et al. 2017).

(Note that this linear regression method belongs to a statistical analysis known as time series analysis. Although time series analysis is relatively unfamiliar to researchers in biological sciences, it is commonly used in economic and business analysis such as GDP growth from year to year and stock performance from quarter to quarter. By emphasizing changes over time, the influence of policies or the efficacy of management regimes behind these changes can be statistically analyzed. As we can see, one distinct strength of time series analysis in addressing behavioral, ecological, and evolutionary issues lies in that it allows researchers to statistically track changes over time, which in turn may lead to the unveiling of the underlying factors that have significant influences on the changes. As such, time series regression is a standard method in the study of social mobility in social sciences.)

To complicate the issue, however, dominance rank can be measured as absolute rank or relative rank. Absolute rank refers to the raw rank data regardless of group size. Relative rank, however, is standardized according to group size. (Note: absolute and relative rank is a way of measuring dominance rank for all members in a group. They have no relationship to, and should not be mistaken as, absolute and relative mobility introduced in Sect. 3.1.) As we have already discussed, the distinction is not meaningful (or even feasible to measure) when a society is very large and the hierarchy is highly nonlinear, such as an industrial society, which tends to show complex forms of strata. However, the distinction becomes important when a society is made of only a few dozen individuals typical of many macaques species. For example, we can see a huge difference between the fifth ranked individual in a group of ten versus the 25th ranked individual in a group of 50 when measured in absolute rank. However, when measured in relative rank, these two individuals are ranked the same, both in the middle of the hierarchies. Clearly, measurement of social mobility will be affected by whether we use absolute or relative rank. Yet, little is known as to how using relative or absolute rank will affect mobility measurements.

In this study, we probed into the critical issue of how group size affects intragenerational mobility by using both absolute and relative rank in Tibetan and Japanese macaques. For Tibetan macaques, data about the hierarchical relationships among all adult females come from our 29-year consecutive observation on one group (YA1) at Mt. Huangshan, Anhui, China. Details about the dataset and how the data were collected are available in Sun et al. (2017). For the Japanese macaque, we obtained data from a published paper, where 11 years of hierarchical relationships among adult females were documented in a group on Koshima Island from 1957 to 1986 (Mori et al. 1989).

Age is well known to affect dominant status in primates (e.g., de Vries 1998; Packer et al. 2000; Alberts et al. 2003; Bayly et al. 2006; Balasubramaniam et al. 2013). Thus, comparing individuals of the same age class is critically important. However, because male and female Tibetan macaques sexually mature at different ages (seven for males and five for females, see Li 1999), we only included adult females that were 7 years or older in our analysis for two reasons. First, it allowed us to compare male and female adults of the same age class in our discussion here and in other places (Sun et al. 2017). Second, the ranks of immature females from the same mother are inversed. That is, the later born are higher in dominance rank than their elder sisters (Li 1999). Therefore, the inclusion of the 5- and 6-year-old females may introduce an unnecessary confounding factor, which may in turn reduce the accuracy in measuring social mobility for adults. The data about rank and rank change in Japanese macaques were collected from adult females of 5 years or older (Mori et al. 1989). Although the research in the Japanese macaque spanned 29 consecutive years, only 11 years of rank data were published.

For the study groups of both species, we used the original rank data for all adult females as their absolute ranks. We then calculated the relative ranks by scaling all absolute rank data into values between 0 (alpha female) and 1 (omega female) using a simple conversion formula:

$$RR = (AR - 1)/(n - 1)$$

where RR and AR refer to relative and absolute rank, respectively, and n is the size of the group. We then measured mobility using the same time series regression method as that detailed in Sun et al. (2017); that is, by using linear regression between rank $R(t)$ in year t and rank $R(t + 1)$ in year $t + 1$ to measure the annual rate of mobility for all adult group members (see Clark 2014) for both Tibetan and Japanese macaques. (Note that sample sizes about yearly rank changes were calculated as individual-years, rather than individuals in most statistics used in behavioral studies. In other words, each independent data point or sampling unit here is individual-year. This is a standard practice in time series analysis.) In our study, none of the datasets used for measuring mobility violated the normality assumption required for linear regression based on Kolmogorov-Smirnov test. As we have discussed in an earlier section, social mobility can be used to quantify the availability of social opportunities (e.g., Breen 2010; Clark 2014; Sun et al. 2017), which then can be used to quantify the benefit and cost of a behavior in terms of gain or loss in opportunity for social advancement. Here, we calculated net social mobility (subtracting the downward mobility from the upward mobility) and then used it to compare the difference between adult females of the two species. For the ease of reading, the statistical methods used are provided as they occur in the next section. As usual, all tests were two-tailed with 0.05 as the a priori level of significance for any statistical difference.

3.5 Results: Social Mobility in Tibetan and Japanese Macaques

From the adult female perspective (which may also be true if all individuals are included, if the number of adults, subadults, and infants is scaled in roughly similar proportions), the group size of Japanese macaques (33.09 ± 7.968SD; range, 21–49; $n = 11$) was more than three times larger than that of Tibetan macaques (7.62 ± 2.624 SD; range, 3–12; $n = 29$; Mann-Whitney U-test, $U = 385$; $n_1 = 11$; $n_2 = 29$; $p = 0.000$; see Fig. 3.1).

Using absolute rank, we found that group members experienced a larger rate of rank change in Japanese macaques than in Tibetan macaques. This was adequately reflected in both the mean ($U = 40{,}821$, $n_1 = 192$, $n_2 = 204$, $p = 0.013$) and variation (Levene test, $L = 65.07$, $df_1 = 1$, $df_2 = 394$, $p = 0.000$, Fig. 3.2) of annual rank change for Japanese macaques. Also, Japanese macaques had a higher rate in both upward (Chi-square test with Yates' correction, $\chi^2 = 13.187$, $df = 1, p < 0.001$) and downward movement ($\chi^2 = 18.705$, $df = 1$, $p = 0.000$) in hierarchy and a lower probability of staying in the same rank ($\chi^2 = 37.436$, $df = 1$, $p = 0.000$) than Tibetan macaques (Fig. 3.3). However, when we compared mobility between the two species using the regression method (rate of mobility, 0.115 for Japanese macaques and 0.144 for Tibetan macaques), the difference was not statistically significant (t-test for the slope, $t = 0.658$, $df = 392$, $p = 0.510$, Fig. 3.4).

For Tibetan macaques, mobility was higher using relative rank than using absolute rank, although the difference was marginally insignificant ($t = 1.706$, $df = 380$, $p = 0.089$). For Japanese macaques, the situation was the opposite. That is, mobility

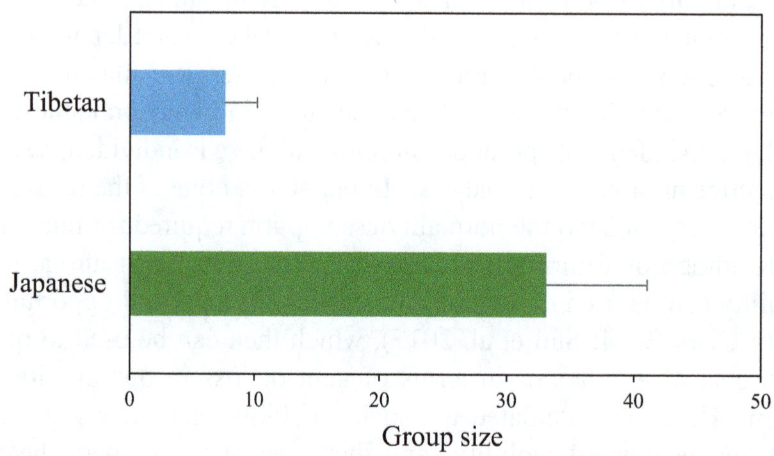

Fig. 3.1 Comparison of the mean group size (only adult females) in Tibetan and Japanese macaques. Error bars are standard deviations

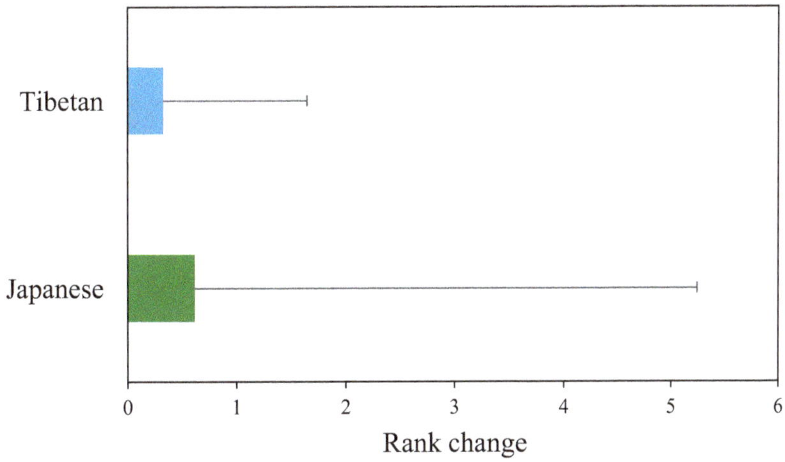

Fig. 3.2 Mean annual rate of absolute rank change in Tibetan and Japanese macaques. Error bars are standard deviations

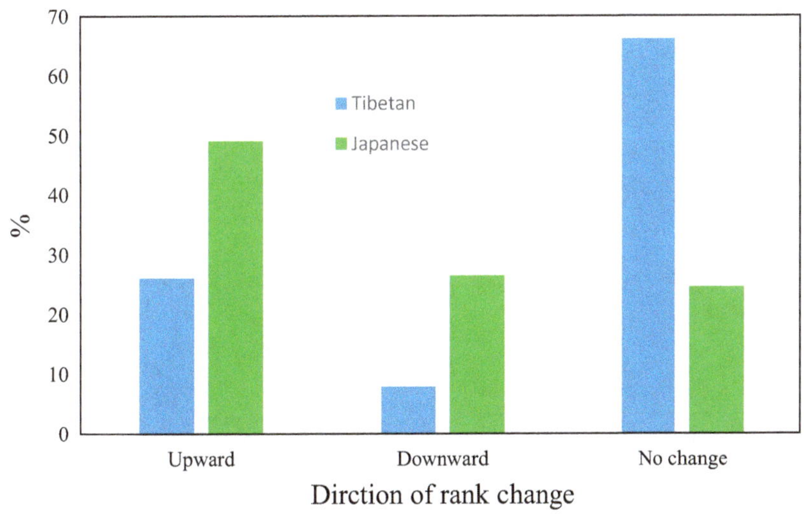

Fig. 3.3 Comparison of the frequency (%) of annual rank change (upward, downward, and no change) between Tibetan and Japanese macaques

was significantly lower when relative rank was used than when absolute rank was used ($t = 2.291$, $df = 404$, $p = 0.037$). Between-species comparison shows that there was no significant difference in absolute rank mobility, but a highly significant difference in relative rank mobility ($t = 3.901$, $df = 392$, $p = 0.0001$; see Fig. 3.5).

3.6 Discussion

This chapter presents a case study of two macaque species to illustrate how social mobility can be used as a new approach to the study of social dynamics as well as an in-depth analysis to explore the effect of group size. Our results show that social

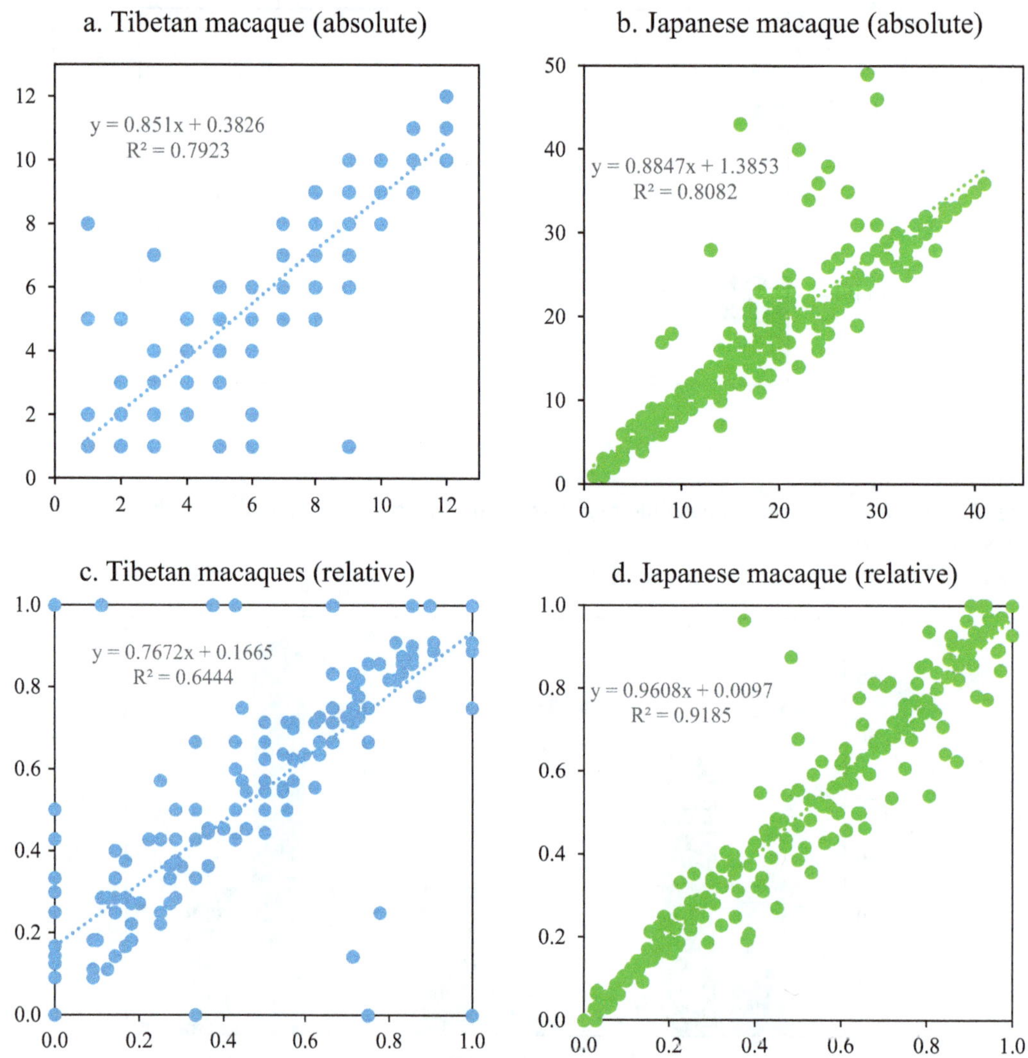

Fig. 3.4 Social mobility as measured by absolute ("absolute") and relative ("relative") rank in Tibetan and Japanese macaques using time series regression analysis. The X-axis indicates rank in year *t*, and the Y-axis indicates rank in the next year, year *t* + 1, for all adult individuals used in the analysis. Note that fewer data points than sample size shown in Panel (**a**) is due to data overlap

mobility can unveil new and unexpected insights into the social structure of primate societies. Our analyses show that, although Japanese macaques appeared to be more dynamic in rank changes, as shown by the higher mean and larger variation than Tibetan macaques, social mobility as measured in absolute rank was virtually the same for the two species. This result was surprising considering that Tibetan macaques and Japanese macaques are two different species with marked difference in so many aspects of their behavior and social organization (see Thierry 2011), despite some shared sociological traits such as matrilineality in social organization and strong despotism (see below).

One factor we particularly examined was group size, which was much larger in Japanese macaques than in Tibetan macaques. (Note that this difference may be

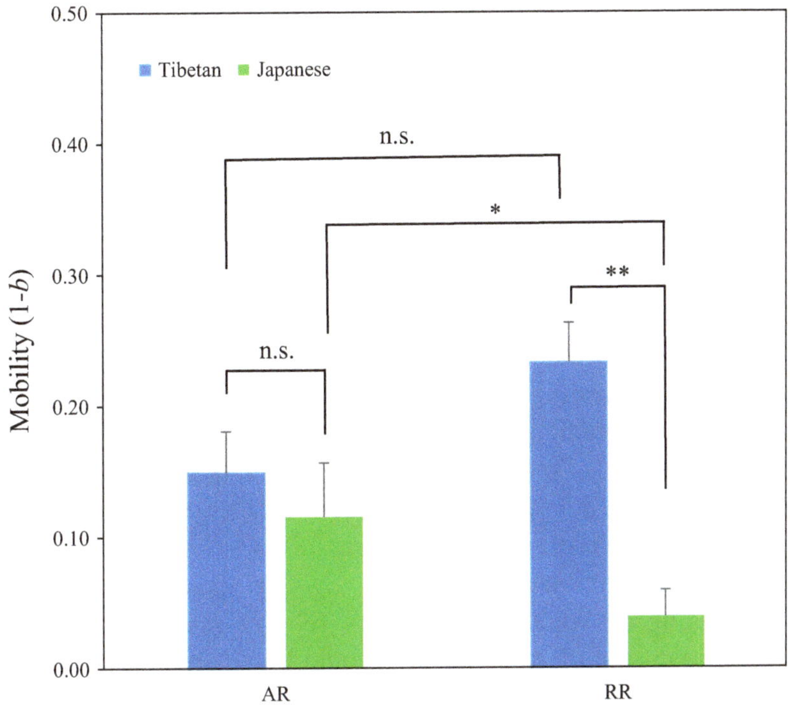

Fig. 3.5 Within- and between-species comparisons of social mobility using absolute ("AR") and relative ("RR") rank for the Tibetan and Japanese macaque. "n.s." stands for not significant, "asterisk" indicates $p < 0.05$, and "double asterisk" represents $p < 0.001$

slightly less prominent had only females of 7 years or older been included in the analysis for Japanese macaques.) This apparent discrepancy between the results from measuring the rate of rank change and from time series regression analysis might be due to two reasons. One is that, although the Japanese macaque troop studied had a larger rate of rank change, much of the significance was offset by a higher level of variation (Fig. 3.2). Also, using a nonparametric test for a non-normal dataset (Mann-Whitney U test here) might also lead to a slightly different statistical result from a parametric linear regression analysis.

Comparing the two species, we found that Tibetan macaque society appeared much more stable than that of the Japanese macaque in absolute rank change, despite no difference in mobility between the two species. Indeed, Tibetan macaques live in a small, simple, matrilineally structured group of strictly linear hierarchy (Li 1999; Berman et al. 2008) and their dominance hierarchy usually remains stable for a prolonged period of time, at least several months (Li 1999). (These behavioral conditions are particularly favorable for observing and determining social status for every adult in the hierarchy with little ambiguity. That is why Tibetan macaques are a highly desirable model species in the evolutionary study of behavior, a point compellingly addressed in the previous two chapters). Even so, social mobility in Tibetan macaques turned out to be similar to that in Japanese macaque. It appears that a larger group (society) may require a higher level of rank change to maintain the same level of mobility as a smaller group (society). Conversely, with the same level of mobility, a larger group would be more dynamic in terms of rank change. From the dynamic stability point of view, therefore, a larger society may require more rank change than a smaller society to maintain its stability in the long run, if the amount of mobility for long-term social stability is assumed to be approximately the same for all societies.

Another major reason for why the two species had similar levels of mobility may lie in the fact that they both form highly despotic societies; Grade 1 (highly despotic) for the Japanese macaque and Grade 2 (despotic) for the Tibetan macaque (Thierry 2000, 2011; Berman et al. 2004, 2006). Apparently, the two species we analyzed here may not be different enough in dominance style to show marked difference in social mobility. In other words, social mobility is likely to be related to dominance style. One logical hypothesis is that despotism may suppress social mobility. If true, this hypothesis would predict that a more relaxed or egalitarian primate society should have a higher level of mobility. This, however, requires a broader comparison with a larger diversity in dominance style, ideally in the genus *Macaca*, which has been well studied for the topic (Thierry 2011).

The most surprising finding in our study was that social mobility could vary a great deal depending on whether absolute or relative rank was used. In Tibetan macaques, the group size of adult females could be so small that rank data, once standardized to be within the range between 0 and 1, appear erratic (see in Fig 3.4c). Since relative rank can change with group size, even though there is no change in absolute rank, small changes can become prominent, which may in turn lead to increment in mobility. Obviously, the smaller the group is, the higher the mobility will become when relative rank is used. In Japanese macaques, on the contrary, larger rank changes (with the extreme records of 9 ranks upward and 27 ranks downward in a year in Japanese macaques compared with up 8 and down 7 in Tibetan macaques) became smaller after these changes were scaled to be within the range between 0 and 1. Clearly, this effect increases with group size. As a result, mobility decreases with increasing group size. These changes led to the counterintuitive result that, despite Japanese macaques appearing to be more dynamic in terms of absolute rank change, their society actually had a lower level of mobility than Tibetan macaques when relative rank was used.

This part of our results leads to an unexpected insight into social evolution. That is, group size may have a major impact on group dynamics. In both macaque species we examined, the net mobility was positive for social advancement. This may be true for most primate societies, in which social rank is more or less related to seniority and/or tenure. Consequently, as higher-ranking individuals become senile or die, younger, lower-ranking individuals can move up over time. So, even without major disruptive events, net mobility tends to be positive (upward). Overall, if everything else is equal, individuals in general should prefer smaller groups to larger groups, not only because absolute ranks go up (as the length of hierarchy becomes shorter) for most members but also because mobility goes up as well in terms of relative rank. Therefore, individuals in a larger society may have more incentive to break off to form smaller societies. This interesting and intriguing difference in mobility when measured in absolute versus relative rank may provide a novel explanation for why group fission is so common whereas group fusion is so rare in primates. It also sheds new light on the balkanization of human societies, happening in so many nations under the name of regional autonomy and sovereignty rather than the other way around. Although it is too early to claim that such events in human and nonhuman primate societies share biological roots, their uncanny similarities should warrant a close and serious examination on such possibilities from the evolutionary perspective.

Acknowledgments We are grateful to Mike Huffman, Peter Kappeler, and an anonymous reviewer for their helpful comments and constructive suggestions, which led to a considerable improvement in the clarity of our chapter. We also thank the Huangshan Garden Forest Bureau for their permission and support of this work and H.B. Cheng's family for their outstanding logistic support to our study. Rose Amrhein proofread the chapter. This work was supported in part by grants from the National Natural Science Foundation of China (No. 31772475; 31672307).

References

Acemoglu D, Egorov G, Sonin K (2018) Social mobility and stability of democracy: re-evaluating de Tocqueville. Q J Econ 133:1041–1105

Alberts SC, Watts HE, Altmann J (2003) Queuing and queue jumping: long term patterns of dominance rank and mating success in male savannah baboons. Anim Behav 65:821–840

Bai Y, Jia R (2016) Elite recruitment and political stability: the impact of the abolition of China's civil service exam system. Econometrica 84:677–733

Balasubramaniam KN, Berman CM, De Marco A et al (2013) Consistency of dominance rank order: a comparison of David's scores with I&SI and Bayesian methods in macaques. Am J Primatol 75:959–971

Bayly KL, Evans CS, Taylor A (2006) Measuring social structure: a comparison of eight dominance indices. Behav Processes 73:1–12

Berman CM, Ionica CS, Li J (2004) Dominance style among *Macaca thibetana* on Mt. Huangshan, China. Int J Primatol 25:1283–1312

Berman CM, Ionica CS, Dorner M et al (2006) Postconflict affiliation between former opponents in *Macaca thibetana* on Mt. Huangshan, China. Int J Primatol 27:827–854

Berman CM, Ogawa H, Ionica CS (2008) Variation in kin bias over time in a group of Tibetan macaques at Huangshan, China: contest competition, time constraints or risk response? Behaviour 145:863–896

Breen R (2010) Social mobility and equality of opportunity Geary lecture spring 2010. Econ Soc Rev 41:413–428

Chetty R, Grusky D, Hell M et al (2017) The fading American dream: trends in absolute income mobility since 1940. Science 356:398–406

Clark G (2014) The son also rises: surnames and the history of social mobility. Princeton University Press, Princeton NJ

Corak M (2013) Income inequality, equality of opportunity, and intergenerational mobility. J Econ Perspect 27:79–102

Cox MW (2012) Myths of rich and poor: why we're better off than we think. Basic Books, New York

De Tocqueville A (2002) Democracy in America. Regnery Publishing, Washington, DC

De Vries H (1998) Finding a dominance order most consistent with linear hierarchy: a new procedure and review. Anim Behav 55:827–843

Erikson R, Goldthorpe JH (1994) The constant flux: a study of class mobility in industrial societies. Clarendon Press, Oxford

Hamilton MJ, Milne BT, Walker RS et al (2007) The complex structure of hunter-gatherer social networks. Proc Biol Sci 274:2195–2202

Heckman JJ, Mosso S (2014) The economics of human development and social mobility. Annu Rev Econ 6:689–733

Lammam C, Karabegović A, Veldhuis N (2012) Measuring income mobility in Canada. Fraser Institute, Social Science Electronic Press, Vancouver, BC

Li J (1999) The Tibetan macaque society: a field study. Anhui University Press, Hefei

Lipset SM, Bendix R (1992) Social mobility in industrial society. Transaction Publishers, Brunswick

Marmot MG, Smith GD, Stansfeld S et al (1991) Health inequalities among British civil servants: the Whitehall II study. Lancet 337:1387–1393

Matthys M (2012) Cultural capital, identity, and social mobility. Routledge, New York

Mori A, Watanabe K, Yamaguchi N (1989) Longitudinal changes of dominance rank among the females of the Koshima group of Japanese monkeys. Primates 30:147–173

Packer C, Collins DA, Eberly LE (2000) Problems with primate sex ratios. Philos Trans R Soc Lond B Biol Sci 355:1627–1635

Piketty T (2014) Capital in the 21st century. Belknap, Cambridge, MA

Price TD, Brown JA (1985) Aspects of hunter-gatherer complexity. In: Price TD, Brown JA (eds) Prehistoric hunter-gatherers: the emergence of cultural complexity. Academic, Orlando, FL, pp 3–20

Sapolsky RM (2005) The influence of social hierarchy on primate health. Science 308:648–652

Seabrook JA, Avison WR (2012) Socioeconomic status and cumulative disadvantage processes across the life course: implications for health outcomes. Can Rev Sociol 49:50–68

Simandan D (2018) Rethinking the health consequences of social class and social mobility. Soc Sci Med 200:258–261

Snyder-Mackler N, Sanz J, Kohn JN et al (2016) Social status alters immune regulation and response to infection. Science 354:1041–1045

Sun L (2013) The fairness instinct: the Robin Hood mentality and our biological nature. Prometheus Books, Amherst, NY

Sun L et al (2017) The prospect of rising in rank is key to long-term stability in Tibetan macaque society. Sci Rep 7(1):7082

Thierry B (2000) Covariation of conflict management patterns across macaque species. In: Aureli F, de Waal FBM (eds) Natural conflict resolution. University of California Press, Berkeley, CA, pp 827–843

Thierry B (2011) The macaques: a double-layered social organization. In: Campbell CJ, Fuentes A, McKinnon KC et al (eds) Primates in perspective. Oxford University Press, New York, pp 229–241

Wilkinson R, Pickett K (2009) The spirit level: why greater equality makes societies stronger. Bloomsbury Press, New York

Zhang P, Wanatabe K, Li B et al (2006) Social organization of Sichuan snub-nosed monkeys (*Rhinopithecus roxellana*) in the Qinling Mountains, Central China. Primates 47:374–382

Chapter 4
Behavioral Exchange and Interchange as Strategies to Facilitate Social Relationships in Tibetan Macaques

Dong-Po Xia, Paul A. Garber, Cédric Sueur, and Jin-Hua Li

4.1 Introduction

Since the 1970s, sociobiologists have focused on two issues when studying social animals. One is the evolutionary processes that have favored group living, and the other is how ecological factors influence the internal structure of groups and the nature of social relationships and how changes in behavior influence the costs and benefits of group living to individuals of different age, sex, and dominance status (Wilson 1975; Barash 1977; Morse 1980; Wrangham 1980; Krebs and Davies 1984; van Schaik 1989). Studies on social animals have argued that the benefits of group living (such as social learning, cooperative hunting, collective resource and mate defense, cooperative infant caregiving, increased foraging efficiency, a reduction in stress associated with social support and reconciliatory behavior, alliance formation,

D.-P. Xia
School of Life Sciences, Anhui University, Hefei, China

International Collaborative Research Center for Huangshan Biodiversity and Tibetan Macaque Behavioral Ecology, Anhui, China

P. A. Garber
Department of Anthropology, Program in Ecology, Evolution, and Conservation Biology, University of Illinois, Urbana, IL, USA
e-mail: p-garber@illinois.edu

C. Sueur
CNRS, IPHC, UMR, Université de Strasbourg, Strasbourg, France
e-mail: cedric.sueur@iphc.cnrs.fr

J.-H. Li (✉)
School of Resources and Environmental Engineering, Anhui University, Hefei, Anhui, China

International Collaborative Research Center for Huangshan Biodiversity and Tibetan Macaque Behavioral Ecology, Anhui, China

School of Life Sciences, Hefei Normal University, Hefei, Anhui, China
e-mail: jhli@ahu.edu.cn

J.-H. Li et al. (eds.), *The Behavioral Ecology of the Tibetan Macaque*, Fascinating Life Sciences, https://doi.org/10.1007/978-3-030-27920-2_4

and partner reliability) serve to counteract the costs and offer fitness benefits to individuals living in a well-functioning and stable social unit (e.g., multilevel selection, Krause and Ruxton 2002; Sussman and Garber 2011).

According to Hinde (1976), social structures are group characteristics based on the patterns of social relationships such as affiliative, cooperative, sexual, and agonistic interactions that characterize a species or members of an established social unit. In order to maintain group cohesion, individuals are required to coordinate their daily activities, including collective movement to feeding and resting sites. Understanding how this is accomplished, whether individuals of a particular age, sex, or rank class consistently lead or direct group movement, and how "leaders" build a consensus in deciding when to leave a feeding or resting site and which location to travel to next, offers critical insights into the dynamics of animal social systems (see also Chap. 5).

A recent paper by Schino and Aureli (2017) describes two alternative processes associated with social relationships. Partner control highlights dyadic interactions between dyads in which the behavior of each subject is dependent exclusively on the previous behavior of a partner (Noë 2006). This assumes that one member of a dyad can control the behavior of its partner with no opportunity of partner switching. Partner control has been used to describe social interactions between individuals living in a social unit. Partner choice assumes that individuals can freely select and change partner preferences based on the assessed benefit each potential partner is expected to provide. For example, in studies of *Sapajus nigritus* (formerly *Cebus apella*) and *Macaca fuscata* (Schino and Aureli 2009, 2017; Schino et al. 2007, 2009), partner choice was found to influence social relationships and group stability.

Biological market theory proposes that social behaviors can be considered as valuable commodities and explains how commodities are exchanged or interchanged among group members (Noë and Hammerstein 1994, 1995; Noë 2006). This theory considers social behaviors as exchangeable commodities to obtain behavioral services as benefits (Noë 2001; Barrett and Henzi 2006; Noë & Voelkl 2013). Individuals compete over access to partners, by offering more valuable services rather than engaging in aggressive or coercive behavior. Therefore, individuals are chosen as partners depending on the services they offer, and the choice is made by comparing the offers of all potential partners depending on the services needed. As such, biological market theory provides an alternative framework to study social relationships and partner preferences and facilitates the understanding of behavioral adaptations for group stability and social cohesiveness in group-living animals.

In most nonhuman primates, behavioral exchange and interchange have been well documented (reviewed in Sánchez-Amaro and Amici 2015). For example, Seyfarth (1977) proposed that lower-ranking females compete to groom higher-ranking females who can offer them more effective aid (agonistic support) in competitive interactions with other group members. In some species such as chacma baboons, *Papio cynocephalus* (Barrett et al. 1999); red-fronted lemurs, *Eulemur fulvus rufus* (Port et al. 2009); and chimpanzees, *Pan troglodytes* (Newton-Fisher and Lee 2013; Kaburu and Newton-Fisher 2015a), grooming is reported to be exchanged for grooming (reciprocal exchange). Within a dyad consisting of individuals A and B, the amount of grooming given by A to B depends on the amount of grooming given by B to A. In some other species, studies show that behavioral services can also be interchanged for different behaviors. For example, grooming can be interchanged for food sharing to maintain social bonds and agonistic support in chimpanzees (de Waal 1997; Kaburu and Newton-Fisher 2015a). In sooty mangabeys (*Cercocebus atys*) and vervet monkeys (*Chlorocebus aethiops*), infant handling can influence dyadic relationships. In this species, grooming is interchanged for opportunities to handle infants in order to establish social relationships (Fruteau et al. 2011). In long-tailed macaques (*M. fascicularis*), grooming is interchanged for tolerance to increase spatial proximity and to reduce aggression from higher-ranking individuals (Gumert and Ho 2008). Additionally, grooming can be interchanged for copulations (Kaburu and Newton-Fisher 2015b). Mating opportunities increase for females who groom a higher-ranking male compared to females who groom a lower-ranking male (Gumert 2007a). Similarly, Gomes and Boesch (2009) found that over a 22-month period, female chimpanzees copulated more with males who shared meat with them compared to males who did not engage in food sharing. In this species, food sharing by adult males provides more mating opportunities.

In this chapter, our goal is to review the evidence for multiple behavioral exchanges and interchanges in Tibetan macaque (see Chap. 2 for more information about Tibetan macaques) groups at Mt. Huangshan, China (Fig. 4.1), and to analyze these dyadic behavioral interactions within the context of behavioral exchange or interchange. We define exchange as the reciprocation of the same behaviors (e.g., grooming for grooming) and interchange as the reciprocation between different

Fig. 4.1 A social group (known as YA1) of Tibetan macaques at Mt. Huangshan, China

behaviors of the same category (e.g., grooming for tolerance, which are different behaviors but both belong to the same category of affiliative relationships) or behaviors of different categories (e.g., agonistic support for copulation). We focus on grooming exchanges, the interchange of grooming for tolerance among female or male intrasexual dyads, the interchange of grooming for infant-handling opportunities among females, and the interchange of male-to-female agonistic support for mating opportunities among intersexual dyads.

4.2 Exchange Between the Same Behaviors

4.2.1 Grooming for Itself

Grooming has been reported to be exchanged for grooming (reciprocal exchange) in primates (such as chacma baboons, Barrett et al. 1999). In Tibetan macaques, similarly, approximately 20% of their daily activity budget is devoted to grooming, regardless of sex, social rank, age, and other factors (see Xia et al. 2012, 2013) (Fig. 4.2 shows grooming in Tibetan macaques). Xia et al. (2012, 2013) examined female and male intrasexual grooming relationships in free-ranging Tibetan macaques and hypothesized that among a broad set of fitness-maximizing strategies, grooming can be used by individuals to enhance social relationships through reciprocity and/or through the interchange of grooming for a different but equivalent service. Overall, they found that social rank played an important role in defining female and/or male social interaction and partner choice in Tibetan macaques. Among female or male intrasexual dyads, the authors found that there

Fig. 4.2 Female-to-female social grooming in Tibetan macaques

were positive correlations between the average frequency of grooming bouts received (A groom B) and reciprocated (B groom A). These patterns were consistent during the mating and non-mating seasons. Similarly, grooming effort or the time the initiator invested in grooming the recipient was positively correlated with grooming received among female and male intrasexual dyads. Furthermore, the longer the initiator groomed her partner (A groom B), the longer her partner was likely to groom her in return (B groom A). This pattern also was consistent during mating and non-mating periods.

In addition, as indicated in Fig. 4.3, in intrasexual dyads composed of individuals of equal ranks, grooming was reciprocally exchanged (equal rates and equal duration) during both the mating and non-mating seasons. This suggests that, within each sex, dyads composed of animals of equal ranks use reciprocal exchange grooming as a social tool to maintain long-term alliances and partner preferences. Grooming frequency and grooming duration were equal within female–female and male–male dyads consisting of high- and high-ranking individuals, middle- and middle-ranking individuals, and low- and low-ranking individuals during both mating and non-mating periods.

The findings that closely ranked females, regardless of their position in the hierarchy, acted as reciprocal traders are similar to the pattern reported in chacma baboons (*P. c. ursinus*, Barrett et al. 1999), Japanese macaques (*M. fuscata*, Ventura et al. 2006), and hamadryas baboons (*P. hamadryas hamadryas*, Leinfelder et al. 2001). Furthermore, the authors also documented a pattern in which the initiator (A) of a bout invested less and received greater benefit than the recipient. Thus, it appears that among female and/or male Tibetan macaques, the act of initiating a grooming bout was a behavioral investment or strategy to obtain a grooming reward or other social dividend. This effect was strongest when the initiator was higher in rank than the receiver.

4.3 Exchange Between Different Behaviors

4.3.1 Grooming for Tolerance

Grooming represents one of the most common commodities traded among primates and has long been considered a reliable indicator of both long-term and short-term social tolerance (Dunbar 2010; Schino and Aureli 2010). Seyfarth (1977) has proposed that there exists a balance between a group member's preference for particular grooming partners and competition among females for access to the most valuable grooming partners. According to Seyfarth's model, lower-ranked individuals should compete to groom higher-ranked individuals, who can offer them more effective aid in competitive interactions with other group members. This is expected to result in higher-ranked individuals receiving more grooming than lower-ranked individuals and grooming partners tending to be of similar or adjacent ranks (Henzi et al. 2003). However, Henzi and colleagues (2003) found no

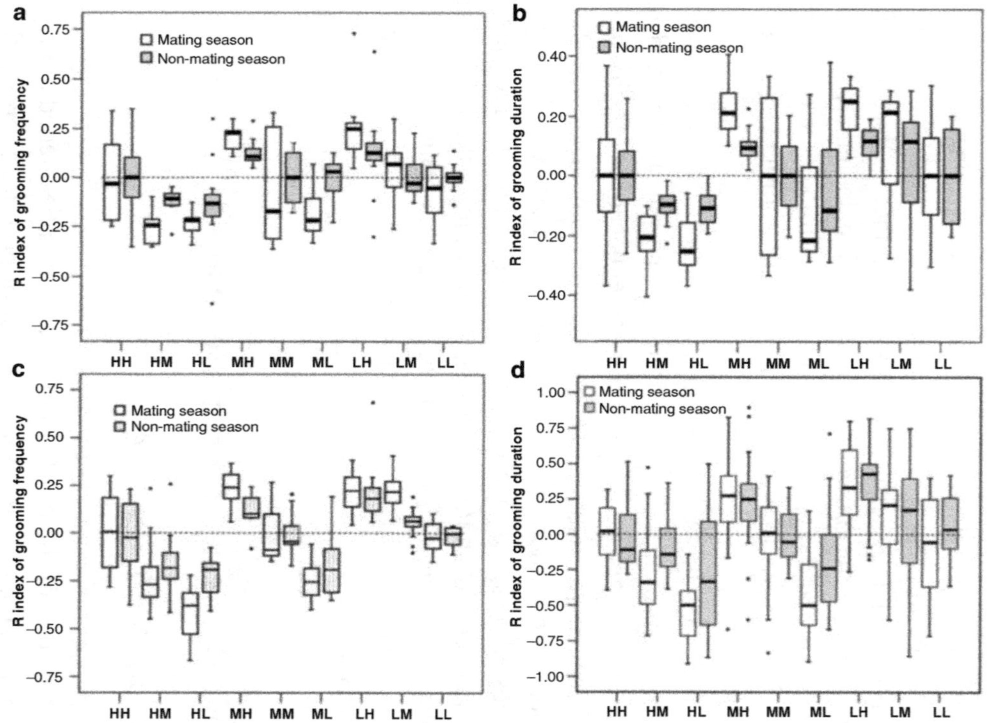

Fig. 4.3 Mean and SE (boxes) of R indices ($R = \frac{G_A - G_B}{G_A + G_B}$, in which G_A is grooming frequency or grooming duration for animal A and G_B is grooming frequency or grooming duration for animal B) within the nine female–female (**a**, grooming frequency; **b**, grooming duration) and male–male (**c**, grooming frequency; **d**,

grooming duration) categories according to rank (H, high-ranking; M, middle-ranking; L, low-ranking). The *R* index can range from -1 to 1, where a value of 0 represents complete reciprocity (the dashed line in each panel), negative values indicate that individual A received more grooming than it gave, and positive values indicate that individual B received more grooming than it gave. Outliers are given as dots. *R* indices were calculated for all grooming bouts and then were assigned each bout to one of three groups according to the level of maternal kin relationships of grooming partners. (Fig. 4.3a, b is from Xia et al. 2012, and Fig. 4.3c, d is from Xia et al. 2013.) *H* represents high-rankings, *M* represents middle-ranking, and *L* represents low-ranking support for Seyfarth's model in their study of chacma baboons. Rather, they argued that in these primates "grooming is unimpeded by restrictions on access to partners" and proposed that baboon females exchanged grooming for increased social tolerance.

In Tibetan macaque society, Xia et al. (2012, 2013) also found that in intrasexual dyads composed of individuals of different social ranks (e.g., high- and middle-ranking females/males, high- and low-ranking females/males, and middle- and low-ranking females/males), initiators received grooming more frequently and for a longer duration than they gave during both mating and non-mating periods. Conversely, in dyads consisting of middle- and high-ranking females, low- and high-ranking females, and low- and middle-ranking females, initiators received grooming less frequently and for a shorter duration than they gave during both mating and non-mating periods. The same results were found among male intrasexual dyads consisting of middle- and high-rankings, low- and higher-rankings, and middle- and high-rankings (see Fig. 4.3).

Xia et al. (2012, 2013) also found that the effect of grooming reciprocity among female and/or male intrasexual dyads was stronger in the non-mating season. This suggests that during periods of reduced social tension (such as the non-mating period), adult female and/or male Tibetan macaques used grooming to maintain and enhance affiliative social relationships. In contrast, during periods when the rate and the threat of male intrasexual aggression increased (such as the mating season), there was a tendency to promote social affiliation by increasing the duration of grooming. The authors found negative relationships between grooming given and aggression received both among male intrasexual dyads (e.g., as showed in Fig. 4.4) and among female intrasexual dyads. This provided supportive evidence that both female and male Tibetan macaques vary their grooming strategies based on the current rates of within-group intrasexual aggression. The results indicated that high-ranking females and/or males who received more grooming from lower-ranking males were less aggressive toward their grooming partners. In other words, lower-ranking females and/or males who devoted more time to grooming higher-ranking individuals received less aggression than did similarly ranked individuals who groomed higher-ranking females and/or males less frequently. Thus, there is evidence that, for female and/or male intrasexual dyads, Tibetan macaques interchanged grooming for social tolerance and possibly competed for access to partners of higher social rank.

Taken together, Xia et al. (2012, 2013) suggest that Tibetan macaques employed an alternative set of social strategies depending on rank relationships. This result highlights the importance of both grooming reciprocal exchange and interchange for tolerance as behavioral mechanisms that regulate female and/or male primate social relationships.

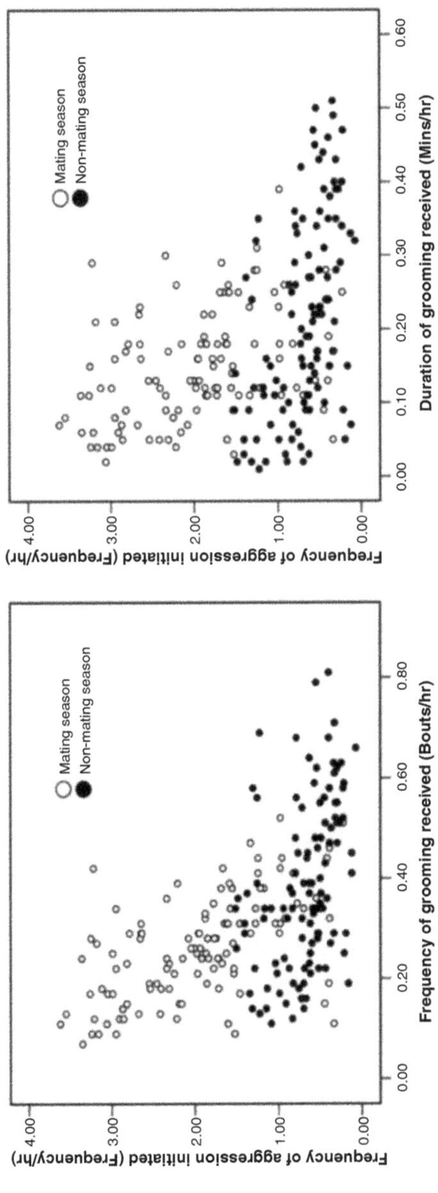

Fig. 4.4 Aggression initiated and grooming received among male intrasexual dyads. Figure is from Xia et al. (2013)

4.3.2 Grooming for Infant Handling

Infants are important components in a primate social group, and alloparental care is widespread among primates to facilitate the survival of newborn infants and increase benefits to caregivers (Maestripieri 1994). While females' attraction to infants represents a common feature of primate species, maternal response to infant handling shows a certain degree of variability (Nicolson 1987; Maestripieri 1994). In some species, such as langurs (*Presbytis pileata*) and vervet monkeys (*C. aethiops sabaeus*), mothers allow other group members to frequently hold and carry their newborn infants for long durations (Stanford 1992; Fairbanks 1990). In other species, such as baboons and macaques, although mothers are more restrictive of their infants (Nicolson 1987; Altmann 2002), group members still have opportunities to contact newborn infants (see Maestripieri 1994). It has been well documented that infants can be social tools to buffer agonistic behavior from higher-ranked individuals and facilitate dyadic social bonds (Silk and Samuels 1984 in bonnet macaques, *M. radiata*; Ogawa 1995 in Tibetan macaques). Grooming for females is an effective way to increase the opportunity for accessing *Sapajus nigritus* infants. For example, in tufted capuchin monkeys (*Sapajus apella nigritus*), Tiddi et al. (2010) proposed that potential handlers have strongly attracted to infants and grooming their mothers. Similar results have also been found in baboons, *P. anubis* (Frank and Silk 2009), long-tailed macaques (Gumert 2007b), and golden snub-nosed monkeys, *Rhinopithecus roxellana* (Wei et al. 2013). These studies imply that grooming might be interchanged for infant handling or vice versa to facilitate dyadic social bonds among females.

In Tibetan macaques, Jiang et al. (2019) found that there were bidirectional relationships between grooming and infant handling among females. Jiang and her colleagues found that there were five patterns of infant handling, including hand touching, mouth licking with teeth chattering, holding, grooming, and bridging. Female Tibetan macaques received more grooming time after giving birth than before parturition, and lactating females invested less grooming time than females without infants. For example, the duration of mothers grooming non-mothers before birth was significantly longer than after birth, whereas the duration of non-mothers grooming mothers after birth was significantly longer than that before birth. Apparently, females with infants are more attractive as grooming partners than females without infants. In a Tibetan macaque social group, adult females with infants received more grooming when this adult female allowed their newborn infants to be handled. For example, the duration of non-mother to mother post-handling-grooming was higher than the baseline values of daily grooming (see left side of Fig. 4.4) and the grooming duration when these two females were in proximity (see left side of Fig. 4.5). Moreover, with the increase in the number of infants, the amount of grooming received from females without infants decreased.

Additionally, in Tibetan macaques, the frequency of non-mother post-grooming-infant handling was higher than the baseline values of daily infant handling (see right side of Fig. 4.4) and the infant handling when these two females are in proximity (see

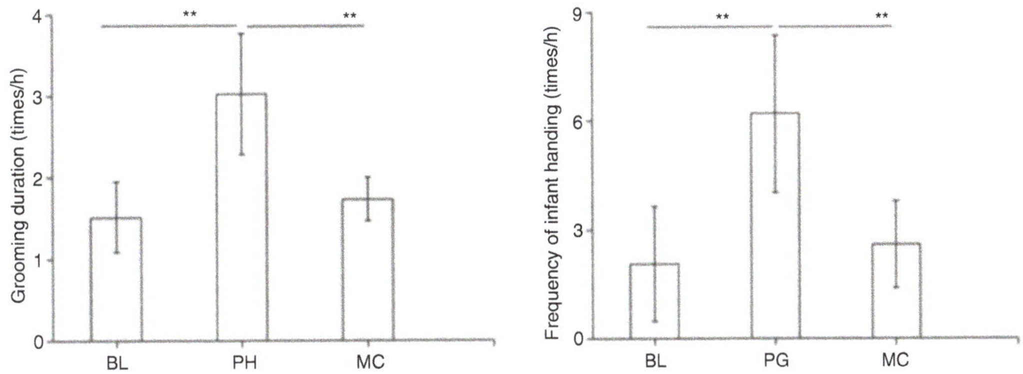

Fig. 4.5 Variation of grooming duration from non-mothers to mothers and frequency of non-mother infant handling. Baseline represents the daily activity, which is from focal sampling data. PH represents post-infant handling-grooming. PG represents post-grooming-infant handling. MC represents the mean from the data where mothers and non-mothers were in proximity. Figure is modified from Jiang et al. (2019)

right side of Fig. 4.5). These results suggested that females obtained more opportunities for infant handling when she groomed more females with infants; thus, grooming was an effective way to gain access to an infant.

4.3.3 Agonistic Support for Copulation

In multi-male–multi-female primates, mating is associated with individual's social rank, with a higher frequency for higher-ranked individuals and lower frequency for lower-ranked individuals (reviewed in Li 1999). Mating is also associated with dyadic affiliative relationships (such as grooming). For example, in a study of Barbary macaques (*M. sylvanus*), Sonnweber et al. (2015) demonstrated that males initiated grooming after copulations with ejaculation in order to keep females from mating with other males. Agonistic support is predicted to increase mating opportunity. Agonistic support has been defined as instances in which an individual joins an ongoing agonistic interaction and supports one animal by directing aggression against its opponent (Seyfarth and Cheney 1984; Hemelrijk 1994), including intrasexual and intersexual agonistic support.

Wang et al. (2013) used a 15-min focal sampling method to collect data on copulations and agonistic support as a random variable in two groups (YA1 and YA2) of Tibetan macaques in Huangshan, China. A total of 216 male-to-female agonistic support events were collected (YA1 32 and YA2 64 in the non-mating season; YA1 43 and YA2 77 in the mating season). Correlation tests showed that male-to-female agonistic support was correlated with the frequency with which the male copulated with that female in both the mating season and the non-mating season. This indicated that male-to-female agonistic support could increase the male mating opportunities with the female he supported in agonistic events.

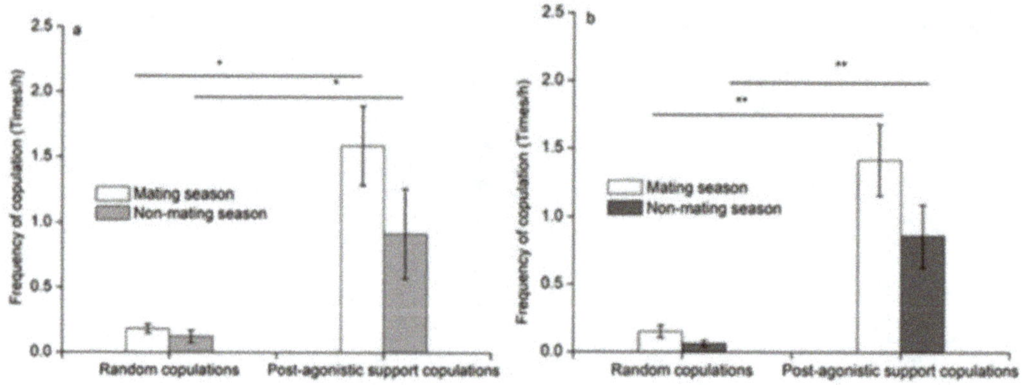

Fig. 4.6 Post-agonistic copulations and post-copulation agonistic supports among female Tibetan macaques. (**a**) from YA1 group. (**b**) from YA2 group. $*P < 0.05$. $**P < 0.01$. Figures are modified from Wang et al. (2013)

In addition, Wang et al. (2013) used a 15-min post-agonistic support focal sampling method to collect copulation data to compare with random values (the average value of copulation from the focal sampling methods). The results showed that, in the YA1 group, copulations in the post-agonistic support observation period (PO) were more frequent, but not significantly so, than random observations (RO) in the non-mating season, and post-agonistic support copulations were significantly more frequent than random observations in the mating season (see Fig. 4.6a). For the YA2 group, copulations in the post-agonistic support period were more frequent than in random observations in both mating and non-mating seasons (see Fig. 4.6b). Taken together, these results indicate that, in Tibetan macaques, male-to-female agonistic support could be interchanged for mating opportunities with females who are participating in an ongoing aggressive event.

Wang et al.'s (2013) study characterized not only the pattern of behavioral exchange in Tibetan macaque social groups but also offers new evidence for how certain mating strategies can increase mating opportunities in males. Male Tibetan macaques copulate under three different circumstances: opportunistic mating, which involves frequent copulation with many different females in the group setting (Li et al. 2015); possessive mating, which involves a single male's attempts to monopolize copulations in spite of the presence of other males (Li et al. 2015); and consortships, in which mating takes place between one male and one female, who travel apart from the rest of the group for several hours or days (Li 1999). Xia et al. (in preparation) found that males preferentially mate with their female grooming partners based on dyadic relationships. This indicates that, similar to the findings in studies in Barbary macaques (Sonnweber et al. 2015) and rhesus macaques, *M. mulatta* (Manson 1992), dyadic social relationships play a vital role in obtaining mating opportunities. Agonistic support is one of the effective interactions for males to form and maintain social relationships with females and to obtain and improve copulatory success. In Tibetan macaques, the relationships between agonistic support and copulatory behavior provide insight into understanding male–male competition and female mate choice in social primates.

4.4 Conclusions

In a Tibetan macaque social group, there are diverse exchange networks consisting of different kinds of behavioral exchanges (see Fig. 4.7). Tibetan macaques provide behavioral services for the same behavioral services in return, such as grooming in exchange for grooming. Among intra-sexual dyads (female–female and male–male dyads), grooming investment is exchanged for itself with equal value, in terms of frequency and/or duration. Secondly, Tibetan macaques can also provide behavioral services that are exchanged for different behavioral services of the same category, such as grooming for tolerance. Here, grooming and tolerance are different behaviors; however, both are friendly interactions for facilitating social relationships. Both male and female Tibetan macaques groom group members of the same sex in exchange for tolerance by decreasing aggressive interactions from higher-ranking individuals. In addition, Tibetan macaques provide behavioral services by interchanging different behavioral services with different categories, such as agonistic support for copulations.

In addition, group members can choose their partners for exchanging different types of behavioral services based on their own social status within a group. For example, grooming is regularly traded reciprocally (for grooming) among female and male intrasexual social dyads consisting of similar social ranks. Grooming can be interchanged for rank-related benefits from higher-ranking individuals, such as for tolerance from. Grooming can also be interchanged for opportunities to access newborn infants to facilitate social relationships with the infants' mothers.

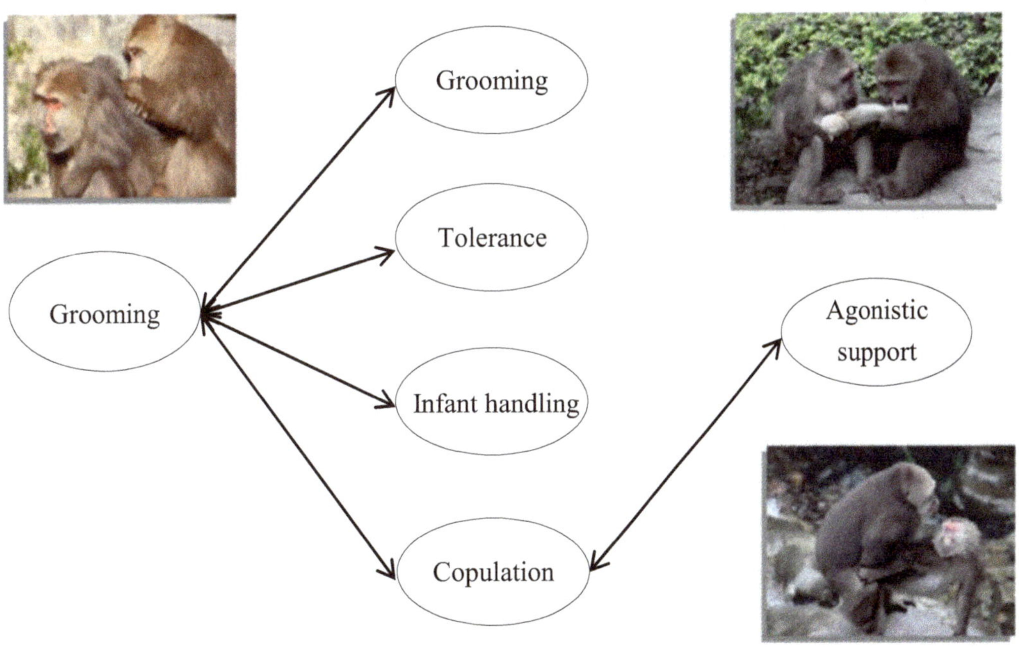

Fig. 4.7 Exchange networks in Tibetan macaque society

Finally, behavioral exchanges facilitate the establishment and maintenance of social relationships and group stability in multiple ways, by obtaining multiple behavioral services in return. For example, grooming exchanges enhance male–male and female–female social bonds. Grooming with higher-ranking group members allows lower-ranking animals to coexist peacefully with more dominant animals. Non-mothers that groom females with infants could increase their opportunity for accessing infants, and accessing newborn infants could facilitate social relationships between the females who join in infant handling. Among intersexual dyads, male Tibetan macaques support certain females against attacks by other animals and subsequently mate more frequently with that female. This implies that males can interchange copulatory behavior by supporting females to improve mating opportunity.

Although we have several case studies of behavior networks in Tibetan macaque society, we need more evidence to draw a clearer picture about behavioral exchange and interchange in this primate species. Apparently, both males and females may establish diverse forms of behavioral exchange or interchange networks with multiple group members. However, we know little about strategies for partner control or partner choice based on more complex measures of individual and social status (such as personality and dominance) to strengthen social bonds, obtain behavioral services, and maximize fitness. Future studies will need to pay closer attention to the variation of behavioral strategies in connection with group dynamics to better understand how behavioral exchange and interchange are related to the maintenance and stability of social bonds and group networks.

Acknowledgments We are very grateful to the Huangshan Garden Forest Bureau for their permission and support of this work. We also gratefully acknowledge Mr. H.B. Cheng's family for their outstanding logistic support to our study. PAG wishes to thank Chrissie, Sara, and Jenni for their support. DPX wishes to thank Randall C. Kyes, professor at the University of Washington, whom DPX worked with as a visiting scholar, Lori K. Sheeran for her encouragement and support, and Lixing Sun and Rose Amrhein for their help. This work was supported in part by grants from the National Natural Science Foundation of China (No. 31772475; 31672307; 31401981) and the China Scholarship Council.

References

Altmann J (2002) Baboon mothers and infants. Afr J Ecol 40(4):419–421

Barash DB (1977) Sociobiology and behaviour. Heinemann, London

Barrett L, Henzi SP (2006) Monkeys, markets and minds: biological markets and primate sociality. In: Kappeler PM, van Schaik CP (eds) Cooperation in primates and humans: mechanisms and evolutions. Springer, New York, pp 209–232

Barrett L, Henzi SP, Weingrill T, Lycett JE, Hill RA (1999) Market forces predict grooming reciprocity in female baboons. Proc R Soc Lond B 266:665–670

de Waal FBM (1997) The Chimpanzee's service economy: food for grooming. Evol Hum Behav 18(6):375–386

Dunbar RIM (2010) The social role of touch in humans and primates: behavioural function and neurobiological mechanisms. Neurosci Biobehav Rev 34(2):260–268

Fairbanks LA (1990) Reciprocal benefits of allomothering for female vervet monkeys. Anim Behav 9:425–441

Frank RE, Silk JB (2009) Impatient traders or contingent reciprocators? Evidence for the extended time-course of grooming exchanges in baboons. Behaviour 146:23–1135

Fruteau C, van de Waal E, van Damme E, Noë R (2011) Infant access and handling in sooty mangabeys and vervet monkeys. Anim Behav 81(1):153–161

Gomes CM, Boesch C (2009) Wild chimpanzees exchange meat for sex on a long-term basis. PLoS One 4:e5116

Gumert MD (2007a) Payment for sex in a macaque mating market. Anim Behav 74(6):1655–1667

Gumert MD (2007b) Grooming and infant handling interchange in *Macaca fascicularis*: the relationship between infant supply and grooming payment. Int J Primatol 28(5):1059–1074

Gumert MD, Ho MR (2008) The trade balance of grooming and its coordination of reciprocation and tolerance in Indonesian long-tailed macaques (*Macaca fascicularis*). Primates 49(3):176–185

Hemelrijk CK (1994) Support for being groomed in long-tailed macaques, *Macaca fascicularis*. Anim Behav 48:479–481

Henzi SP, Barrett L, Gaynor D, Greeff J, Weingrill T, Hill RA (2003) Effect of resource competition on the long-term allocation of grooming by female baboons: evaluating Seyfarth's model. Anim Behav 66(5):931–938

Hinde RA (1976) Interactions, relationships and social-structure. Man 11:1–17

Jiang Q, Xia DP, Wang X, Zhang D, Sun BH, Li JH (2019) Interchange between grooming and infant handling in female Tibetan macaques (*Macaca thibetana*). Zool Res 40:139–145

Kaburu SSK, Newton-Fisher NE (2015a) Egalitarian despots: hierarchy steepness, reciprocity and the grooming-trading model in wild chimpanzees, *Pan troglodytes*. Anim Behav 99:61–71

Kaburu SSK, Newton-Fisher NE (2015b) Trading or coercion? Variation in male mating strategies between two communities of East African chimpanzees. Behav Ecol Sociobiol 69:1039–1052

Krause J, Ruxton G (2002) Living in groups. Oxford University Press, Oxford

Krebs JR, Davies NB (1984) Behavioral ecology: an evolutionary approach. Blackwell Scientific, Oxford

Leinfelder I, de Vries H, Deleu R, Nelissen M (2001) Rank and grooming reciprocity among females in a mixed-sex group of captive hamadryas baboons. Am J Primatol 55(1):25–42

Li JH (1999) The Tibetan macaque society: a field study. Anhui University Press, Hefei. (In Chinese)

Li ZP, Li JH, Xia DP, Zhu Y, Wang X, Zhang D (2015) Mating strategies of subordinate males in Tibetan macaqus (*Macaca thibetana*) at Mt. Huangshan, China. Acta Theriologica Sinica 35 (1):29–39. (In Chinese)

Maestripieri D (1994) Social structure, infant-handling, and mother styles in group-living Old World monkeys. Int J Primatol 15:531–553

Manson JH (1992) Measuring female mate choice in Cayo Santiago rhesus macaques. Animl Behav 44:405–416

Morse DH (1980) Behavioral mechanisms in ecology. Harvard University Press, Cambridge, MA

Newton-Fisher NE, Lee PC (2013) Grooming reciprocity in wild male chimpanzees. Anim Behav 81(2):439–446

Nicolson N (1987) Infants, mothers, and other females. In: Smuts BB, Cheney DL, Seyfarth RM, Wrangham RM, Struhsaker TT (eds) Primate societies. University of Chicago Press, Chicago, IL, pp 330–342

Noë R (2001) Biological markets: partner choice as the driving force behind the evolution of mutualisms. In: Noe R, van Hooff JARAM, Hammerstein P (eds) Economics in nature: social dilemmas, mate choice and biological markets. Cambridge University Press, Cambridge, pp 93–118

Noë R (2006) Digging for the roots of trading. In: Kappeler PM, van Schaik CP (eds) Cooperation in primates and humans: mechanisms and evolution. Springer, Berlin, pp 223–251

Noë R, Hammerstein P (1994) Biological market: supply and demand determine the effect of partner choice in cooperation, mutualism and mating. Behav Ecol Sociobiol 35:1–11

Noë R, Hammerstein P (1995) Biological markets. Trends Ecol Evol 10:336–339

Noë R, Voelkl B (2013) Cooperation and biological markets: the power of partner choice. In: Sterelny K, Joyce R, Calcott B, Fraser B (eds) Cooperation and its evolution. MIT Press, Cambridge, MA, pp 131–152

Ogawa H (1995) Bridging behavior and other affiliative interactions among male Tibetan macaques (*Macaca thibetana*). Int J Primatol 16:707–729

Port M, Clough D, Kappeler PM (2009) Market effects offset the reciprocation of grooming in free-ranging redfronted lemurs, *Eulemur fulvus rufus*. Anim Behav 77(1):29–36

Sánchez-Amaro A, Amici F (2015) Are primates out of the market? Anim Behav 110:51–60

Schino G, Aureli F (2009) Reciprocal altruism in primates: partner choice, cognition, and emotions. Adv Study Behav 39:45–69

Schino G, Aureli F (2010) Primate reciprocity and its cognitive requirements. Evol Anthropol Issues News Rev 19(4):130–135

Schino G, Aureli F (2017) Reciprocity in group-living animals: partner control *versus* partner choice. Biol Rev 92:665–672

Schino G, Polizzi di Sorrentino EP, Tiddi B (2007) Grooming and coalitions in Japanese macaques (*Macaca fuscata*): partner choice and the time frame reciprocation. J Comp Psychol 121:181–188

Schino G, di Giuseppe F, Visalberghi E (2009) The time frame of partner choice in the grooming reciprocation of *Cebus paella*. Ethology 115:70–76

Seyfarth RM (1977) A model of social grooming among adult female monkeys. J Theor Biol 65:671–698

Seyfarth RM, Cheney DL (1984) Grooming, alliances and reciprocal altruism in vervet monkeys. Nature 308:541–542

Silk JB, Samuels A (1984) Triadic interactions among *Macaca radiata*: passports and buffers. Am J Primatol 6:373–376

Sonnweber RS, Massen JJM, Fitch WT (2015) Post-copulatory grooming: a conditional mating strategy? Behav Ecol Sociobiol 69:1749–1759

Stanford C (1992) The costs and benefits of allomothering in wild capped langurs (*Presbytis pileata*). Behav Ecol Sociobiol 30:29–34

Sussman RW, Garber PA (2011) Cooperation, collective action, and competition in primate social interactions. In: Campbell CJ, Fuentes A, MacKinno KC, Bearder S, Stumpf R (eds) Primates in perspective, vol 2. Oxford University Press, New York, pp 587–599

Tiddi B, Aureli F, Schino G (2010) Grooming for infant handling in tufted capuchin monkeys: a reappraisal of the primate infant market. Anim Behav 79:1115–1123

van Schaik CP (1989) The ecology of social relationships amongst female primates. In: Standen V, Foley RA (eds) Comparative socioecology: the behavioural ecology of humans and other animals. Blackwell Scientific, Oxford, pp 195–218

Ventura R, Majolo B, Koyama NF, Hardie S, Schino G (2006) Reciprocation and interchange in wild Japanese macaques: grooming, cofeeding, and agonistic support. Am J Primatol 68 (12):1138–1149

Wang S, Li JH, Xia DP, Zhu Y, Sun BH, Wang X, Zhu L (2013) Male-to-female agonistic support for copulation in Tibetan macaques (*Macaca thibetana*) at Huangshan, China. Zool Res 34:139–144

Wei W, Qi XG, Garber PA, Guo ST, Zhang P, Li BG (2013) Supply and demand determine the market value of access to infants in the golden snub-nosed monkey (*Rhinopithecus roxellana*). PLoS One 8(6):e65962

Wilson EO (1975) Sociobiology. Harvard University Press, Cambridge, MA

Wrangham RW (1980) An ecological model of female-bonded primate groups. Behaviour 75:262–299

Xia DP, Li JH, Garber PA, Sun L, Zhu Y, Sun BH (2012) Grooming reciprocity in female Tibetan macaques *Macaca thibetana*. Am J Primatol 74:569–579

Xia DP, Li JH, Garber PA, Matheson MD, Sun BH, Zhu Y (2013) Grooming reciprocity in male Tibetan macaques. Am J Primatol 75:1009–1020

Chapter 5
Social Relationships Impact Collective Decision-Making in Tibetan Macaques

Xi Wang, Claudia Fichtel, Lixing Sun, and Jin-Hua Li

5.1 Introduction

Group-living offers many benefits related to survival and reproduction for animals (Bertram 1978; van Schaik 1983; Zemel and Lubin 1995; Krause and Ruxton 2002). Nevertheless, it also involves some unavoidable costs, such as mate competition or interindividual conflict (Kappeler et al. 2015). Therefore, animals need to coordinate their actions and maintain group cohesion to gain the benefits of group-living (Conradt and Roper 2003; Fichtel et al. 2011). Group coordination is often difficult to achieve because individuals may differ in their needs and interests. When these differences cannot be reconciled, cohesiveness can be at risk (Rands et al. 2003,

X. Wang
School of Resources and Environmental Engineering, Anhui University, Hefei, Anhui, China

International Collaborative Research Center for Huangshan Biodiversity and Tibetan Macaque Behavioral Ecology, Anhui, China
e-mail: xwang@ahu.edu.cn

C. Fichtel
Behavioral Ecology and Sociobiology Unit, German Primate Center, Leibniz Institute for Primate Research, Göttingen, Germany
e-mail: Claudia.Fichtel@gwdg.de

L. Sun
Department of Biological Sciences, Primate Behavior and Ecology Program, Central Washington University, Ellensburg, WA, USA
e-mail: Lixing@cwu.edu

J.-H. Li (✉)
School of Resources and Environmental Engineering, Anhui University, Hefei, Anhui, China

International Collaborative Research Center for Huangshan Biodiversity and Tibetan Macaque Behavioral Ecology, Anhui, China

School of Life Sciences, Hefei Normal University, Hefei, Anhui, China
e-mail: jhli@ahu.edu.cn

J.-H. Li et al. (eds.), *The Behavioral Ecology of the Tibetan Macaque*, Fascinating Life Sciences, https://doi.org/10.1007/978-3-030-27920-2_5

2008). How consensus is achieved and implemented at the behavioral level is often studied during natural group movements to and from specific resources and/or locations (e.g., sleeping and foraging sites). Such studies can provide an ecologically relevant context to probe into fundamental mechanisms of social coordination of collective actions (Boinski and Garber 2000; Fichtel et al. 2011).

Social interactions could affect the process of group movements. For example, in large anonymous groups, such as fish schools and bird flocks, in which members do not know each other individually, direction and action during movements are regulated by self-coordination by individuals following the simple rule of keeping a certain distance to the nearest neighbor (Parrish and Edelstein-Keshet 1999; Couzin et al. 2002; Hemelrijk 2002). In contrast, in primate groups, where members know each other individually, certain dominant or affiliated individuals may play specific roles in the context of collective behavior, such as initiating or terminating a group movement (Boinski and Garber 2000; King and Cowlishaw 2009; King and Sueur 2011).

In primate species, initiation of a group movements can be accompanied by notifying behaviors (Kummer 1968) or preliminary behaviors (Sueur and Petit 2008a) that are exhibited in the pre-departure period, directly preceeding the group movement. This recruitment process often includes visual and acoustic communication, and which can influence the recruitment success of an initiation, i.e., whether the initiator is followed and, if so, by how many group members and how quickly (Sueur and Petit 2010; Seltmann et al. 2016; Sperber et al. 2017).

A successful collective movement can be driven by an unshared decision-making mechanism, i.e., one individual leading all group movements and other members following it all the time (Conradt and Roper 2005). In this decision-making process, the highest-ranking male usually plays a major role in leadership in several species of Old World monkey (Sueur and Petit 2008a). Alternatively, shared or partially shared decision-making mechanism can also result in a collective movement. That is, all group members or a subgroup can lead the movement of the entire group on different occasions (Pyritz et al. 2011). In this case, individuals with better social connections enjoy higher rates of initiating group movements (Sueur and Petit 2008b; Strandburg-Peshkin et al. 2015; Fratellone et al. 2018).

A major focus in the study of collective decisions in primates is the joining process, which occurs once an individual has initiated a collective movement. Often, primates do not decide independently on activity changes, but, rather, base their choices on the actions of their group mates. This form of joining rule, when one individual taking an action makes it more likely for another to do so as well, has been termed mimetism (Deneubourg and Goss 1989). Mimetism can be further categorized as anonymous mimetism and selective mimetism. Anonymous mimetism refers to individuals being more likely to take the actions of other individuals irrespective of their identity, and thus group movements can simply depend on the number of individuals who have already left or performed a certain behavior (Petit et al. 2009). Selective mimetism refers to joining decisions based on some other factors, such as distance, with individuals being more likely to join a movement when in close proximity (Ramseyer et al. 2009; Ward et al. 2013), and affiliation, with individuals being more likely to follow those group members with whom they have strong social bonds (King et al. 2011; Sueur et al. 2009, 2011; Seltmann et al. 2013; Strandburg-Peshkin et al. 2015; Farine et al. 2016).

Individuals may also engage another joining process based on a quorum rule. This rule states that once a minimum number of group members joins a movement, group movement will occur all the time (Conradt and Roper 2003; Wang et al. 2015; Rowe et al. 2018). A response to a quorum is observed when the probability of members exhibiting a vote by joining a movement depends on the number of individuals already performing the voting behavior (Pratt et al. 2002; Seeley and Visscher 2004; Sumpter 2006; Ward et al. 2008).

Interestingly in macaques, social style appears to influence the organization of group movements (Sueur and Petit 2008a). Social styles of the species of *Macaca* have been divided into four grades (Thierry 2000), ranging from grade 1 (the most intolerant) to grade 4 (the most tolerant). These styles appear to influence the initiation and joining process of group movements (Sueur and Petit 2008a). In species with a more despotic dominance style, decision-making is more likely to be unshared and social rank determines leadership (Sueur and Petit 2008a, b). In contrast, in species exhibiting a more egalitarian style, decision-making is more likely to be shared and social relationships determine leadership (Sueur and Petit 2008a). For instance, in rhesus (*Macaca mulatta*) and Japanese macaques (*M. fuscata*), both of which have a despotic social style, movements are mainly initiated by dominant individuals, and joining processes are also determined by dominance order (Sueur et al. 2009; Jacobs et al. 2011). In contrast, in macaques with a more egalitarian social style, such as Tonkean (*M. tonkeana*) and Barbary macaques (*M. sylvanus*), decision-making is equally or partially shared, and joining processes are determined by affiliation (Sueur et al. 2009; Jacobs et al. 2011; Seltmann et al. 2013).

Currently, the relationship between social style and leadership and other important aspects of collective decision-making has been investigated only in a limited number of macaque species. Therefore, detailed studies of collective movement in relatively less known species, such as the Tibetan macaque (*M. thibetana*), are of special interest for a better understanding of the link between social relationships and collective decision-making. We therefore take this opportunity to review and synthesize information based largely on our studies of Tibetan macaques. We hope that some of the findings and insights (including those that have not been fully developed in our publications) will enrich our understanding of decision-making processes in collective movement in primates in general and macaques in particular.

5.2 Collective Decision-Making in Tibetan Macaques

5.2.1 A Macaque Species for Studying Decision-Making

Because *Macaca* species vary in dominance style (Thierry et al. 2004), they serve as an interesting taxon to study how social relationship influences the process of group movements (Jacobs et al. 2011). Tibetan macaques are highly gregarious and live in cohesive groups (Li 1999). Similar to other macaques, Tibetan macaques show female philopatry, male dispersal, and linear dominance hierarchies (Berman et al. 2004).

The wild group of macaques we studied (YA1, see Chap. 2 for detailed information about the history, demography, and habitat of the study group) engaged in social activities in nearby forest during most of the day without any restriction on their home range. For the convenience of viewing by tourists, they were supplied with 3–4 kg of corn daily (Berman and Li 2002; Berman et al. 2008; Xia et al. 2012). After corn feeding, they regularly switched locations from the feeding site to the nearby forest. Collective movements often occurred at the time of the switch. We therefore investigated decision-making processes during group movements in this group from August to December of 2012.

5.2.2 Decision-Making During the Initiation Process of Group Movements

It is still debated whether Tibetan macaques exhibit a despotic or a tolerant dominance style (Thierry 2000; Berman et al. 2004). In accordance with a despotic dominance style (Berman et al. 2004), Tibetan macaques should be expected to show an unshared decision-making process when initiating group movements, i.e., a single, highest-ranking individual leads most movements. In reality, however, Tibetan macaques demonstrate a shared decision-making process with affiliative individuals more often initiating group movements than less sociable group members (Wang et al. 2016). Considering the above contradiction about dominance style in Tibetan macaques, we assume that there may be a potential connection between social rank/affiliative relationship and the initiation of group movement.

We observed initiation processes when Tibetan macaques returned from the feeding site to the nearby forest. Thus, an initiator was defined as the first individual that moved more than 10 m in less than 30 s from the provisioning area to the forest. Any individual walking more than 5 m and within 45° of the direction taken by the initiator was considered as a follower (Sueur and Petit 2008a). We used the criterion of 5 min for each successive follower who joined the movement after the first mover or previous follower (Wang et al. 2016). In our study, only those movements including at least two-thirds of group members were counted as successful group movements. To quantify leadership in group movements, we standardized initiation

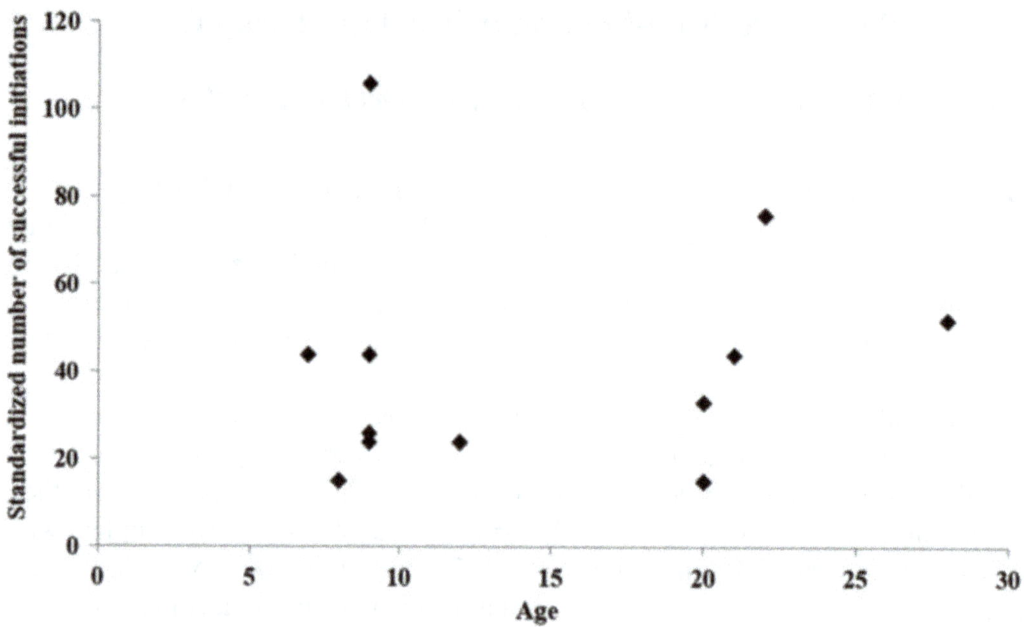

Fig. 5.1 Age of adults and their successful initiations of group movements in Tibetan macaques. There was no correlation between the two variables

data on the number of initiations of each individual by the number of times in which this individual was identified at the provisioning area (Wang et al. 2016).

During the 5 months of our observation period, we recorded more than 200 initiation attempts, all of them by adult members. Two-thirds of these initiations were considered as successful group movements. We found that all adults could initiate group movements, but that they differed significantly in the standardized number of successful initiations. This result clearly showed that decision-making during the initiation process of group movement was shared among adults.

To explore which factors might affect collective decision-making, we analyzed the relationship between several key biological/social attributes and the initiation of group movement. Interestingly, there was no significant difference in the standardized number of successful initiations between adult males and females. Second, there was neither a correlation between social rank and the standardized number of successful initiations nor with the success ratio of initiations. Also, age of adults was not correlated with the successful initiation of group movement (Fig. 5.1).

To evaluate the relationship between social affiliation of an adult and its leadership in group movements, we related the number and ratio of successful initiations of every subject in the provisioning area to its eigenvector centrality coefficient based on proximity relations among group members when they were in the forest (in comparison with the situations when they were in the feeding site). We used focal animal sampling method to collect proximity data for assessing affiliative relationships among group members (Altmann 1974; Li 1999; Berman et al. 2008). We found a positive correlation between the eigenvector centrality coefficient (based on proximity relations) and the standardized number of successful initiations

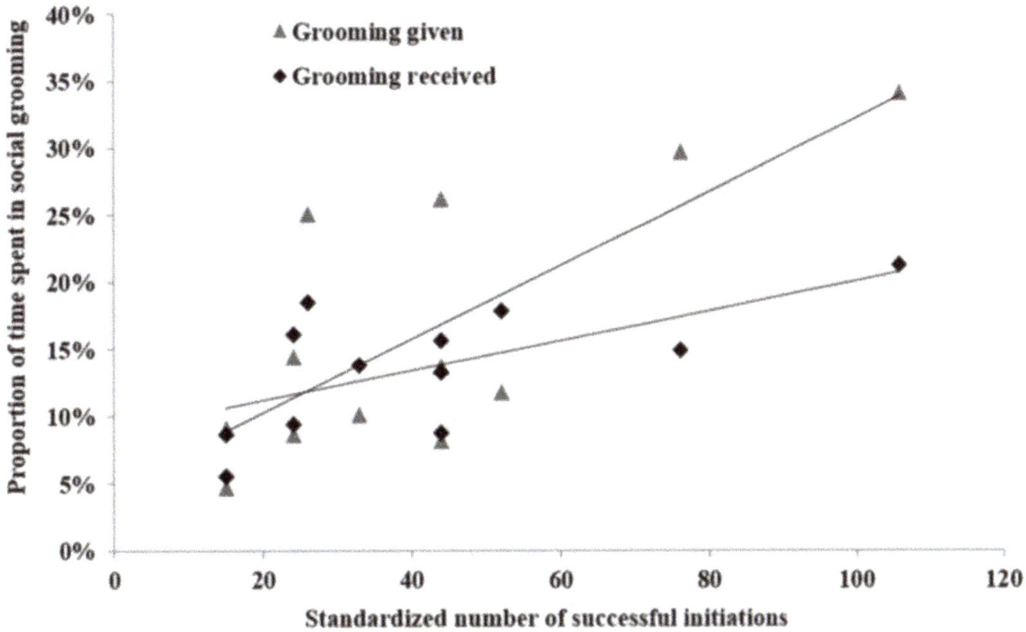

Fig. 5.2 Correlation between successful initiations of group movements and social grooming in Tibetan macaques. Proportion of time spent in social grooming indicates the duration of grooming of every subject divided by its total focal observation time. The two lines respectively represent positive correlations between grooming given/grooming received and successful initiations of group movements

across adults. Eigenvector centrality was also correlated positively with the success ratio of initiations.

Moreover, to further analyze the relationship between affiliated behavior and leadership of group movement, we correlated the initiation of group movement with social grooming among adults. Results showed that the standardized number of successful initiations was positively correlated with the duration of social grooming, including grooming given and grooming received (Fig. 5.2).

5.2.3 Decision-Making During the Joining Process of Group Movements

Joining processes were also observed when Tibetan macaques returned from the feeding site, where they were regularly provisioned, to the nearby forest. We tested whether joining occurs according to a quorum decision or mimetism as shown in a simple schematic to illustrate how the two responses would differ (Fig. 5.3).

During our preliminary observation of group YA1 (August 1–14, 2012), 5 min were used as the minimum duration of initiating a successful group movement. Therefore, an early joiner was defined as an individual that moved in the first 5 min after the initiator departed (Wang et al. 2015). According to this definition, an

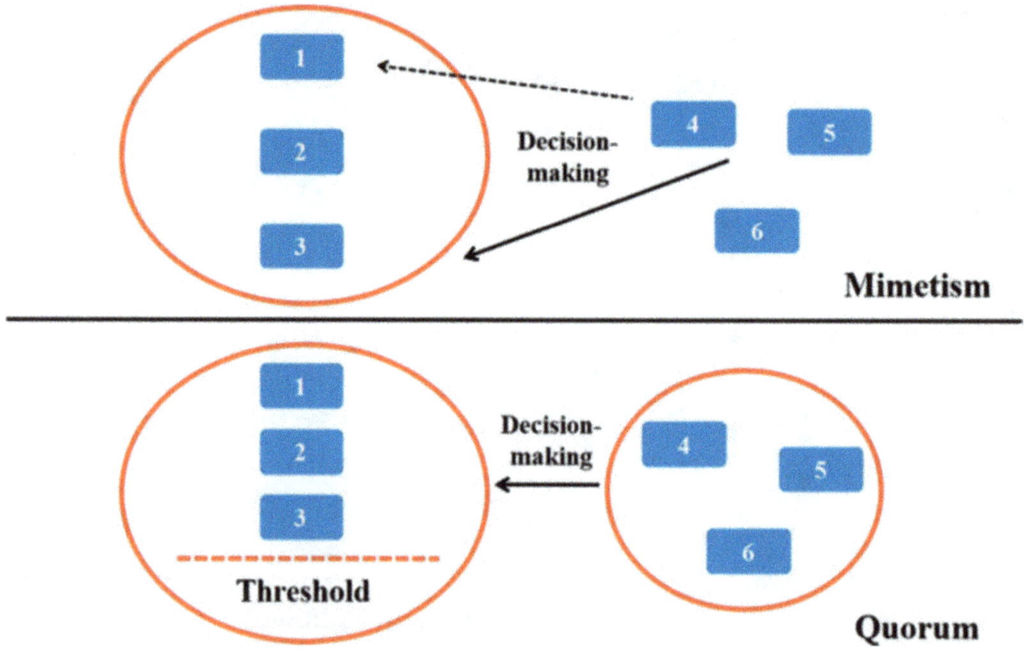

Fig. 5.3 A schematic depiction of mimetism and quorum. Nos. 1, 2, and 3 in red circles represent individuals who have joined the movement. Nos. 4, 5, and 6 are individuals waiting to join the movement. Dotted line of upper figure indicates selective mimetism: Nos. 4, 5, and 6 would join the movement based on the choice of specific members (e.g., No. 1 is the highest-ranking male). Solid line of upper figure indicates anonymous mimetism: Nos. 4, 5, and 6 would join the movement based on the number of joiners by linearly. Threshold of lower figure indicates quorum: for example, once half of the members (i.e., three individuals) have joined the movement, other individuals would follow the collective action all the time

initiator in our study was also considered as an early joiner because he/she left in the first 5 min.

We assessed the relationship between the number of adult early joiners and the probability of successful group movement. The results revealed that group movements were not successful unless three or more early joiners participated in the movement. When three to six early joiners participated, successful group movements occurred without a consistent pattern, showing some fluctuations in the probability of successful group movement. Nonetheless, once more than half of the early joiners participated in the movement, other group members followed the collective actions all the time.

To further study the role of early joiners in group movements, we performed a correlation analysis between the mean joining order and eigenvector centrality coefficient for adults in group movements (Wang et al. 2015). Results showed that the earlier an individual joined the movement, the higher its centrality was (Fig. 5.4). We then explored key attributes of early joiners in the social network of those in group movements. Our results showed that early joiners differed significantly in eigenvector centrality coefficient based on the half-weight index (HWI: co-occurrence index in group movements, Wang et al. 2015), but there was no difference between adult males and females. Also, age and eigenvector centrality

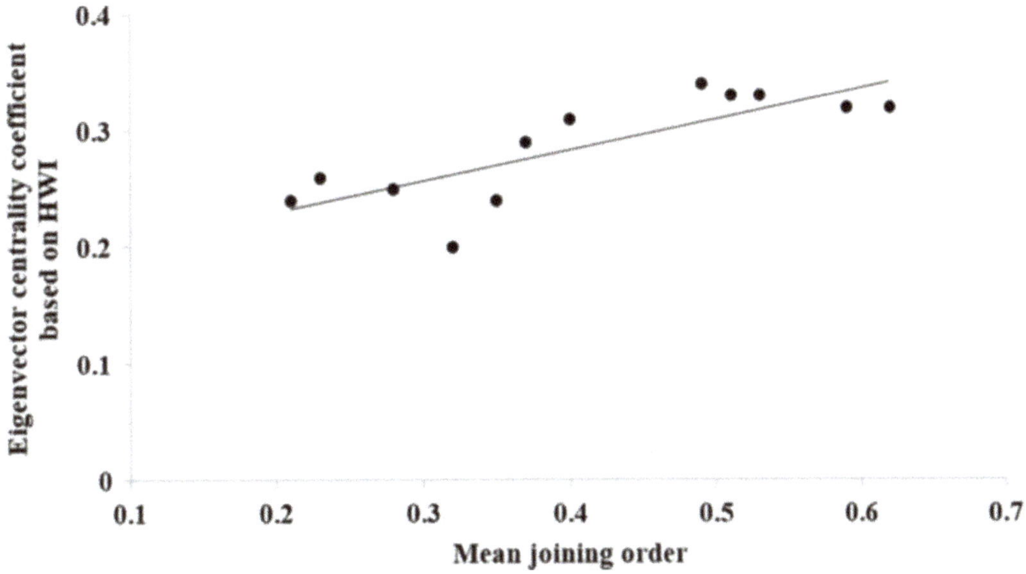

Fig. 5.4 Relationship between individuals' joining order of group movements and eigenvector centrality coefficient. HWI: co-occurrence index in group movements. The bigger the value of mean joining order was, the earlier an individual joined the movement

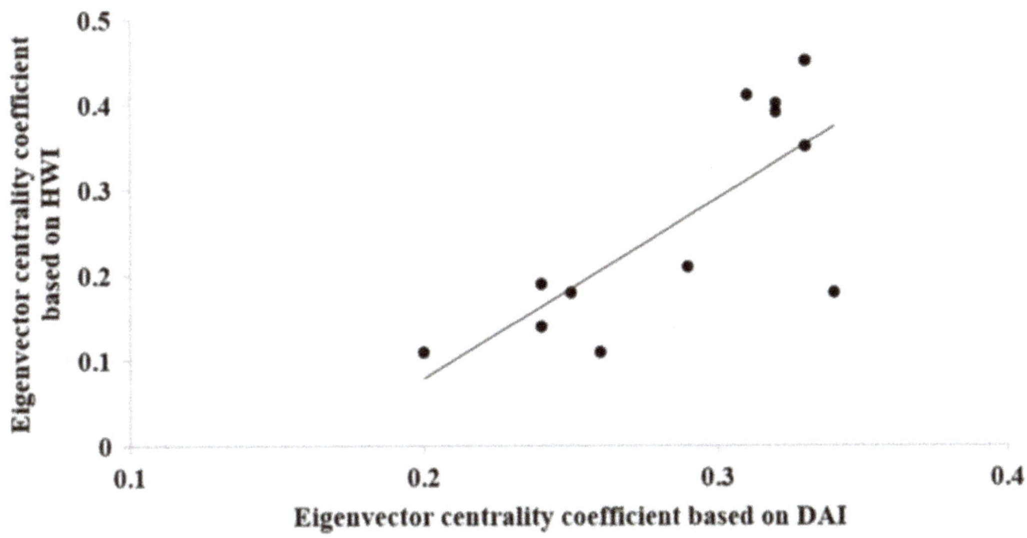

Fig. 5.5 Eigenvector centrality coefficients of individuals based on HWI and DAI. HWI: co-occurrence index in group movements. DAI: dyadic association index in proximity relations

coefficients were not correlated. However, social rank was positively correlated with eigenvector centrality coefficients in both adult males and females.

Finally, we compared two eigenvector centrality coefficients of individuals in group movements and in proximity relations (Wang et al. 2015). We found a positive correlation between the two coefficients (Fig. 5.5). This result indicates a close relation between affiliative behavior and the joining process of group movements.

5.3 Social Relationship and Collective Decision-Making

Our collection of studies revealed that all adult Tibetan macaques had opportunities to successfully initiate group movements. This result is consistent with studies in Barbary macaques (Seltmann et al. 2013) and Tonkean macaques (Sueur and Petit 2008a), both of which have a tolerant social style. However, different adults had varying times of success in initiating group movements across the study period. Eigenvector centrality coefficients were positively correlated with the number of successful initiation of group movements.

Our data demonstrate that more affiliative individuals were more likely to assume leadership roles. The importance of social relationship is consistent with the results of several other studies (e.g., Tonkean macaques: Sueur and Petit 2008a, b; Sueur et al. 2009). King et al. (2011) suggested that chacma baboons (*Papio ursinus*) with higher eigenvector centrality coefficients in their social network were more likely to attract partners to follow their initiations. We assume that individuals preferentially followed "friends" because following friends may be associated with benefits. For instance, female baboons (*P. cynocephalus*) that had closer bonds with others lived longer and their offspring also had a higher probability to survive (Silk et al. 2010), but in Tibetan macaques neither these friendships nor their potential consequences have been studied yet.

We found a positive correlation between the number of successful initiations and the amount of time spent grooming given/received, suggesting that the motivation to move may lay in staying away from the provisioning area and moving to the relatively calm forest, where animals could engage in social grooming. We carried out our study at a tourist site (Wang et al. 2016), and adult Tibetan macaques at our study site had indeed higher rates of aggression in the provisioning area than in the forest (Berman et al. 2007). Moreover, as no large predators were present at the study site in recent decades (Li 1999), the forest was supposed to be a safer and less stressful location for monkeys to engage in affiliative behaviors such as grooming. We see the location change from the provisioning area to forest as an adaptive response to the socioecological conditions experienced by our study group.

Our work on the joining process of group movements has provided several important insights into which rules might be used by Tibetan macaques in decision-making during collective movement. First, we found that the more central a group member was in the social network, the earlier it participated in a movement, as shown by the positive correlation between the mean joining order of every joiner and its eigenvector centrality coefficient (Wang et al. 2015). This result is comparable with situations in black howler monkeys (*Alouatta pigra*), where females at the front of a group movement have the highest centrality eigenvectors among the adult group members (Belle et al. 2013). Our results demonstrate the importance of early joiners in the decision-making process and indicate that the initiator was not always the only decision-maker. Other joiners in the first 5 min might also play a key role in decision-making.

Second, we found that higher-ranking early joiners tended to have higher eigenvector centrality coefficients. Because the eigenvector centrality coefficient can quantify the attraction of early joiners to other members during the joining process (Newman 2004), our data (Wang et al. 2015) showed the role of social rank in early joiners. That is, higher-ranking joiners could have more companions in group movements. A similar influence of social rank on decision-making has also been reported in other species. For instance, alpha males have been reported to be the consistent decision-makers in group movements in mountain gorillas, *Gorilla gorilla* (Watts 2000).

Third, we found that early joiners who had higher centrality coefficients in proximity activities also had higher centrality coefficients during collective movements. This means that early joiners with frequent social interactions could also attract more members during the joining process. This result is comparable to findings in chacma baboons and red-fronted lemurs, *Eulemur rufifrons* (Stueckle and Zinner 2008; King et al. 2011; Sperber et al. 2019).

Our data also showed that when the number of early joiners in a group was below seven, group members preferred to follow higher-ranking or affiliative early joiners during the joining process of group movements. By this time, selective mimetism had most likely been used as the joining rule. Apart from our study, selective mimetism has also been suggested in several other studies (Detrain et al. 1999; Camazine et al. 2001; Couzin and Krause 2003; Sumpter 2006; Gautrais et al. 2007; Sperber et al. 2019). In Tonkean macaques, for instance, how an individual decides to join a collective movement depends on whether it is strongly affiliated to departing individuals (Sueur et al. 2009).

We can explain selective mimetism in Tibetan macaques with respect to their social style. For example, Tonkean macaques exhibit an egalitarian social structure (Sueur and Petit 2008b). In this species, individuals decide when to move via a quorum. The lack of centrality for dominant or old Tonkean macaques suggests that all individuals may have equal weight in the voting process and interactions are not constrained by individual status (Sueur and Petit 2008b). However, rhesus macaques, with a despotic style, prefer to join high-ranking or related individuals during collective movements, showing selective mimetism (Sueur and Petit 2008a, b; Sueur et al. 2010). In Tibetan macaques, higher-ranking early joiners were socially connected to more individuals than lower-ranking group members, both in collective movements and during proximity activities. As our results showed that high-ranking individuals had high centrality coefficients based on the co-occurrence index in group movements, they were more attractive to other members than low-ranking individuals during the joining process.

Our data also showed that when the number of early joiners had accumulated to more than half of the adults in our study group, all group members would participate in group movement all the time. Clearly, this threshold of ">50% adult members" indicates the existence of another joining rule, the quorum rule, during collective movements in Tibetan macaques. The voting process in the group of Tibetan macaques we studied may be explained by the reduction in the risk of being left behind from the group (Wang et al. 2015). A similar joining rule has also been found in white-faced capuchins (*Cebus capucinus*), rhesus macaques, and Tonkean macaques, all of which use the threshold of four in collective movements (Petit et al. 2009; Sueur and Petit 2010).

5.4 Conclusions

In this chapter, we reviewed and synthesized studies of collective movement and decision-making in Tibetan macaques. We suggest that leadership of group movements in Tibetan macaques was distributed among adults rather than exclusively taken by a single, high-ranking individual. Different members led the group on different occasions, and social relationships were more related to leadership than social rank, age, or sex. Performing group movement presumably produced opportunities to switch locations for adults to participate in other social activities, including social grooming. Moreover, social relationship also mattered for the joining process. Tibetan macaques used selective mimetism and a quorum process in collective decision-making, and early joiners with closer affiliation played a critical role as to which rule was used. Thus, our study provided further evidence for the link between social relationships and collective decision-making in a little known macaque species. Future studies can examine whether social relationships affect the decision-making process in Tibetan macaques at the same level when group size varies. This will lead us to a better understanding as to whether a general pattern exists for group coordination and social cohesion through collective decision-making.

Acknowledgments We thank the Huangshan Monkey Management Center and the Huangshan Garden Forest Bureau for their permission for us to conduct research at the field site. We are also grateful to Haibin Cheng's family for their outstanding logistic support to our study. This work was supported by grants from the National Natural Science Foundation of China (No. 31801983; 31772475; 31672307; 31401981; 31372215), the Initial Foundation of Doctoral Scientific Research in Anhui University (J01003268), the Training Program for Excellent Young Teachers (J05011709) and China Scholarship Council.

References

Altmann J (1974) Observational study of behavior: sampling methods. Behaviour 49:227–267

Belle SV, Estrada A, Garber PA (2013) Collective group movement and leadership in wild black howler monkeys (*Alouatta pigra*). Behav Ecol Sociobiol 67:31–41

Berman CM, Li JH (2002) Impact of translocation, provisioning and range restriction on a group of *Macaca thibetana*. Int J Primatol 23:383–397

Berman CM, Ionica C, Li JH (2004) Dominance style among *Macaca thibetana* on Mt. Huangshan, China. Int J Primatol 25:1283–1312

Berman CM, Li JH, Ogawa H, Ionica C, Yin HB (2007) Primate tourism, range restriction, and infant risk among *Macaca thibetana* at Mt. Huangshan, China. Int J Primatol 28:1123–1141

Berman CM, Ogawa H, Ionica C, Yin HB, Li JH (2008) Variation in kin bias over time in a group of Tibetan macaques at Huangshan, China: contest competition, time constraints or risk response? Behaviour 145:863–896

Bertram BCR (1978) Living in groups: predators and prey. In: Krebs JR, Davies JB (eds) Behavioural ecology. Blackwell, Oxford, pp 64–96

Boinski S, Garber PA (2000) On the move: how and why animals travel in groups. University of Chicago Press, Chicago, IL

Camazine S, Deneubourg JL, Franks NR, Sneyd J, Theraulaz G, Bonabeau E (2001) Self-organization in biological systems. Princeton University Press, Princeton, NJ

Conradt L, Roper TJ (2003) Group decision-making in animals. Nature 421:155–158

Conradt L, Roper TJ (2005) Consensus decision making in animals. Trends Ecol Evol 20:449–456

Couzin ID, Krause J (2003) Self-organization and collective behavior in vertebrates. Adv Stud Behav 32:1–75

Couzin ID, Krause J, James R, Ruxton GD, Franks NR (2002) Collective memory and spatial sorting in animal groups. J Theor Biol 218:1–11

Deneubourg JL, Goss S (1989) Collective patterns and decision-making. Ital J Zool 1:295–311

Detrain C, Deneubourg JL, Pasteels JM (1999) Decision-making in foraging by social insects. In: Detrain C, Deneubourg JL, Pasteels JM (eds) Information processing in social insects. Birkhaüser Verlag, Basel, pp 331–354

Farine DR, Strandburg-Peshkin A, Berger-Wolf T, Ziebart B, Brugere I, Li J, Crofoot MC (2016) Both nearest neighbours and long-term affiliates predict individual locations during collective movement in wild baboons. Sci Rep 6:27704

Fichtel C, Pyritz L, Kappeler PM (2011) Coordination of group movements in non-human primates. In: Boos M, Kolbe M, Kappeler P, Ellwart T (eds) Coordination in human and primate groups. Springer, Heidelberg, pp 37–56

Fratellone GP, Li JH, Sheeran LK, Wagner RS, Wang X, Sun L (2018) Social connectivity among female Tibetan macaques (*Macaca thibetana*) increases the speed of collective movements. Primates 60(3):183–189. https://doi.org/10.1007/s10329-018-0691-6

Gautrais J, Michelena P, Sibbald A, Bon R, Deneubourg JL (2007) Allelomimetic synchronization in merino sheep. Anim Behav 74:1443–1454

Hemelrijk CK (2002) Understanding social behaviour with the help of complexity science. Ethology 108:655–671

Jacobs A, Watanabe K, Petit O (2011) Social structure affects initiations of group movements but not recruitment success in Japanese macaques (*Macaca fuscata*). Int J Primatol 32:1311–1324

Kappeler PM, Cremer S, Nunn CL (2015) Sociality and health: impacts of sociality on disease susceptibility and transmission in animal and human societies. Philos Trans R Soc Lond B Biol Sci 370:20140116

King AJ, Cowlishaw G (2009) Leaders, followers and group decision-making. Commun Integr Biol 2:147–150

King AJ, Sueur C (2011) Where next? Group coordination and collective decision making by primates. Int J Primatol 32:1245–1267

King AJ, Sueur C, Huchard E, Cowlishaw G (2011) A rule-of-thumb based on social affiliation explains collective movements in desert baboons. Anim Behav 82:1337–1345

Krause J, Ruxton GD (2002) Living in groups. Oxford University Press, Oxford

Kummer H (1968) Social organisation of hamadryas baboons. University of Chicago Press, Chicago, IL

Li JH (1999) The Tibetan macaque society: a field study. Anhui University Press, Hefei. (In Chinese)

Newman MEJ (2004) Analysis of weighted networks. Phys Rev E 70:056131

Parrish JK, Edelstein-Keshet L (1999) Complexity, pattern, and evolutionary trade-offs in animal aggregation. Science 284:99–101

Petit O, Gautrais J, Leca JB, Theraulaz G, Deneubourg JL (2009) Collective decision-making in white-faced capuchin monkeys. Proc R Soc B 276:3495–3503

Pratt SC, Mallon EB, Sumpter DJT, Franks NR (2002) Quorum sensing, recruitment, and collective decision-making during colony emigration by the ant *Leptothorax albipennis*. Behav Ecol Sociobiol 52:117–127

Pyritz LW, Kappeler PM, Fichtel C (2011) Coordination of group movements in wild red-fronted lemurs: processes and influence of ecological and reproductive seasonality. Int J Primatol 32:1325–1347

Ramseyer A, Boissy A, Dumont B, Thierry B (2009) Decision making in group departures of sheep is a continuous process. Anim Behav 78:71–78

Rands SA, Cowlishaw G, Pettifor RA, Rowcliffe JM, Johnstone RA (2003) Spontaneous emergence of leaders and followers in foraging pairs. Nature 423:432–434

Rands SA, Cowlishaw G, Pettifor RA, Rowcliffe JM, Johnstone RA (2008) The emergence of leaders and followers in foraging pairs when the qualities of individuals differ. BMC Evol Biol 8:51

Rowe AK, Li JH, Sun L, Sheeran LK, Wagner RS, Xia DP, Uhey DA, Chen R (2018) Collective decision making in Tibetan macaques: how followers affect the rules and speed of group movement. Anim Behav 146:51–61

Seeley TD, Visscher PK (2004) Quorum sensing during nest-site selection by honeybee swarms. Behav Ecol Sociobiol 56:594–601

Seltmann A, Majolo B, Schülke O, Ostner J (2013) The organization of collective group movements in wild Barbary macaques (*Macaca sylvanus*): dominance style drives processes of group coordination in macaques. PLoS One 8:e67285

Seltmann A, Franz M, Majolo B, Qarro M, Ostner J, Schülke O (2016) Recruitment and monitoring behaviors by leaders predict following in wild Barbary macaques (*Macaca sylvanus*). Primate Biol 3:23–31

Silk JB, Beehner JC, Bergman TJ, Crockford C, Engh AL, Moscovice LR, Wittig RM, Seyfarth RM, Cheney DL (2010) Strong and consistent social bonds enhance the longevity of female baboons. Curr Biol 20:1359–1361

Sperber AL, Werner LM, Kappeler PM, Fichtel C (2017) Grunt to go – vocal coordination of group movements in redfronted lemurs. Ethology 123:894–905

Sperber AL, Kappeler PM, Fichtel C (2019) Should I stay or should I go? Individual movement decisions during group departures in redfronted lemurs. Proc R Soc Open Sci 6(3). https://doi.org/10.1098/rsos.180991

Strandburg-Peshkin A, Farine DR, Couzin ID, Crofoot MC (2015) Group decisions. Shared decision-making drives collective movement in wild baboons. Science 348:1358–1361

Stueckle S, Zinner D (2008) To follow or not to follow: decision making and leadership during the morning departure in chacma baboons. Anim Behav 75:1995–2004

Sueur C, Petit O (2008a) Shared or unshared consensus decision in macaques? Behav Process 78:84–92

Sueur C, Petit O (2008b) Organization of group members at departure is driven by dominance style in *Macaca*. Int J Primatol 29:1085–1098

Sueur C, Petit O (2010) Signals use by leaders in *Macaca tonkeana* and *Macaca mulatta*: groupmate recruitment and behaviour monitoring. Anim Cogn 13:239–248

Sueur C, Petit O, Deneubourg JL (2009) Selective mimetism at departure in collective movements of *Macaca tonkeana*: an experimental and theoretical approach. Anim Behav 78:1087–1095

Sueur C, Deneubourg JL, Petit O (2010) Sequence of quorums during collective decision making in macaques. Behav Ecol Sociobiol 64:1875–1885

Sueur C, Deneubourg J-L, Petit O (2011) From the first intention movement to the last joiner: macaques combine mimetic rules to optimize their collective decisions. Proc R Soc B 278:1697–1704

Sumpter DJT (2006) The principles of collective animal behaviour. Philos Trans R Soc Lond B Biol Sci 361:5–22

Thierry B (2000) Covariation of conflict management patterns across macaque species. In: Aureli F, de Waal FBM (eds) Natural conflict resolution. University of California Press, Berkeley, CA, pp 106–128

Thierry B, Singh M, Kaumanns W (2004) Macaque societies: a model for the study of social organization. Cambridge University Press, Cambridge

van Schaik CP (1983) Why are diurnal primates living in groups? Behaviour 87:120–144

Wang X, Sun L, Li JH, Xia DP, Sun BH, Zhang D (2015) Collective movement in the Tibetan macaques (*Macaca thibetana*): early joiners write the rule of the game. PLoS One 10:e0127459

Wang X, Sun L, Sheeran LK, Sun BH, Zhang QX, Zhang D, Xia DP, Li JH (2016) Social rank versus affiliation: which is more closely related to leadership of group movements in Tibetan macaques (*Macaca thibetana*)? Am J Primatol 78:816–824

Ward AJW, Sumpter DJT, Couzin ID, Hart PJB, Krause J (2008) Quorum decision-making facilitates information transfer in fish shoals. Proc Natl Acad Sci U S A 105:6948–6953

Ward AJW, Herbert-Read JE, Jordan LA, James R, Krause J, Ma Q, Rubenstein DI, Sumpter DJT, Morrell LJ (2013) Initiators, leaders, and recruitment mechanisms in the collective movements of damselfish. Am Nat 181:748–760

Watts D (2000) Mountain gorilla habitat use strategies and group movements. In: Boinski S, Garber PA (eds) On the move. University of Chicago Press, Chicago, IL, pp 351–374

Xia DP, Li JH, Garber PA, Sun LX, Zhu Y, Sun BH (2012) Grooming reciprocity in female Tibetan macaques *Macaca thibetana*. Am J Primatol 74:569–579

Zemel A, Lubin Y (1995) Inter-group competition and stable group sizes. Anim Behav 50:485–488

Chapter 6
Considering Social Play in Primates: A Case Study in Juvenile Tibetan Macaques (*Macaca thibetana*)

Jessica A. Mayhew, Jake A. Funkhouser, and Kaitlin R. Wright

6.1 Introduction

The evolutionary origins and adaptive value of animal play behavior have long been contemplated. Social play in gregarious animals is a multidimensional topic that has long been debated, insufficiently investigated, and a source of enigmatic questions regarding its development, relationship to cognition, and adaptive value. Play has been characterized as the most sophisticated manifestation of communication (Bekoff 1972; Bekoff and Allen 1997; Burghardt 2005; Fagen 1981; Pellis and Pellis 2009), which is partially why it is challenging to study. Beyond its incorporation into species' activity budgets, researchers have encountered multiple logistical and theoretical stumbling blocks, including difficulty in operationalizing definitions, identifying multiple juveniles and their varied social relationships, keeping pace with changes in interaction tempo and player composition, quantifying the behaviors observed and their short- and long-term costs and benefits, and determining an individual's motivation and intention to engage with others in this specific manner. Nevertheless, social play warrants attention because of its inherent complexity,

J. A. Mayhew (✉)
Department of Anthropology and Museum Studies, Central Washington University, Ellensburg, WA, USA

Primate Behavior and Ecology Program, Central Washington University, Ellensburg, WA, USA
e-mail: MayhewJ@cwu.edu

J. A. Funkhouser
Primate Behavior and Ecology Program, Central Washington University, Ellensburg, WA, USA

Department of Anthropology, Washington University in St. Louis, St. Louis, MO, USA
e-mail: jakefunkhouser@wustl.edu

K. R. Wright
Primate Behavior and Ecology Program, Central Washington University, Ellensburg, WA, USA
e-mail: wrightkr@uw.edu

J.-H. Li et al. (eds.), *The Behavioral Ecology of the Tibetan Macaque*, Fascinating Life Sciences, https://doi.org/10.1007/978-3-030-27920-2_6

perceived contributions to an individual's fitness, communicative content, and interspecies variability.

In this chapter, we provide a brief overview of play behavior, with special attention to macaques, and propose some considerations to others interested in studying play. To emphasize some of these points, we provide an example where we consider the multiple factors characterizing the social play behavior of the 2017 Yulingkeng A1 infant and juvenile Tibetan macaques (*Macaca thibetana*) at Mt. Huangshan, China. Additionally, we map their positions in a social play network and use this foundation to generate hypotheses regarding their future network positions. We encourage future studies to address how the construction of these juvenile relationships contributes to an individual's overall group position and the formation of adult relationships and the potential adaptive advantage play provides.

6.2 Play Behavior: An Overview

The diversity in the form and content of play has generated discussion about the proximate and ultimate costs/benefits and whether generalizations about this behavior can be made across taxa. There are multiple definitions of play emphasizing either structure or function (e.g., Bekoff and Allen 1997; Fagen 1981; Martin and Caro 1985), but the development of five descriptive criteria by Burghardt (2005) has provided researchers the parameters to identify play as distinct from other common behaviors in a repertoire. For a behavior to be labeled "playful," Burghardt (2005) proposes it should (1) have a limited immediate function, (2) be endogenous, (3) have structural or temporal properties that are different from "serious" behaviors, (4) be flexibly exercised and not stereotypical, and (5) be performed in a relaxed field (i.e., free of stress or social/physical pressures). Guided by these criteria, researchers now have a foundation to tackle more complex questions of how and why animals play.

Play is often broadly categorized as solitary, object, and social (Bekoff and Byers 1981; Fagen 1981), but these categories are not mutually exclusive. Solitary play typically includes intense or sustained locomotor movements performed alone, e.g., the running and gamboling of young ungulates. Object play can be solitary or social and includes the manipulation of an object for no immediate benefit. Object play is commonly observed in carnivores, in which predatory movements, such as grabbing and shaking, are performed on a non-prey item (Burghardt 2005). Primates also engage in object play from stick carrying in chimpanzees (*Pan troglodytes*) (Kahlenberg and Wrangham 2010), stone handling in Japanese macaques (*M. fuscata*, Nahallage et al. 2016; Shimada 2006, 2010), to the manipulation of one's environment in functional object substitution, sparking discussion about pretense, imagination, and theory of mind (Gómez and Martín-Andrade 2002). Social play is identified as being interactive and occurs between two or more conspecifics who may influence one another's actions (Thompson 1996). Social play often includes aggressive behaviors, such as biting and wrestling (Burghardt 2005), but tends to be reciprocal between partners (Fagen 1981).

The type and exhibition of play depends on the species observed. Typically, the occurrence of play increases throughout juvenilehood but then tapers off at sexual maturity where activity budgets and individual priorities shift toward competition and reproductive resources. In birds, orders with more altricial species tend to engage in more play than orders with mostly precocial species (Ortega and Bekoff 1986). Raptors (e.g., eagles and hawks) engage in object play (Ortega and Bekoff 1986), young herring gulls (*Larus argentatus*) have been observed performing "drop-catch" behavior with clams and nonfood objects over hard substrates (Gamble and Cristol 2002), and ravens (*Corvus corax*) slide down snowy inclines and hang upside down from tree branches (Heinrich and Smolker 1998). Members of the family Canidae, including wolves, coyotes, foxes, and jackals, engage in solitary, object, and social play. Of this group, domestic dogs (*Canis familiaris*) are the most familiar play partners to humans and even use play signals that can be readily understood and responded to (e.g., the play bow) (Bekoff 1974). Cetacean play also takes on multiple forms, including the creation of play objects in the form of bubbles (Hill et al. 2017; Jones and Kuczaj 2014). For elephants (*Loxodonta africana*), play occurs throughout the life course on land and in water, and object play can be observed by individuals of all ages (Lee and Moss 2014). Guided by Burghardt's criteria, play can also be identified in animals that are not commonly regarded as being playful, including cichlid fish (*Tropheus duboisi*, Burghardt et al. 2014), Nile softshell turtles (*Trionyx triunguis*, Burghardt et al. 1996), octopus (*Octopus dofleini*, Mather and Anderson 1999), and even spiders (Pruitt et al. 2012).

Social play intrinsically involves partner cooperation, complex communication, and learning, and these are critical variables to investigate if we are to understand the cognitive and social development in young individuals (Bekoff and Allen 1997; Palagi et al. 2007). The functional significance of this suite of "nonserious" social behaviors remains elusive yet intriguing to ethologists; nevertheless, play has been noted to nourish the physical, social, and cognitive aspects of an individual (see Palagi 2018 for review). Namely, play functions to increase and maintain physical fitness or motor performance (Byers and Walker 1995; Fontaine 1994; Martin and Caro 1985), refines social skills and increases behavioral flexibility (Baldwin and Baldwin 1974; Brown 1988; Fagen 1984), and may be an important context that enhances cognitive skills in which an individual learns to identify and respond to the intentions of others through social cues (Bekoff 1972; Bekoff and Allen 1997; Palagi et al. 2007). These types of interactions serve as an opportunity to accrue and refine adult social skills and enhance behavioral flexibility (Baldwin and Baldwin 1974; Brown 1988; Fagen 1984). The unpredictable nature of this intimate social exchange challenges the participants to literally "think on their feet" as the interaction occurs, which imposes some inherent risk, but sharpens social tactics and reactions. In general, play is expected to occur more frequently with a partner that is evenly matched in skills and ability (*M. fuscata*, Kulik et al. 2015) as well as with individuals that are likely to be frequently encountered in adulthood (e.g., Maestripieri and Ross 2004). Pellis and Pellis (1996) hypothesize that rough-and-tumble play influences the developmental of dominance relationships in postpubertal juveniles (i.e., subadult), especially in male-male play bouts, through testing a play partner's strength. Intense or more aggressive behaviors, such as bite or chase, could be used to create or maintain a competitive edge in a play bout (e.g., *Gorilla gorilla*, van Leeuwen et al. 2011) that serves as practice for future, more aggressive interactions requiring the defense or acquisition of resources. Therefore, playing at increased rates and with a diversity of partners is likely an adaptive strategy during juvenile development to enhance one's success in "serious" behaviors later in life. However, with variable social systems (between or within species), it can be expected that juvenile play rates, relationships, and partner diversity depend on the life stage (i.e., the priorities of an individual at different ages) and group composition (i.e., social partner diversity and range restrict group demography).

In addition to potentially preparing a juvenile for future social experiences (Burghardt 2005; Fagen 1981), researchers can study play in attempt to predict the context, quantity, and dyadic relationship quality of future social relationships between group members (e.g., Pellis et al. 2019). Play partner choice tends to reflect the adult social partners of cooperative, competitive, and reproductive relationships experienced in adulthood (e.g., Maestripieri and Ross 2004). At its core, play builds upon reciprocity and turn-taking between players; without reciprocity and clear communication it is difficult to maintain the interaction and playful context. Partner selection is thus important, as regularly playing against larger, stronger individuals may mean being disadvantaged more frequently. Similarly, selecting a smaller, less skilled, or younger opponent might mean you possess a more frequent advantage but may require more self-handicapping (e.g., *Cebus apella*, Lutz and Judge 2017; *Macaca mulatta*, Yanagi and Berman 2014b). Interactions with larger age disparities may require increased communication, i.e., play signaling, to maintain a playful context. For example, self-handicapping behaviors (e.g., adopting a supine position) are positively associated with play signaling in domestic dogs (*Canis lupus familiaris*, Ward et al. 2008), and gorillas often pair an open-mouth face with more intense play behaviors, such as chasing, potentially to maintain an equitable level of cooperation between play partners (van Leeuwen et al. 2011).

In the order Primates, there is marked interspecific variation in play, and even within a species, the frequency in which an individual engages in play is not uniform. Teasing apart the contributing individual- and group-level variables for these differences is no small task, and these intrinsic and external factors can be multidimensional. For example, young infants may be physically handicapped when engaged in play with larger juveniles due to their small size, limited motor coordination, and unrefined social skills. However, young infants may also be physically restricted or actively discouraged by their mothers from participating in play (e.g., the mother has a restrictive rearing style or is low-ranking in the group). Other variables, including group social organization, structure, style, and status (Ciani et al. 2012; Fagen 1981; Maestripieri 2004; Thierry 2007), an individual's opportunity (Panksepp and Beatty 1980), prior experience (Cloutier et al. 2013), personality (Lampe et al. 2017; Pellis and McKenna 1992), and brain chemistry (Siviy et al. 2011) may also influence an individual's participation in and style of play. Although it has been acknowledged that play typically declines in frequency with age, some primates continue to engage in play beyond sexual maturity, maintaining playful relationships with juveniles and even playing with other adults (Pellis and Iwaniuk 2000; e.g., *Lemur catta*, Palagi 2009; *Macaca tonkeana*, Ciani et al. 2012; *Pan paniscus*, Palagi 2006; *Pan troglodytes*, Yamanashi et al. 2018; *Theropithecus gelada*, Mancini and Palagi 2009). The motivation to continue playing into adulthood likely differs depending on a variety of factors, including the degree of social tolerance in the species (e.g., despotic vs. egalitarian society), species-specific affiliative patterns of behavior, and the evolved communicative mechanisms to maintain a playful context.

6.3 Macaque Play

Macaca spp. vary widely in their geographical distribution, but the genus shares certain foundational similarities in social organization; for example, they can be found living in female philopatric, multi-male, multi-female groups with overlapping home ranges. However, the differing geographic distribution and phylogeny of the genus has resulted in interspecific variation in patterns of affiliation, reconciliation, dominance, aggression, nepotism, and temperament (Thierry 1985, 1990; Thierry et al. 2000). From this variation, the social organization of macaques is typically regarded as a continuous, four-grade scale of dominance style: species occupying the first grade are considered hierarchical and nepotistic and those occupying the fourth grade are considered more tolerant or egalitarian (Thierry et al. 2000). Dominance style can be defined as the dominance relationships, categorized by agonistic interactions, within dyads in a social group (Thierry et al. 2000). A difference in dominance style between primate taxa can be the result of environmental variables, such as contest over food resources (Matsumura 1999), but may also be context-dependent (Funkhouser et al. 2018b). Grade one despotic species are generally marked by dominant individuals that show intense and highly asymmetrical patterns of aggression, little tolerance around resources, and infrequent reconciliation (e.g., *M. mulatta*, *M. fuscata*, and *M. cyclopis*). Species with a grade four dominance style show the opposite tendencies, with low or moderate levels of kin bias in affiliation, tolerant and supportive interactions with group members, strong group cohesion, and maternal tolerance for infant handling, for example, *M. maura*, *M. nigra*, *M. ochreata*, and *M. tonkeana*, all endemic to Sulawesi (Thierry et al. 2000). Although the above species fit easily onto this graded scale, other macaques are more difficult to categorize based on inconsistent or a lack of behavioral data. The classification of species as a grade two or three can often rely on relative comparisons: grade two macaques possess behavioral traits that are more similar to grade one species, and grade three macaques are more similar to grade four species.

Regardless of their place on this graded scale, the variation in social tolerance observed across the 23 extant macaque species provides an opportunity to directly compare different facets of behavior, including play structure and content, across multiple, differing social structures. Currently, much of the play research in macaques derives from species occupying opposite ends of the dominance scale, and the differences in observed play style, rate of play, and play signaling have been suggested to reflect this social dominance style (Petit et al. 2008; Reinhart et al. 2010; Yanagi and Berman 2014b). The social play of despotic species, such as rhesus (*M. mulatta*) or Japanese macaques (*M. fuscata*), can be characterized as competitive, whereas more tolerant species, such as Tonkean (*M. tonkeana*) or Sulawesi crested macaques (*M. nigra*), engage in a more cooperative play style (Petit et al. 2008; Reinhart et al. 2010). These differences are reflected in the targets attacked during play fighting as well as the type and frequency of play signals utilized throughout a play bout (Reinhart et al. 2010; Scopa and Palagi 2016; Yanagi and Berman 2014b). A more competitive or risky play style, one in which sensitive targets like the face are attacked, may generate miscommunication between partners and risk ending the play contact. Therefore, using play signals to indicate imminent play (Yanagi and Berman 2014a) or to reinforce playful intent (Wright et al. 2018) could be used to mitigate potential aggression and prolong the interaction (Scopa and Palagi 2016). These signals may be more specific and less interchangeable in more aggressive or competitive species to minimize the risk of misinterpretation (Scopa and Palagi 2016; Thierry et al. 2000; Yanagi and Berman 2014b) whereas in more tolerant species, these signals may be used redundantly and interchangeably to initiate/terminate play (Pellis et al. 2011; Scopa and Palagi 2016). Thus, investigating grade two and grade three macaque species, such as Tibetan macaques (*M. thibetana*), can help to determine the degree of overlap in species play patterns and play signaling.

6.4 Tibetan Macaques

Tibetan macaques are female philopatric and live in multi-male, multi-female social groups of 15–50 individuals (Berman et al. 2004; Thierry 2011; Thierry et al. 2000). This species has a strong kin bias with linear male and female dominance hierarchies, in which males generally occupy the top ranks of the hierarchy although females can outrank them (Berman et al. 2004). Despite being originally designated as having a grade three dominance style (Thierry et al. 2000), Tibetan macaques have been re-established as grade two, showing some despotism and low conciliatory tendencies, especially for female-female interactions (Berman et al. 2004). However, some female behavior is also inconsistent with a more despotic style, including that female individuals display a markedly high preference for female kin in proximity relationships and maternal tolerance for infant handling (Berman et al. 2004).

Female *M. thibetana* rank is based on matrilines with a daughter occupying the rank directly below her mother and above her older siblings (Berman et al. 2004; Thierry 2011; Zhao 1997). This hierarchy influences intergroup competition among females and generates preferential bonds between kin (Thierry 2011). Tibetan macaques groom at symmetric rates (exchange grooming for grooming received) and prefer female kin grooming partners, and females prefer to groom higher-ranking females (even if unrelated) (Xia et al. 2012). These investigations illuminate the generally despotic nature of Tibetan macaque social organization (Berman et al. 2004; Thierry et al. 2000), bias for female kin across a number of social contexts (viz., coalitionary support, grooming, and infant handling; e.g., Berman et al. 2004), and the overall value of grooming in this species (e.g., Xia et al. 2012, 2013). Tibetan macaques and other species with intermediate, graded behavioral variability may demonstrate behavioral nuances that are more likely a difference in degree rather than kind; therefore, it is important to determine the extent of these differences and how they manifest and compare across species.

Similar to other macaque species, Tibetan macaque juveniles engage in dyadic and polyadic play (Wright et al. 2018) that is fast-paced and rough and tumble. Few studies have examined social play in juvenile Tibetan macaques, primarily because few studies have been conducted on this species compared to other species of macaques. The play studies that have been performed have occurred at Valley of the Wild Monkeys in Anhui Province, Huangshan, China, specifically with the Yulingkeng A1 (YA1) group. Juveniles in this group have been noted as preferring similarly aged partners for both social play bouts and affiliation (Batts 2012). The type of play engaged in, whether solitary or social, also appears to depend on the age of the individual: infants tend to engage in more solitary play and juvenile males engage in the most social play (Batts 2012). The majority of juveniles in this group participate in play, but participation frequency, duration, and the rate of play signaling is variable (Wright et al. 2018). Nine candidate play signals, such as crouch and stare, have been observed, six of which overlap with the play signal repertoire of rhesus macaques (Yanagi and Berman 2014a). In this group, the play face is an important signal used throughout bouts but is not a reliable indicator that play will be initiated (see Wright et al. 2018 for a more detailed discussion). Having other individuals in proximity to the play bout (an audience) also seems to impact the play signaling of individuals involved, i.e., the data represent a negative parabolic trend. Play signaling in these juveniles increased as the number of audience members increased from zero to two but decreased with additional individuals beyond this threshold suggesting that the communication value of the play signal may degrade as the complexity and size of the play group increases (Wright et al. 2018). However, additional comparative research on this topic is necessary.

In the following example, we build upon previous social play studies of the YA1 juvenile Tibetan macaques of Mt. Huangshan, China, using social network analyses to supplement our current understanding of the social dynamics of this group. Although the majority of research performed on these macaques highlights the relationships between the adults, such analyses can help construct a more complete picture of group social dynamics. Additionally, this information can be used to generate future hypothesis-driven research about potential adult relationships and social position within the group as the juveniles age and become integrated into the adult social network.

6.5 Study Subjects and Data Collection

The YA1 group of Tibetan macaques resides at Valley of the Wild Monkeys in the Huangshan Scenic District, Anhui Province, China. This group has been habituated to human presence since 1986 (see Berman and Li 2002; Berman et al. 2004) and is free-ranging but regularly provisioned with corn multiple times per day in supplement to their natural diet. All group members are individually recognized, and records of the group structure, including name, sex, birth date, and matriline, are maintained by Anhui University researchers. Additional adult female grooming data, collected July 7–August 28, 2016, between 06:30 and 17:30, was used to supplement some of the following analyses. To generate maternal dominance rank, all-occurrence sampling (Altmann 1974) was used to collect agonistic data from July 14 to August 27, 2016, from 7:00 to 12:00 and 14:00 to 17:00 daily (data contributed by Lori K. Sheeran). All research herein was approved by the Central Washington University Institutional Animal Care and Use Committee (protocols: #A051602, #M061603, #A051702), and all protocols adhered to the legal requirements of the People's Republic of China and the American Society of Primatologists' Principles for the Ethical Treatment of Primates.

6.5.1 *Maternal Allogrooming and Dominance Rank*

In summer 2016, the YA1 group was composed of 47 individuals (19 males, 28 females): eight adult males, 13 adult females, and 26 infants and juveniles (between the ages of approximately 30 days and 6 years old) (see Table 6.1). All adult females were randomly sampled, and 10-min focal follows were conducted to collect all instances of auto- and allogrooming. This generated approximately 400 min of observation time per focal individual. Data were collected on actor/recipient identities, rank, sex, matriline, and duration. An allogrooming bout was initiated with physical touch between partners, and the bout ended when all grooming partners ceased to groom for >10 s. If a grooming bout was polyadic (more than two individuals), the identity and durations of all partners were recorded, and the interactions were treated as dyadic.

To determine maternal rank, agonistic data were collected. Agonism consisted of fear grin, scream, flee, displace, threat, lunge, chase, grab, slap, and bite (as defined by Berman et al. 2004). We coded unambiguously directed *agonistic interactions* (or "competitions") in a 1:0 dichotomous fashion, where 1 indicated the "actor" who "won" the interaction and 0 indicated the "recipient" who "lost" the interaction. For these reasons, submissive behaviors (*lack of agonism*) were reverse-coded, where the actor was said to have lost (0) to the winning (1) recipient. We then derived Elo scores for each individual using methods similar to Neumann et al. (2011) in R.

Table 6.1 Yulingkeng A1 player demographics in 2017

Name	Age (days)	Age (years)	Sex	Mother
TouQiuGuo (TQG)	30	<1	F	TouXiaHua
YeXiaYun (YXYun)	49	<1	F	YeHong
YeXiaDuo (YXD)	65	<1	F	YeChunYu
YeXiaMing (YXM)	102	<1	M	YeChunLan
HuaXiaYun (HXYun)	126	<1	F	HuaHong
TouHuaLi (THL)	399	1	F	TouRongYu
TouFuHua (TFH)	451	1	F	TouHuaYu
TouQiuYing (TQY)	474	1	F	TouXiaXue
YeXiaYue (YXYue)	794	2	F	YeHong
TouQiuSong (TQS)	802	2	M	TouXiaHua
TouHuaNan (THN)	823	2	M	TouRui
HuaXiaYue (HXYue)	852	2	F	HuaHong
YeChunHua (YCH)	1461	4	F	Unknown
YeRongLan (YRL)	1514	4	F	YeZhen
TouRongXi (TRX)	1537	4	F	TouTai
TouQiuLan (TQL)	1549	4	F	TouXiaXue
HuaXiaWei (HXW)	1579	4	F	HuaHong
YeXiaKun (YXK)	1588	4	M	YeHong
HuangYu (HY)	1643	5	M	Unknown
YeChunLan (YCL)[a]	1740	5	F	YeMai
TouXiaLong (TXL)	1817	5	M	TouHong
TouHuaXue (THX)[a,b]	1889	5	F	TouRui
TouRongYu (TRY2)	1969	5	M	TouTai
YeChunLong (YCLong)[a,b]	2557	7	M	YeMai
YeRongQiang (YRQ)[a]	2586	7	M	YeZhen
TouRongGong (TRG)[a]	2635	7	M	TouTai
HuaXiaMing (HXM)[a]	2675	7	M	HuaHong

Age was calculated from the recorded birth date to 6-21-2017, which was when Anhui University researchers prepared the demographic information for the field season. In instances where the birth date and matriline were unknown (e.g., YeChunHua and HuangYu), age was estimated based on the individual's size and sex characteristics

[a]The six individuals classified as adults during 2017, including two 5-year-old females and four 7-year-old males

[b]The two adult individuals of the six who participated in play with younger individuals (THX and YCLong)

6.5.2 *Juvenile Play Behavior*

In summer 2017, the YA1 group was composed of 46 individuals (17 males, 29 females): 10 adult males, 15 adult females, and 21 infants and juveniles (between the ages of approximately 30 days and 5 years old). Data were collected from July 6 to August 5, 2017, between 06:30 and 17:30. A complete list of player demographics, including age, sex, and matriline, are shown in Table 6.1.

Video footage of social play behavior was collected using all-occurrence sampling and a Sony Handycam camcorder. Regular scans of infant and juvenile interactions occurred until social play was initiated. If a second play bout occurred while recording the first, the video frame was widened to include both bouts. If this was not possible, the first bout was recorded to completion, and then the second bout was recorded. The juveniles were followed until play concluded and the players ceased to engage in play for >10 s. All play videos were coded using VLC media player (version 3.0.3, Vetinari) and Microsoft Excel (version 16.14) for player identity, demographic information (e.g., age, sex, and matriline), duration of the play bout, and the total number of players involved. All ages were recorded in days, and all durations were recorded in seconds.

A coding distinction was made between the longer encompassing *play bout* and the smaller component units designated here as *play periods* (see Mayhew 2013 for a discussion of play structure, including play periods and vigilance periods). Here, only play periods were examined. Guided by Burghardt (2005) and Fagen (1981), the start of a play bout was marked by the exchange of playful behavior between two or more juveniles. A play bout was considered to have concluded when (1) *a player engaged in non-playful behavior* (e.g., aggression), (2) *a non-player directed non-playful behavior toward the players* (e.g., the bout was interrupted by an adult initiating a grooming session), or (3) *player(s) withdrew from the interaction, thus dissolving the play group*. Because play dynamics change quickly, it was noted each time a player withdrew from or was added to the interaction even if play continued following this change in player composition. Each partner interchange marked the transition to a new play period; for example, players A and B are playing, and player C approaches and joins; therefore the play period between A and B terminates and a new period containing players A, B, and C begins (see example in Fig. 6.1).

6.5.3 Statistical Analyses

As indicated by its frequent appearance in recent animal behavior studies, social network analysis (SNA) has become a valuable tool for understanding the role and

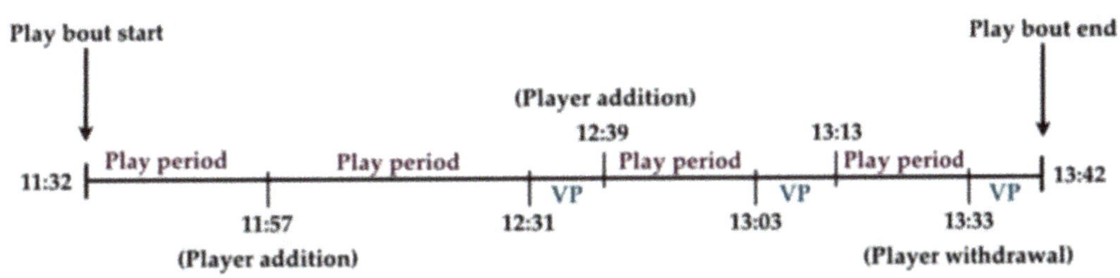

Fig. 6.1 An example of a play bout divided into its smaller components: play periods and vigilance periods (VP). A change in player composition (addition or withdrawal) is marked and indicates a transition into a new play period (adapted from Mayhew 2013)

positioning of individuals within a larger social group. SNA uses matrix-based data to analyze individual social interactions, where individuals can be depicted as nodes and network-based descriptive and statistical analyses are generated to describe the relationship between nodes (Sueur et al. 2011a; Whitehead 2008). A comprehensive overview of the utility of SNA, its terminology, methods, and analyses can be found elsewhere in comprehensive texts (see Borgatti et al. 2013 or Whitehead 2008).

SNA has been used to characterize these relationships in primate social groups, as well as identify clusters of individuals, subgroups, isolated group members, and, overall, diagram of the group's social network (e.g., Clark 2011; Farine and Whitehead 2015; Funkhouser et al. 2018a; Sueur et al. 2011a). For example, using SNA, researchers have examined the correlation between juvenile play network positions, ontogenetic social development, and later-life social connectedness (*P. troglodytes*, Shimada and Sueur 2014; *M. fuscata*, Shimada and Sueur 2017; *Macaca* spp., Sueur et al. 2011b). Animals that are strongly connected to one another (i.e., having strong bonds) tend to have relationships on multiple independent measures of relation; thus, it can be expected that playing preferentially with certain partners will correlate with other measures, such as grooming, copulation, and proximity (Whitehead 2008).

In this study, all statistical analyses were performed using IBM SPSS Statistics (Version 23) software ($\alpha = 0.05$). UCINET (Borgatti et al. 2002) was used to calculate network statistics, and NetDraw was used to construct sociograms. Elo scores (calculating position in the dominance hierarchy) were derived using R. To investigate correlations between independent social networks (e.g., durations of play and difference in age) we used QAP correlation analyses in UCINET; this test analyzes whole matrices against one another. The following node-by-node statistics were calculated: (1) *degree* (the sum of each node's ties with all other nodes, also known as strength), (2) *eigenvector centrality* (how well an individual is associated with others and how well the associates are associated), (3) *closeness* (how close nodes are within a network, often discussed in terms of the time it takes for information to spread from one individual to others), and (4) *betweenness* (a value used to assess a node's ability to control flow through a network). We also constructed a principal coordinate sociogram to illustrate the observed social play relationships between individuals. This sociogram plots individuals with strong associations near one another. To define the minimum edge value in this diagram, we used the mean of all directional dyadic indexes plus one standard deviation (mean + SD). Detailed information about these node-by-node statistics can be found in Borgatti et al. (2013) and Whitehead (2008).

6.6 Results

In total, 256 play bouts (12,458 s total) were observed containing 965 play periods. Twenty-three macaques (eight males, 15 females) participated in play, ranging in age from approximately 30 days (TQG) to 7 years old (YCLong) (Table 6.1). *Play bouts* ranged in duration from one to 895 s (nearly 15 min), but the mean was considerably shorter (48.99 ± 80.33 s). *Play periods* ranged in duration from 1 to 227 s with a mean of 12.91 ± 16.70 s.

6.6.1 Player Age

Play was observed more frequently between players of the same birth cohort (i.e., individuals born in the same year) (QAP, $r = 0.289$, $p = 0.001$) and age class (QAP, $r = 0.219$, $p < 0.001$). The mean age of a player was 867.66 ± 496.55 days old (approximately 2.4 years old), and the mean age of male players (977.75 ± 488.22 days) was slightly older than female players (736.84 ± 476.56 days). A Pearson correlation was used to determine whether the age of the player (in days) correlated with the total amount of time spent playing with any other player(s) during the study period. Player age was significantly correlated with play duration ($r = -0.246$, $p \leq 0.001$), and this negative relationship suggests that as age increased, the time spent involved in play decreased. Notably, when only dyadic play periods were examined, the ages of the players were positively correlated ($r = 0.467$, $p \leq 0.001$), indicating that as the age of Player 1 increased, the age of Player 2 also increased (Fig. 6.2).

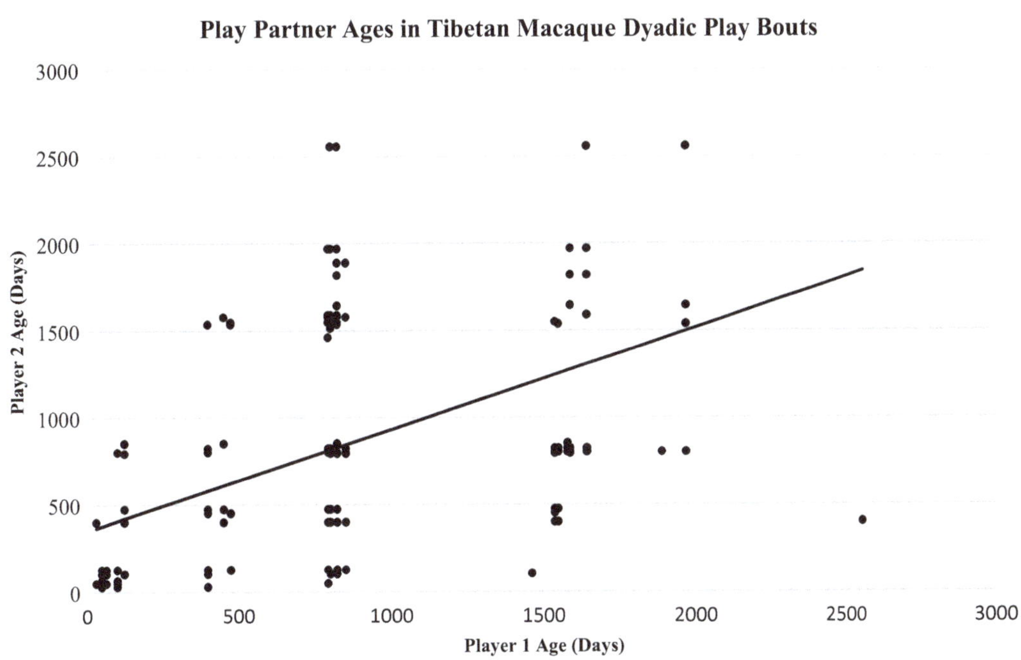

Fig. 6.2 A scatterplot of player ages in dyadic play bouts (two individuals) indicating that as the age of one partner increased, the age of the second partner also tended to increase ($r = 0.467$)

6.6.2 Number of Players

Consistent with Wright et al. (2018), polyadic play periods were observed in this group of macaques (four players, $n = 66$; five players, $n = 10$) but occurred less frequently than dyadic ($n = 628$) or triadic ($n = 261$) interactions; the mean number of players in a play period was 2.44 ± 0.67. Similarly, the total number of players in a play period was significantly correlated with the duration of a period ($r = -0.138$, $p \leq 0.001$), indicating that playing with more individuals decreased the duration of play. Additionally, the total number of players in a play period was significantly correlated with player age, indicating that as player age increased, the number of players involved decreased ($r = -0.152$, $p \leq 0.001$).

6.6.3 Player Composition

An analysis of player composition indicated that mixed sex play ($n = 533$ periods, 55.2%) represented more than half of all play periods (all-male play, $n = 273$, 28.3%; all-female play, $n = 159$, 16.5%), and this was significantly different from the expected values ($\chi^2(2) = 228.47$, $p \leq 0.001$). Therefore, juvenile males remained active participants despite the presence of more juvenile females in the group.

6.6.4 Matrilineal Relatedness and Rank

YA1 matrilines are known and documented annually by Anhui University researchers; therefore it was possible to determine and rank the degree of maternal relatedness between juveniles. There was no significant relationship between maternal difference in Elo score (dominance) and the duration of play in offspring (QAP, $r = 0.00$, $p = 0.025$). Similarly, there was no significant relationship between maternal dominance status (dominant or subordinate) and the duration of play in offspring (QAP, $r = -0.025$, $p = 0.411$).

6.6.5 Maternal Social Relationships

Adult female grooming duration data from 2016 was factored in as a proxy for maternal social dyadic relationships, but there was no significant relationship between maternal grooming (total seconds) and the duration of play in offspring (QAP, $r = 0.018$, $p = 0.274$).

6.6.6 Individual Playfulness

In addition to the above demographic variables, individualistic patterns emerged for the juveniles in this group. Predictably, some juveniles played more frequently and longer than others (see Figs. 6.3 and 6.4), and overall, TouQiuSong (TQS), TouHuaNan (THN), and YeXiaYue (YXYue) were the three most active participants (appearing in 56.58, 40.70, and 32.40% of play periods observed, respectively). All three of these players were born within 29 days of one another and belonged to the 2015 birth cohort. Interestingly, the oldest individual in this cohort, HuaXiaYue (HXYue), was born 29 days before THN, but did not participate in play nearly as much as her three peers.

Network statistics for each player were calculated, including degree, eigenvector centrality, closeness, and betweenness (Table 6.2). To better visualize these play relationships, a sociogram was constructed (minimum edge weights of mean ± SD) for all dyadic interactions (Fig. 6.5). Examining the network statistics and sociogram, TQS, THN, and YXYue appear to unite three clusters of players.

Based on a cursory examination of the play frequency and duration data as well as from personal observation, it is unsurprising that TQS, THN, and YXYue were the

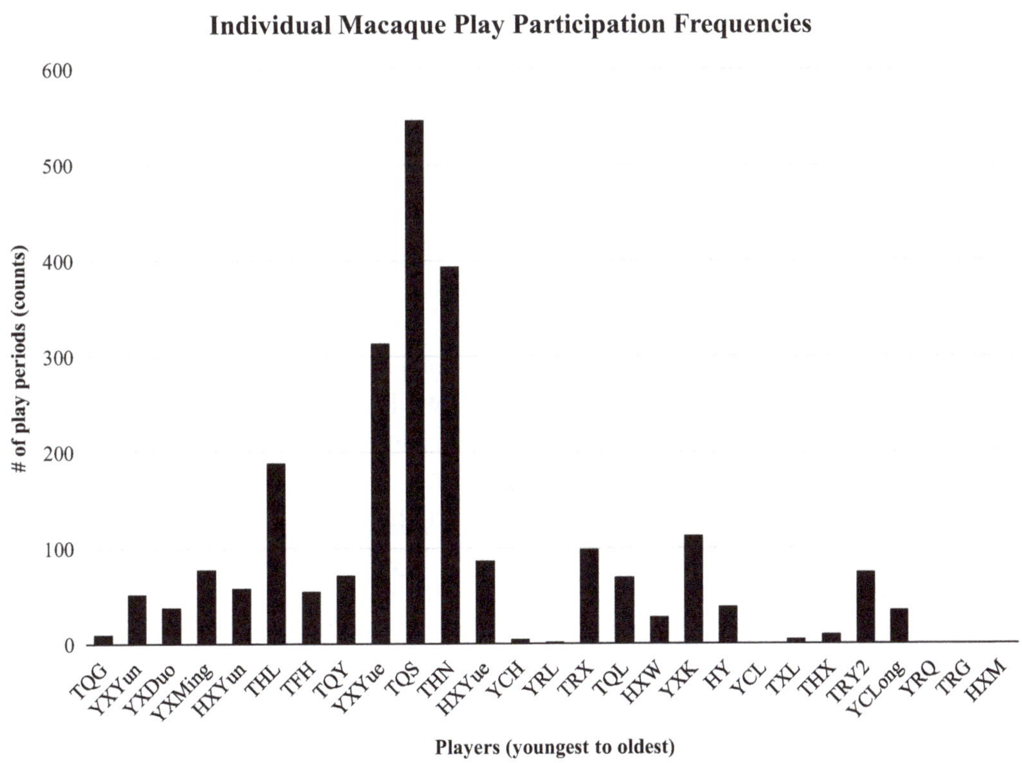

Fig. 6.3 The total individual counts of play participation for all group members ≤7 years old in the YA1 group during 2017. TouQiuGuo (TQG) on the left is the youngest player at 30 days old and HuaXiaMing (HXM) is the oldest at 2675 days. Some individuals were not observed to engage in play at all (e.g., HuaXiaMing), whereas others (e.g., YXYue, TQS, and THN) were frequent participants

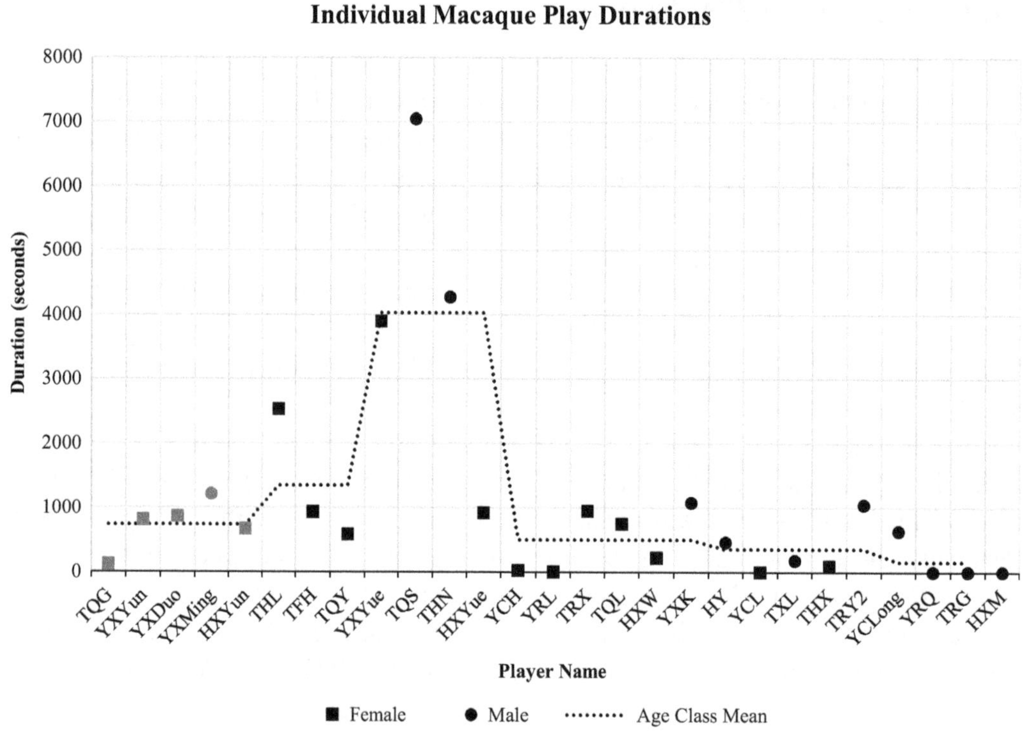

Fig. 6.4 The total individual durations of play (in seconds) for all male and female individuals ≤7 years old. Males are depicted as circles, females as squares, infants as gray, and juveniles as black. Individuals are organized left to right by age in days (TQG is the youngest, HXM is the oldest). The dotted line represents the mean play duration for each age class, including all individuals of that age class even if they were not observed playing (e.g., YCL, YRQ, TRG, and HXM)

top three ranking individuals for all four calculated network measures. Overall, TQS has the highest strength (6896) and betweenness (47.905) values, indicating that he is highly gregarious and links discrete clusters of juvenile players. His high eigenvector centrality of 1.0 indicates that he is well connected to others that are themselves well connected within the network. He also has the lowest closeness value (24), which suggests that social information that originates at a random node in the network would reach him quickly and with high fidelity, again emphasizing his central position in the network. The results are very similar for THN and YXYue, and these three juveniles are found clustered in the center of the sociogram.

The individuals in the three clusters evident in Fig. 6.5 are notably comprised of individuals with similar demographic characteristics. Cluster A contains the group's older and larger males (ages four to seven), including YeChunLong (YCLong), TouRongYu (TRY2), HuangYu (HY), and YeXiaKun (YXK). Cluster B contains nearly all remaining juveniles (ages two to four), who are predominantly female, including TouRongXi (TRX), TouQiuLan (TQL), TouQiuYing (TQY), TouHuaLi (THL), and TouFuHua (TFH). Cluster C contains HXYue (age three) and the majority of the group's infants: YeXiaYun (YXY), YeXiaDuo (YXD), and YeXiaMing (YXM). Clusters A and C appear to be connected via cluster B,

Table 6.2 2017 Yulingkeng A1 player network measures sorted by degree

Player	Degree	Eigenvector centrality	Closeness	Betweenness
TQS	**6896**	**1**	**24**	**47.905**
THN	**4448**	**0.74**	**26**	**28.6**
YXYue	**4108**	**0.791**	**25**	**28.807**
THL	3393	0.579	30	6.211
YXMing	1351	0.062	30	13.186
TRX	1266	0.153	34	1.5
TQL	1237	0.145	35	0.25
YXK	1215	0.216	37	1.715
TRY2	1201	0.15	36	1.394
TQY	1187	0.155	32	1.313
YXDuo	1027	0.016	46	0.2
YXYun	1023	0.054	35	3.781
TFH	1005	0.124	32	1.153
HXYue	879	0.144	32	1.416
HXYun	722	0.07	32	7.197
YCLong	719	0.056	39	0.966
HY	539	0.088	36	2.276
HXW	291	0.045	34	0.862
TXL	188	0.014	43	0.2
TQG	161	0.006	40	1.069
YCH	36	0.008	40	0
YRL	4	0.001	44	0

Bold indicates the top three individuals with the highest network measures

suggesting that certain individuals or individuals of intermediate age more generally act as scaffolding for infants into the juvenile social network. One juvenile was a notable social isolate with the lowest degree, eigenvector, and betweenness values and highest closeness value: YeRongLan (YRL), who only participated in one play bout for four seconds. Similarly, other individuals had few or weak social network connections, reflected by their low degree, eigenvector, and betweenness values and high closeness scores: YeChunHua (YCH), TouQiuGuo (TQG), TouXiaLong (TXL), and HuaXiaWei (HXW). These isolated individuals were both males and females and ranged from infancy to nearly 5 years old.

6.7 Discussion

The adult macaques in the YA1 group are well known and studied, but juvenile interactions have not been well documented. The results outlined in this chapter provide a window into the social lives of these juveniles, and the results presented here build upon previous Tibetan macaque social play research performed at this site (viz., Wright et al. 2018). Importantly, the findings in this work aim to establish these

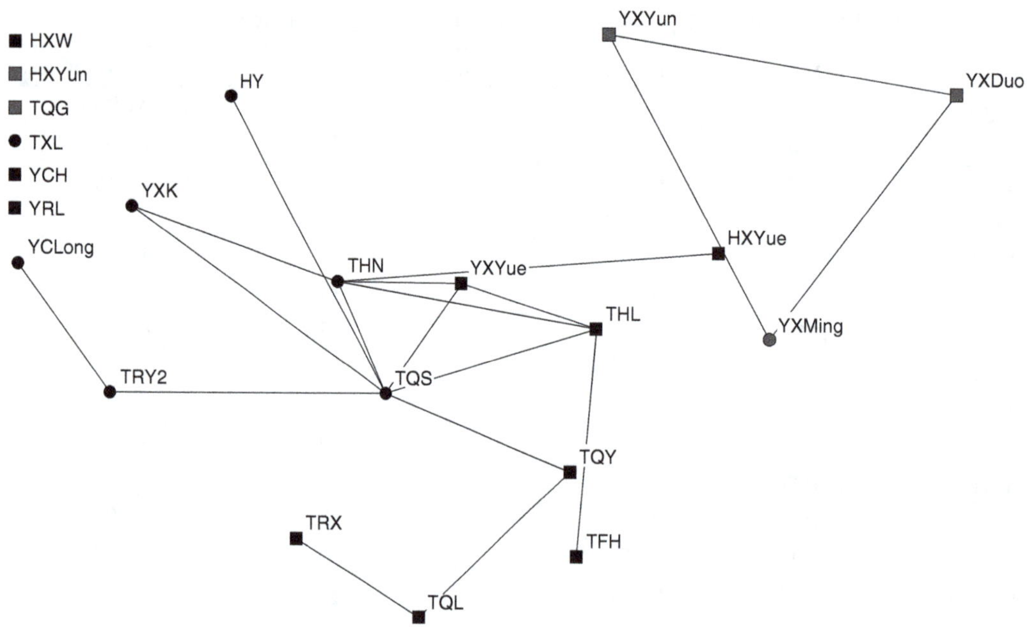

Fig. 6.5 Sociogram for the 2017 juvenile and infant play network based on duration data (mean $\pm$ SD $=$ 242). This network highlights those juveniles who participated more frequently in social play. Males are depicted as circles, females as squares, juveniles as black, and infants (<1 years old) as gray. YCLong is included in this network as a juvenile but was categorized as an adult at the time of the study

juveniles as important entities within the group that have their own identities and the potential to drive the direction of group social dynamics as they age and continue to interact.

The frequency of social play in these juvenile macaques followed a similar trajectory to other primate species: play increased as age increased, peaked at approximately 2.5 years old, and then declined. Adult macaques in this group rarely participated in play, which supports previous literature indicating a general lack of adult primate play in more despotic species. Additionally, the number of players involved in a play period was fewer when the participating players were older. The QAP results indicated that juveniles had a preference for certain social partners and played more frequently with peers of their birth cohort. In dyadic play bouts, partner ages were positively correlated, which is a further indication of a preference for similarly aged partners. Taken together, partner preferences may not only be indicative of the strong social bonds that result from growing up together in the same age cohort but also reflect a preference for equity in social competition (Kulik et al. 2015; van Leeuwen et al. 2011). As juveniles age, especially males, physical strength and prowess build and are increasingly put to the test in competition with other individuals. This partner equity may be particularly important for more despotic macaque species in attempt to ensure that play occurs with an appropriately matched opponent in skills, abilities, and size in attempt to reduce the risk of play devolving into aggression by interacting with a mismatched partner. Tibetan macaques also engage in polyadic play, which can quickly become complicated for the participants as well as the researchers observing the interaction. Typically, the duration of these polyadic play events is shorter than when play occurs between dyads. This is reasonable considering that when more players become involved in an interaction, it becomes increasingly difficult for each player to monitor their own behavior as well as track the behaviors of all the other partners simultaneously. This inability to effectively monitor multiple partners may contribute to the brevity of polyadic play. Such events also open the door for future research analyzing the structure and composition of these bouts in comparison to dyadic events.

Although there may be some general conclusions to be made about play in macaques depending on their degree of social despotism or tolerance, the influence of individual identity and personality on play participation warrants further exploration. Some individuals, simply based on personality traits (e.g., boldness, tendency to explore, or sociability), may be more inclined to participate in play regardless of their partner's age, sex, or relationship. Although personality data were not collected for these juvenile macaques, and the dataset here only represents a snapshot of the relationships of this group, predictions could be made about which juveniles likely possess which traits and to what degree. This is undoubtedly a large undertaking; however, such data could generate some understanding of how personality contributes to social play participation and can aid in generating predictions about how certain individuals will behave in other realms of social behavior.

6.7.1 Does Social Position Matter?

The introduction of SNA into social play analyses is useful in that it not only literally illustrates the complex relationships shared by juveniles but also provides a method to track individual relationships and compare them across years, events, and other social networks (e.g., Shimada and Sueur 2017). Specifically, SNA allows researchers to follow an individual across time and various social networks to investigate the effects of an individual's characteristics during play as a juvenile (e.g., frequency, partner choice and diversity, and play style) on their position in adult social networks. Such a longitudinal, multilayer social network framework would provide a pointed investigation of the adaptive mechanisms present during the juvenile play period that may impact reproductive fitness later in life (e.g., dominance status, reproductive success, infant survival). The incorporation of such information into future social play studies may be of interest to researchers performing long-term studies, where data-driven predictions about the role or tenure of future juveniles within the social group is invaluable. For the YA1 group, this type of information may be helpful to future researchers interested in the development and maintenance of Tibetan macaque relationships.

In 2017, it appears that TQS, THN, and YXYue provided the foundation for this social play network in 2017, acting as social glue and connecting juveniles across multiple birth cohorts and matrilines. Interestingly, the grandmothers of THN and TQS were central in the 2016 female allogrooming network (TouTai and TouHong, respectively, unpublished data), so it is possible that well-connected maternal relatives influence the social freedom or social position of kin across years. Whether the central positioning of this trio is an artifact of their particular age (nearly 3 years old), their early cohort experience (close dates of birth), or previous sibling or kin scaffolding into the social network is currently unknown, but additional SNA data from 2015 and 2016 can help to clarify this picture.

The central nature of TQS, THN, and YXYue can also be used to make predictions about the transition of future dominance status and mating opportunities in this group. The YA1 group has a history of some resident males remaining in the group longer than expected and rising through the ranks to alpha position (e.g., YeRongBing, HuaXiaMing, and TouGui). This longer group tenure might also be expected of THN and TQS, and they may not be immediately pressured by group males to disperse. As a female, YXYue is expected to remain in YA1, and once she reaches sexual maturity, she will likely fall into place at the top of the hierarchy with the rest of her Ye female relatives, achieving this high rank namely because her mother, YeHong, is the alpha female.

Social predictions can also be made about individuals that occupy more peripheral positions in the juvenile network. For some, such as the infants of the group, this position will most likely change with age and increased social exposure to their peers or may be mediated by the infant's matriline. For others, this window of social opportunity may have closed, and this peripheral social position may persist into adulthood. One important advantage of reliably identifying unique infants and juveniles is that it allows for young individuals to be tracked across development to test for changes in the status of their social lives. Further, reliably identifying young individuals ensures control for bias in data collection, whereby ensuring that nongregarious or cryptic individuals (that might be easily overlooked because they play less) can be accurately represented in a dataset.

Although the infants in this group all had low eigenvector values (less central), they comprised their own cluster and the majority were not social isolates (peripheral). The isolation of TouQiuGuo (TQG) from the juvenile network can be explained by her very young age (a newborn). The clustering of the remaining newborn peer group may have occurred because of limited social access and ability. For infants, having access to social partners beyond their mother can be restricted because macaque mothers often carefully control their offspring's experience during their first vulnerable months. For Tibetan macaques, this can be a physical restriction, where the mother will not release the infant from her grasp, or a social restriction, where the mother actively monitors and interferes in the social interactions between her offspring and other group members for at least the initial weeks of life. Indeed, Tibetan macaque mothers often affiliate or associate with one another, thereby providing limited opportunities for their infants to socially engage while still being heavily monitored. Regardless, this mothering behavior limits the infant's ability to socially engage relative to the rate of older (juvenile) conspecifics. Aside from the influence of maternal rearing behavior, the infant itself has limited physical abilities that prohibit it from keeping up with older players. Growth and development bring physical, mental, and emotional strength, flexibility, endurance, and knowledge of the natural environment, but this is inherently a slow, prolonged process, one that requires time as well as practice and repeated experience. A newborn infant is also a novel object in the environment that others must familiarize themselves with; therefore, it is not simply that the infant must gain physical and social competence to enter into this new social world; they must also gain acceptance from members of the social group and be treated accordingly by older play partners.

A handful of juveniles were social isolates. Both YCH and YRL were rarely seen interacting with any members of the YA1 group, juvenile or otherwise. YCH is thought to be a juvenile immigrant (rare), and therefore she lacks kin relations on which to base her social behaviors; this could explain her peripheral social position. YRL's mother, YeZhen, occupied the lowest position in the social hierarchy during 2017, and this low rank could have impacted YRL's social position within the juvenile network by limiting her opportunities for social engagement. It was surprising to uncover the weak connection of TouXiaLong (TXL) to the juvenile play network because his mother, TouHong, occupied a central, well-connected position within the group's 2016 grooming network (unpublished data). Further, TXL's maternal cousin is TQS, a highly playful and central individual (see above). Although the position of TXL was wholly different from TQS, a possible explanation for his position revolves around his age: TXL was roughly 5 years old and possibly in a liminal developmental stage where he was transitioning out of the juvenile period and therefore more concerned with establishing connections to the adult social network.

6.7.2 Future Considerations

One major challenge to studying social play in primates is that there are typically a large number of participants to track. When studying a large group, especially one containing many young individuals who look quite similar, it is considerably easier to use broad age categories to ease any comparisons (e.g., infant, juvenile, subadult, and adult). However, this generalization of individuals into broad age classes can generate statistical and conceptual issues because the boundaries between age categories differ widely between sexes, species, and studies. In primates, the majority of individuals involved in play are juveniles, which often means they are between the ages of weaning and sexual maturity. This, however, is a large and variable span of developmental time, and the acquisition of physical, social, and cognitive skills may occur at different stages and paces. Not only might there be marked differences in the physical and social abilities among the age classes, there are considerable differences in abilities within age classes, for example, between a 2-year-old juvenile who is 731 days old and a 2-year-old juvenile who is 1094 days old. Therefore, it is our recommendation to use age in days to achieve a fine-grain perspective and to emphasize the individual nature and developmental trajectory of play in any group examined.

Much like any study on social play, more questions emerge and remain open for future exploration than are typically answered. In particular, the idea of social scaffolding—acting as a bridge and supporting or facilitating a connection between groups of individuals—is of particular interest. Who are the individuals that fall into this particular role? Are they kin? Do older siblings help assimilate younger siblings into the network? What is the influence of having a small familial network to begin with? These kinds of questions require longitudinal comparisons, that not only incorporate social play data but also multiple independent measures of association (Shimada and Sueur 2017), such as proximity and grooming.

Furthermore, questions of evolutionary importance surface in the study of social play. Particularly, because play is often regarded as a context for learning and practice for later serious adult interactions (Lutz et al. 2019), juvenile play is rich in opportunities to study the contributions of juvenile behavior to adult reproductive success. Do individuals who play at high frequencies when young go on to be more successful in dominance and mating interactions (relative to juveniles who rarely play)? What about those that demonstrate high play partner diversity? What is the effect of maternal or paternal social network connectedness on the juvenile's frequency or diversity of play? Do observed rates of play in different contexts (e.g., social or object) predict an individual's success in corresponding contexts later in life (e.g., mating or extractive foraging)? Researchers could test such predictions using unidimensional social network analyses (as we illustrate here) or expand on multidimensional social networks. Multiplex social network analyses are particularly useful in studying complex social systems where individuals participate in various social contexts (Smith-Aguilar et al. 2018). With multiplex statistics, researchers could ask questions of variation across multiple behaviorally defined layers (e.g., play, grooming, agonism, and mating), temporal sequences (e.g., years or developmental periods), or biologically relevant connections (e.g., kinship) to test predictions of inclusive fitness and reproductive success.

Studying play could also help to place the four grades of macaque species into clearer context: the differences between these species, especially those considered to be grades two and three, most likely reflect a nuanced degree rather than overt differences. Tibetan macaques are considered a grade two species, meaning that observers should see some evidence of despotism, hierarchical behavior, and matrilineal preference, even in juvenile play. However, researchers should expect to observe some behavioral inconsistencies when compared with grade one macaques (e.g., *M. mulatta* or *M. fuscata*). For example, contrary to the expectation for more despotic species, social play in this group is not often interrupted by an adult; instead, juveniles are fairly self-sufficient at self-regulating play bouts and choose when to participate rather than being controlled by the adults of the group (Wright et al. 2018). Additionally, neither kinship nor maternal rank appears to have a significant impact on the duration of play or partner choice (this study). These are unexpected results that warrant further inquiry and require researchers to ask themselves: just how much variation is reasonable to expect for grade two or three species?

Acknowledgments We are so grateful for the continued support and guidance of Dr. Lori K. Sheeran. Without your collaboration, this work would not be possible. Many thanks to Dr. Crickette Sanz for stimulating discussions on the utility of social networks in an evolutionary context. We are excited for the future research realized from these conversations. We are also appreciative of our collaborators at Anhui University and the helpful and positive feedback provided by two reviewers of this piece. Finally, to all of the individuals of the YA1 group, thank you for letting us spend some time in your world.

References

Altmann J (1974) Observational study of behavior: sampling methods. Behav 49(3/4):227–267

Baldwin JD, Baldwin JI (1974) Exploration and social play in squirrel monkeys (*Saimiri*). Am Zool 14(1):303–315

Batts C (2012) The impact of eco-tourism on infant and juvenile play behaviors in Tibetan macaques (*Macaca thibetana*). Thesis, Central Washington University, Ellensburg, WA

Bekoff M (1972) The development of social interaction, play and metacommunication in mammals: an ethological perspective. Q Rev Biol 47(4):412–434

Bekoff M (1974) Social play and play-soliciting by infant canids. Am Zool 14:323–340

Bekoff M, Allen C (1997) Intentional communication and social play: how and why animals negotiate and agree to play. In: Bekoff M, Byers JA (eds) Animal play: evolutionary, comparative, and ecological perspectives. Cambridge University Press, New York, pp 97–114

Bekoff M, Byers JA (1981) A critical reanalysis of the ontogeny of mammalian social and locomotor play, an ethological hornet's nest. In: Immelmann K, Barlow GW, Petrinovich L, Main M (eds) Behavioral development, the bielefeld interdisciplinary project. Cambridge University Press, New York, pp 296–337

Berman CM, Li JH (2002) The impact of translocation, provisioning and range restriction on a group of *Macaca thibetana*. Int J Primatol 23(2):383–397

Berman CM, Ionica CS, Li JH (2004) Dominance style among *Macaca thibetana* on Mt. Huangshan, China. Int J Primatol 25(6):1283–1312

Borgatti SP, Everett MG, Freeman LC (2002) UCINET for Windows: software for social network analysis. Version 6.627 [software]. 2016 Dec 14. Available from https://sites.google.com/site/ucinetsoftware/home

Borgatti SP, Everett MG, Johnson JC (2013) Analyzing social networks. Sage, Los Angeles, CA

Brown SG (1988) Play behaviour in lowland gorillas: Age differences, sex differences, and possible functions. Primates 29(2):219–228

Burghardt G (2005) The genesis of animal play: Testing the limits. MIT Press, Ann Arbor, MI

Burghardt G, Ward B, Rosscoe R (1996) Problem of reptile play: Environmental enrichment and play behavior in a captive Nile soft-shelled turtle, *Trionyx triunguis*. Zoo Biol 15:223–238

Burghardt G, Dinets V, Murphy JB (2014) Highly repetitive object play in a cichlid fish (*Tropheus duboisi*). Ethology 121:38–44

Byers JA, Walker C (1995) Refining the motor training hypothesis for the evolution of play. Am Nat 146(1):25–40

Ciani F, Dall'Olio S, Stanyon R et al (2012) Social tolerance and adult play in macaque societies: a comparison with different human cultures. Anim Behav 84:1313–1322

Clark FE (2011) Great ape cognition and captive care: Can cognitive challenges enhance well-being? Appl Anim Behav Sci 135:1–12

Cloutier S, Baker C, Wahl K et al (2013) Playful handling as social enrichment for individually- and group-housed laboratory rats. Appl Anim Behav Sci 143:85–95

Fagen R (1981) Animal play behavior. Oxford University Press, New York

Fagen R (1984) Play and behavioural flexibility. In: Smith PK (ed) Play in animals and humans. Basil Blackwell, Oxford, pp 159–173

Farine DR, Whitehead H (2015) Constructing, conducting and interpreting animal social network analysis. J Anim Ecol 84:1144–1163

Fontaine RP (1994) Play as physical flexibility training in five ceboid primates. J Comp Psychol 108 (3):203–212

Funkhouser JA, Mayhew JA, Mulcahy JB (2018a) Social network and dominance hierarchy analyses at chimpanzee sanctuary northwest. PLoS One 13(2):e0191898

Funkhouser JA, Mayhew JA, Sheeran LK, Mulcahy JB, Li JH (2018b) Comparative investigations of social context-dependent dominance in captive chimpanzees (*Pan troglodytes*) and wild Tibetan macaques (*Macaca thibetana*). Sci Rep 8(1):e13909

Gamble JR, Cristol DA (2002) Drop-catch behaviour is play in herring gulls, *Larus argentatus*. Anim Behav 63:339–345

Gómez JC, Martín-Andrade B (2002) Possible precursors of pretend play in nonpretend actions of captive gorillas (*Gorilla gorilla*). In: Mitchell R (ed) Pretending and imagination in animals and children. Cambridge University Press, Cambridge, pp 255–268

Heinrich B, Smolker R (1998) Play in common ravens (*Corvus corax*). In: Bekoff M, Byers JA (eds) Animal play: evolutionary, comparative, and ecological perspectives. Cambridge University Press, New York, pp 27–44

Hill HM, Dietrich S, Cappiello B (2017) Learning to play: A review and theoretical investigation of the developmental mechanisms and functions of cetacean play. Learn Behav 45(4):335–354

Jones BL, Kuczaj SA (2014) Beluga (*Delphinapterus leucas*) novel bubble helix play behavior. Anim Behav Cogn 1(2):206–214

Kahlenberg SM, Wrangham RW (2010) Sex differences in chimpanzees' use of sticks as play objects resemble those of children. Curr Biol 20(24):R1067–R1068

Kulik L, Amici F, Langos D et al (2015) Sex differences in the development of social relationships in rhesus macaques (*Macaca mulatta*). Int J Primatol 36:353–376

Lampe JF, Burman O, Würbel H et al (2017) Context-dependent individual differences in playfulness in male rats. Dev Psychobiol 59:460–472

Lee PC, Moss CJ (2014) African elephant play, competence and social complexity. Anim Behav Cogn 1(2):144–156

Lutz MC, Judge PG (2017) Self-handicapping during play fighting in capuchin monkeys (*Cebus apella*). Behaviour 154:909–938

Lutz MC, Ratsimbazafy J, Judge PG (2019) Use of social network models to understand play partner choice strategies in three primate species. Primates 60(3):1–14

Maestripieri D (2004) Maternal behavior, infant handling, and socialization. In: Thierry B et al (eds) Macaque societies: a model for the study of social organization. Cambridge University Press, Cambridge, pp 231–234

Maestripieri D, Ross SR (2004) Sex differences in play among western lowland gorilla (*Gorilla gorilla gorilla*) infants: Implications for adult behavior and social structure. Am J Phys Anthropol 123:52–61

Mancini G, Palagi E (2009) Play and social dynamics in a captive herd of gelada baboons (*Theropithecus gelada*). Behav Process 82:286–292

Martin P, Caro TM (1985) On the functions of play and its role in behavioral development. In: Rosenblatt JS, Beer C, Busnel MC, Slater PJB (eds) Advances in the study of behavior, vol 15. Academic, New York, pp 59–103

Mather JA, Anderson RC (1999) Exploration, play and habituation in octopuses (*Octopus dofleini*). J Comp Psychol 113(3):333–338

Matsumura S (1999) The evolution of "egalitarian" and "despotic" social systems among macaques. Primates 40(1):23–31

Mayhew JA (2013) Attention cues in apes and their role in social play behavior of western lowland gorillas (Gorilla gorilla gorilla). PhD Thesis, The University of St Andrews, St Andrews, Scotland

Nahallage CAD, Leca JB, Huffman MA (2016) Stone handling, an object play behaviour in macaques: welfare and neurological health implications of a bio-culturally driven tradition. Behaviour 153(6–7):845–869

Neumann C, Duboscq J, Dubuc C et al (2011) Assessing dominance hierarchies: validation and advantages of progressive evaluation with Elo-rating. Anim Behav 82:911–921

Ortega JC, Bekoff M (1986) Avian play: comparative evolutionary and developmental trends. Auk 104:338–341

Palagi E (2006) Social play in bonobos (*Pan paniscus*) and chimpanzees (*Pan troglodytes*): implications for natural social systems and interindividual relationships. Am J Phys Anthropol 129:418–426

Palagi E (2009) Adult play fighting and potential role of tail signals in ringtailed lemurs (*Lemur catta*). J Comp Psychol 123(1):1–9

Palagi E (2018) Not just for fun! Social play as a springboard for adult social competence in human and non-human primates. Behav Ecol Sociobiol 72:90

Palagi E, Antonacci D, Cordoni G (2007) Fine-tuning of social play in juvenile lowland gorillas (*Gorilla gorilla gorilla*). Dev Psychobiol 49(4):443–445

Panksepp J, Beatty WW (1980) Social deprivation and play in rats. Behav Neural Biol 30 (2):197–206

Pellis SM, Iwaniuk AN (2000) Comparative analyses of the role of postnatal development on the expression of play fighting. Ethology 106(12):1083–1104

Pellis SM, McKenna M (1992) What do rats find rewarding in play fighting? An analysis using drug-induced non-playful partners. Behav Brain Res 68(1):65–73

Pellis SM, Pellis VC (1996) On knowing it's only play: the role of play signals in play fighting. Aggress Violent Behav 1(3):249–268

Pellis SM, Pellis VC (2009) The playful brain. OneWorld Publications, Oxford

Pellis SM, Pellis VC, Reinhart CJ et al (2011) The use of the bared-teeth display during play fighting in Tonkean macaques (*Macaca tonkeana*): sometimes it is all about oneself. J Comp Psychol 125(4):393–403

Pellis SM, Pellis VC, Pelletier A et al (2019) Is play a behavior system, and, if so, what kind? Behav Process 160:1–9

Petit O, Bertrand F, Thierry B (2008) Social play in crested and Japanese macaques: testing the covariation hypothesis. Dev Psychobiol 50(4):399–407

Pruitt JN, Burghardt GM, Riechert SE (2012) Non-conceptive sexual behavior in spiders: a form of play associated with body condition, personality type, and male. Ethology 118:33–40

Reinhart CJ, Pellis V, Thierry B et al (2010) Targets and tactics of play fighting: competitive versus cooperative styles of play in Japanese and Tonkean macaques. Int J Comp Psychol 23 (2):166–200

Scopa C, Palagi E (2016) Mimic me while playing! Social tolerance and rapid facial mimicry in macaques (*Macaca tonkeana* and *Macaca fuscata*). J Comp Psychol 130(2):153–161

Shimada M (2006) Social object play among young Japanese macaques (*Macaca fuscata*) in Arashiyama, Japan. Primates 47:342–349

Shimada M (2010) Social object play among juvenile Japanese macaques. In: Nakagawa N et al (eds) The Japanese macaques. Springer, New York, pp 375–385

Shimada M, Sueur C (2014) The importance of social play network for infant or juvenile wild chimpanzees at Mahale Mountains National Park, Tanzania. Am J Primatol 76(11):1025–1036

Shimada M, Sueur C (2017) Social play among juvenile wild Japanese macaques (*Macaca fuscata*) strengthens their social bonds. Am J Primatol 80(1):e22728

Siviy S, Deron L, Kasten C (2011) Serotonin, motivation, and playfulness in the juvenile rat. Dev Cogn Neurosci 1:606–616

Smith-Aguilar SE, Aureli F, Busia L et al (2018) Using multiplex networks to capture the multidimensional nature of social structure. Primates 60(3):1–19

Sueur C, Jacobs A, Amblard A et al (2011a) How can social network analysis improve the study of primate behavior? Am J Primatol 3:703–719

Sueur C, Petit O, De Marco A et al (2011b) A comparative network analysis of social style in macaques. Anim Behav 82:845–852

Thierry B (1985) Patterns of agonistic interactions in three species of macaque (*Macaca mulatta*, *M. fascicularis*, *M. tonkeana*). Aggress Behav 11(3):223–233

Thierry B (1990) Feedback loop between kinship and dominance: the macaque model. J Theor Biol 145(4):511–521

Thierry B (2007) Unity in diversity: lessons from macaque societies. Evol Anthropol 16 (6):224–238

Thierry B (2011) The macaques: a double-layered social organization. In: Campbell CJ et al (eds) Primates in perspective, 2nd edn. Oxford University Press, New York, pp 229–241

Thierry B, Iwaniuk AN, Pellis SM (2000) The influence of phylogeny on the social behaviour of macaques (Primates: Cercopithecidae, genus *Macaca*). Ethology 106(8):713–728

Thompson KV (1996) Play-partner preferences and the function of social play in infant sable antelope, *Hippotragus niger*. Anim Behav 52:1143–1155

van Leeuwen EJC, Zimmerman E, Davila Ross M (2011) Responding to inequities: gorillas try to maintain their competitive advantage during play fights. Biol Lett 7:39–42

Ward C, Bauer E, Smuts B (2008) Partner preferences and asymmetries in social play among domestic dog, *Canis lupus familiaris*, littermates. Anim Behav 76(4):1187–1199

Whitehead H (2008) Analyzing animal societies: quantitative methods for vertebrate social analyses. University of Chicago Press, Chicago, IL

Wright KR, Mayhew JA, Sheeran LK et al (2018) Playing it cool: characterizing social play, bout termination, and candidate play signals of juvenile and infant Tibetan macaques (*Macaca thibetana*). Zool Res 39(4):1–13

Xia DP, Li JH, Garber P et al (2012) Grooming reciprocity in female Tibetan macaques *Macaca thibetana*. Am J Primatol 74(6):569–579

Xia DP, Li JH, Garber P et al (2013) Grooming reciprocity in male Tibetan macaques. Am J Primatol 75:1009–1020

Yamanashi Y, Nogami E, Teramoto M et al (2018) Adult-adult social play in captive chimpanzees: is it indicative of positive animal welfare? Appl Anim Behav Sci 199:75–83

Yanagi A, Berman CM (2014a) Functions of multiple play signals in free-ranging juvenile rhesus macaques (*Macaca mulatta*). Behaviour 151:1983–2014

Yanagi A, Berman CM (2014b) Body signals during social play in free-ranging rhesus macaques (*Macaca mulatta*): a systematic analysis. Am J Primatol 76:168–179

Zhao QK (1997) Intergroup interactions in Tibetan macaques at Mt. Emei, China. Am J Phys Anthropol 104(4):459–470

Chapter 7
The Vocal Repertoire of Tibetan Macaques (*Macaca thibetana*) and Congeneric Comparisons

Sofia K. Blue

7.1 Introduction

The vocal repertoires of mammals usually consist of a fixed number of calls, some of which are closely linked to particular contexts. Though not to the same extent as humans, other mammals have been documented to emit highly modifiable calls with a cognitively rich set of meanings (Seyfarth and Cheney 2010). Fixed vocal production coupled with modifiable context emission and comprehension may have been homologous traits present in our prelinguistic ancestor, since they appear to be present in a wide array of taxonomic groups. Seyfarth and Cheney (2012) suggest that the common ancestor of Old World monkeys, apes, and humans had limited vocal production and open-ended comprehension, and by making comparisons with our closest living relatives, we can further illustrate the implications of theories concerning language evolution.

Our limitless repertoire of sound combinations and capacity for vocal learning is unmatched in the animal kingdom. To date, songbirds have been the focus of vocal learning studies in non-human animals, and evidence in mammals has been restricted to species of cetaceans, pinnipeds, elephants, and bats (Lattenkamp and Vernes 2018). Surprisingly, our closest living relatives, the non-human primates, are deficient in their ability to learn new vocalizations. This inability may be a result of a lack of neuronal potential required for vocal learning although the vocal tract is speech-ready (Fitch et al. 2016). Vocal production is, for the most part, highly constrained in non-human animals, and mammalian repertoires usually consist of a variety of grunts, threatening vocalizations, alarm calls, and screams (Seyfarth and Cheney 2012). Comparing the diverse array of vocal repertoires and communication across taxa is one way to identify the selective pressures behind vocal complexity

S. K. Blue (✉)
Primate Behavior and Ecology Program, Department of Anthropology and Museum Studies, Central Washington University, Ellensburg, WA, USA
e-mail: Sofia.Bernstein@cwu.edu

J.-H. Li et al. (eds.), *The Behavioral Ecology of the Tibetan Macaque*, Fascinating Life Sciences, https://doi.org/10.1007/978-3-030-27920-2_7

and the biological underpinnings of language-specific traits. Finding an appropriate measure of vocal complexity, however, is challenging when there are no standard methods for quantifying, classifying, or describing vocal repertoires.

7.2 Measuring Vocal Complexity

7.2.1 Vocal Repertoire Size

Although communication is undoubtedly multimodal in nature, across the modalities the number of distinct signals or signaling units can be used as a representative of communicative complexity (Peckre et al. 2019). For studies investigating the modality of acoustic communication, one possible metric of vocal complexity is repertoire size. Repertoire size is defined by the number of call types produced by a species or population (Peckre et al. 2019). Therefore, repertoire size is a strictly numeric measure unable to provide any information concerning the function or usage of the calls that constitute them. Variable data collection methods among studies constrain the ability to make exact comparisons across species and genera, further limiting the informative value of repertoire size alone. In addition, comparing repertoire size leaves out the identification of species-specific calls that may have evolved.

Hohmann (1991) carried out a comparative analysis of the vocal repertoires of four species of Old World monkey (*Macaca radiata*, *Macaca silenus*, *Presbytis johnii*, *Presbytis entellus*). For the species under Hohmann's (1991) investigation, previous reports of their vocalizations were very fragmented. To bypass this challenge and make accurate comparisons, Hohmann (1991) recorded calls from these four species and used the same methods to analyze the call recordings. Even though this study involved an intensive classification of the vocal repertoire of four different species, emphasis was placed on comparisons of the frequencies of call type emission and vocal activity among sex/age classes, rather than identifying species-specific calls or call usage. Therefore, the extent to which vocal complexity among the four species was investigated is limited.

7.2.2 Identifying Homologous and Derived Calls

Gustison et al. (2012) proposed that the identification of homologous (acoustically similar calls shared between species) and derived vocalizations (acoustically unique to a species) among closely related species can be used as a measure of vocal complexity. By focusing on the identification of acoustically homologous calls (Hohmann 1991; Gustison et al. 2012) and identifying species-specific derived calls, researchers are not constrained by the variable methods used across studies to classify vocal repertoires. Furthermore, comparing homologous and derived vocalizations among closely related species is essential in the identification of phylogenetic, social, and ecological factors influencing varying degrees of vocal complexity.

To date, few studies in non-human primates have approached investigations on vocal complexity through the identification of derived calls. Kudo (1987) compared the vocal behavior of mandrills (*Mandrillus sphinx*) to savannah baboons (*Papio* spp.) and geladas (*Theropithecus gelada*) and found that differences among the three species' vocal behavior were due to the ecological pressures associated with the attenuation of sound in forest versus savannah habitats. More recently, Gustison et al. (2012) identified the homologous and derived vocalizations in two closely related species, geladas and chacma baboons (*Papio ursinus*). In this comparison, derived vocalizations were uttered in social and reproductive contexts unique to a species, suggesting that differences in sociality and reproductive ecology were the driving factors in the evolution of these species-specific vocalizations.

7.3 Understanding the Evolution of Vocal Complexity

To understand the evolution of vocal complexity, it is necessary to make comparisons across closely related species. Although broader interspecies comparisons do exist and reveal some interesting patterns (McComb and Semple 2005), in-depth investigations among closely related species allow researchers to tease apart which factors have driven the evolution of communication (Bouchet et al. 2013). Macaques may be an ideal genus for investigations of vocal complexity because of several characteristics. Vocalizations in the *Macaca* genus have garnered a considerable amount of attention over the years, and the literature is rich enough to make detailed comparisons across many species. In addition, the genus is the most geographically widespread and behaviorally diverse genus of non-human primate because macaques display a high range of interspecific variation and inhabit the greatest range of habitats (Thierry et al. 2000). By documenting the diversity across the macaques through a comparative perspective and the identification of derived calls, I can investigate how potential selective pressures, like phylogeny, sociality, and ecology, shape which calls are conserved or vary in the vocal repertoires of the genus.

The effects of phylogeny, sociality, and ecology are essential in the explorations on the evolution of vocal complexity (Freeberg et al. 2012). Phylogeny may act as a starting point for mapping the evolutionary convergence of vocal characters. For example, complex oropendola bird songs are conserved and relatively invariant among the three genera (Price and Lanyon 2002), and primate vocalizations appear to be largely genetically predetermined (Newman and Symes 1982). It is widely accepted in the literature that sociality may be an essential driving factor in species with high degrees of vocal complexity (Blumstein and Armitage 1997; Wilkinson 2003; Freeberg 2006; Furrer and Manser 2009). In their investigation on Tonkean macaques (*Macaca tonkeana*), Masataka and Thierry (1993) concluded that sociality determines the vocal repertoire of a species as strongly as phylogenetic constraints. Lastly, since macaques inhabit a wide range of habitats, interspecific variation in vocal repertoires may be the result of these ecological differences (Masataka and Thierry 1993). It is likely, however, that any one of these factors is insufficient to explain vocal diversity and that a complex interplay of phylogenetic, social, and ecological factors influences degrees of vocal complexity.

In this contribution, I explore vocal homologs and derived calls in the *Macaca* genus through the:

1. Identification of the main categories of call production
2. Selection of call types from each category of call production to compare across the genus
3. Identification of homologs based on acoustic characteristics
4. Identification of species-specific derived calls and main differences across the genus
5. Comparisons with the Tibetan macaque vocal repertoire
6. Exploration of phylogenetic, social, and ecological factors that may influence homologous and derived calls

7.4 Methods

7.4.1 Categories of Call Production

Two previous studies designated categories for comparison across different species of Old World monkeys (*M. radiata*, *M. silenus*, *P. johnii*, *P. entellus*: Hohmann 1991; *Papio ursinus*, *T. gelada*, Gustison et al. 2012). I followed Gustison et al. (2012) and identified calls in macaques within three main categories and included the subcategories differentiated for each: allospecific (*alarm* and *food calls*), social (long-distance and close-range calls (competitive, distress, and contact)), and other (see Fig. 7.1). Allospecific calls are elicited by external stimuli and include *alarm* and *food calls*. The majority of vocalizations in mammals are emitted during social interactions with conspecifics (Gustison et al. 2012). Such social calls are further categorized into two subclasses: long-distance and close-range calls. For close-range calls, a further subdivision is necessary; so I classified close-range calls into three additional subcategories to cover the many different contexts in which these calls are emitted: contact, competition, and distress. The last category, other, includes calls that are not strictly emitted in a particular context or the context is unknown.

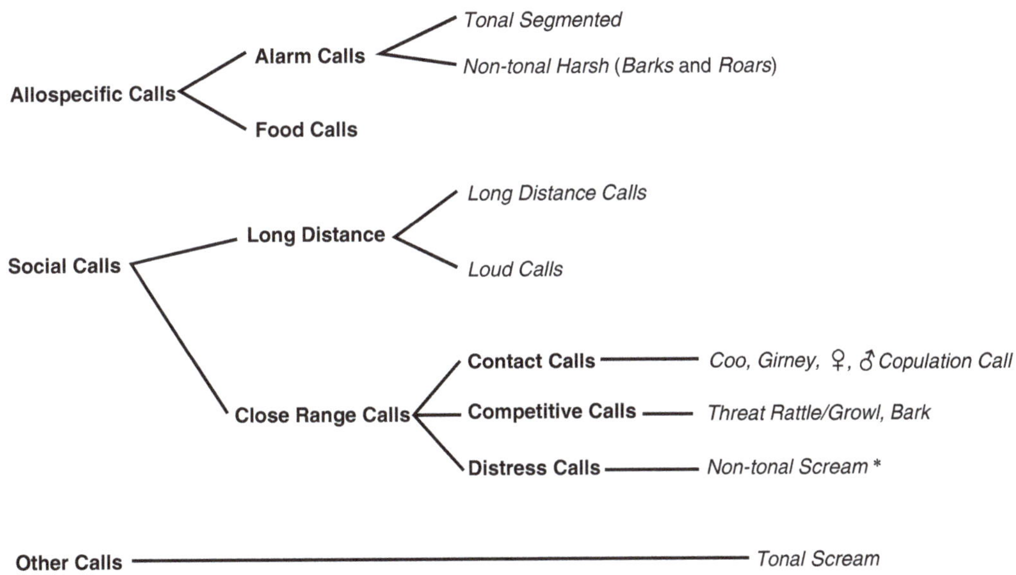

Fig. 7.1 Call categories and call types investigated in the genus *Macaca*. The main categories and subcategories are in bold, and calls in italics indicate the call types chosen for homologous and derived comparisons. (*) Non-tonal screams are not strictly distress calls since they are also emitted in feeding contexts

7.4.2 Exploring Vocal Homologs and Derived Calls in the Genus Macaca

The homologous call types investigated come from a review of the macaque literature and the identification of calls that were similar in acoustic structure alone, even if they were designated different names in the literature or were emitted in different contexts. Derived calls are defined as acoustically unique to a species and were also identified via visual inspection of spectrograms. Similar to my previous study (Bernstein et al. 2016), I followed Rowell and Hinde (1962) and first identified calls as either clear or harsh. Clear calls are tonal with energy concentrated in harmonic frequency bands, while harsh calls are atonal and acoustically unstructured with underlying harmonic frequency bands obscured by broadband noise and a distribution of energy across a wide frequency spectrum (Palombit 1992; Rowell and Hinde 1962). Once calls were classified as either clear or harsh, I then focused on temporal and frequency characteristics to identify homologous calls. These characteristics included the presence or absence of harmonic frequency bands, the modulation of frequency bands, spectral shifts (e.g., abrupt changes like the presence of a fast-rise transient from high pitch to low pitch), the concentration of spectral energy, whether or not calls were attached to or superimposed onto a noisy portion, and the duration of the call.

In order to make comparisons in the macaque genus, I used reports from which the entire repertoire was systematically investigated and excluded investigations that described only particular call types. This method is similar to the one adopted by McComb and Semple (2005) in their investigation of the coevolution of vocal communication and repertoire size as a measure of vocal complexity. To date there are repertoire studies from 11 different macaque species, including my study on Tibetan macaques (Tibetan macaques, *Macaca thibetana*, Bernstein et al. 2016; Formosan macaques, *M. cyclopis*, Hsu et al. 2005; Barbary macaques, *M. sylvanus*, Hammerschmidt and Fischer 1998; Tonkean macaques, *M. tonkeana*, Masataka and Thierry 1993; long-tailed macaques, *M. fascicularis*, Palombit 1992; bonnet macaques, *M. radiata*, Hohmann 1989; lion-tailed macaques, *M. silenus*, Hohmann and Herzog 1985; Sugiyama 1968; stump-tailed macaques, *M. arctoides*, Lillehei and Snowdon 1978; Japanese macaques, *M. fuscata*, Green 1975; Itani 1963; rhesus macaques, *M. mulatta*, Rowell 1962; Rowell and Hinde 1962). Comparison of call emission frequency and in which contexts calls were uttered was analyzed when the results were available across the published repertoires in the genus.

7.4.3 Comparisons with Tibetan Macaques and Phylogenetic, Social, and Ecological Factors

I included a wider array of species for comparisons with the Tibetan macaque vocal repertoire based on phylogenetic closeness, even if their repertoires have not been systematically analyzed and reported. I made this attempt to incorporate studies on specific calls for a more robust comparison with particular Tibetan macaque calls that were not included in the genus comparison. I chose interspecific call types based on their acoustic similarity and not context (for context comparisons, see Table 7.2).

After I identified the homologous and derived calls in the genus, I explored potential phylogenetic, social, and ecological factors. Depending on the source and the type of genetic study, the phylogenetic classification of macaques can vary. One of the most recent studies by Li et al. (2009) investigated the phylogeny of the macaques based on Alu elements. I used their classification for the comparisons investigated here (for each species' designation in the phylogeny, see Table 7.1). For the social component, I compared the contexts of emission and, if available, the rate at which the call types are emitted in a species' repertoire. In addition, I used reviews of the *Macaca* genus (Thierry et al. 1996, 2000; van Schaik et al. 1999; Maestripieri and Roney 2005; Pradhan et al. 2006) and the life history traits provided by Singh and Sinha (2004) to explore social and ecological factors.

Table 7.1 Shared call types in the macaque genus

Group	Species	Call types											
		Coo	Threat rattle/ growl	Non-tonal scream	Girney	Tonal scream	Squeak	Food call	Alarm call	♀ Cop call	♂ Cop call	Bark	Loud call
Sylvanus	*sylvanus*	x[a]	–	x	–	x	x	–	NT	x	–	x	–
Silenus	*silenus*	x	x	x	x	x	x	–	NT	x	x	–	x
	nemestrina	x	x	x	–	–	x	–	NT	x	–	x	–
	tonkeana	x	x	x	x	x	x	–	TS	x	–	x	x
Sinica	*radiata*	x	x	x	x	x	x	–	TS	x	x	x	x
	thibetana	x	x	x	–	x	x	–	TS	x	x	x	x
	arctoides	x	x	x	x	x	x	–	NT	x	x	x	x
Fascicularis	*fascicularis*	x	x	x	–	x	x	–	NT, TS	x	x	x	–
	fuscata	x	x	x	x	x	x	–	TS	x	–	–	x
	mulatta	x	x	x	x	x	x	x	TS	–	x	x	–
	cyclopis	x	x	x	x	x	x	x	TS	x	x	x	–

An x indicates that the call type is present in the repertoire of that species
NT non-tonal harsh alarm calls including barks and roars, *TS* tonal segmented alarm calls
[a]*M. sylvanus* have a limited use of coo calls in their repertoire (Bruce and Estep 1992)

7.5 Results

7.5.1 Homologous and Derived Calls in the Genus

I selected call types from the three main categories (allospecific, social, and other) of call production across the genus. I identified the following as the shared calls in the 11 macaque species investigated: *coo*, *threat rattle/growl*, *non-tonal scream*, *girney/greeting call*, *tonal scream*, *squeak*, *food call*, *alarm call* (*non-tonal harsh* and *tonal segmented* types), *female* and *male copulation call*, *bark*, and *loud call* (Table 7.1). The main category allospecific calls includes the *tonal segmented* and *non-tonal harsh alarm calls* and the *food call*. *Threat rattle/growl*, *bark*, *non-tonal scream*, *coo*, *girney*, *female* and *male copulation call*, *long distance*, and *loud call* are in the social category, and the *tonal screams* are in the other category since they are emitted in feeding and distress contexts (for subcategories, see Fig. 7.1).

For the 11 species investigated, I identified 9 derived calls based solely on acoustic structure: the *krahoo* (*M. fascicularis*); *food yell*, *atonal greeting*, and *tonal girney* (*M. cyclopis*); *warble*, *harmonic arch*, *chirp*, and *male copulation scream* (*M. mulatta*); *male copulation grunt* (*M. arctoides*); *tonal estrus call* (*M. fuscata*); and *female copulation call* (*M. thibetana*).

The main differences found in the homologous calls concerned the frequency of emission and the range of call types within a specific context. For example, *M. arctoides* have a limited use of *coo* calls with considerably more harsh, noisy calls in a purely graded signal system (Bruce and Estep 1992). *Macaca tonkeana* and *M. cyclopis* macaques have a wider range of agonistic and *girney/greeting* calls, respectively, that is unparalleled in the rest of the genus (Masataka and Thierry 1993; Hsu et al. 2005).

7.5.2 Comparisons with the Vocal Repertoire of Tibetan Macaques

The Tibetan macaque vocal repertoire consists of five clear calls (*coo*, *squawk*, *leap coo*, *weeping*, *modulated tonal scream*) and seven harsh calls (*squeal*, *noisy scream*, *growl*, *bark*, *compound squeak*, *pant*, *female copulation call*) (for a quantitative analysis of the vocal repertoire, see Bernstein et al. (2016); see Fig. 7.2). In order to make congeneric comparisons with the Tibetan macaque vocal repertoire, I included a wider array of species based on phylogenetic closeness, even if their repertoires have not been systematically analyzed and reported. I made this attempt to incorporate studies on specific calls for a more robust comparison with particular Tibetan macaque calls that were not included in the genus comparison above (i.e., *squawk*, *squeal*, *leap coo*, *weeping*, and *pant*). I chose interspecific call types based on their acoustic similarity and not context (for context comparisons see Table 7.2).

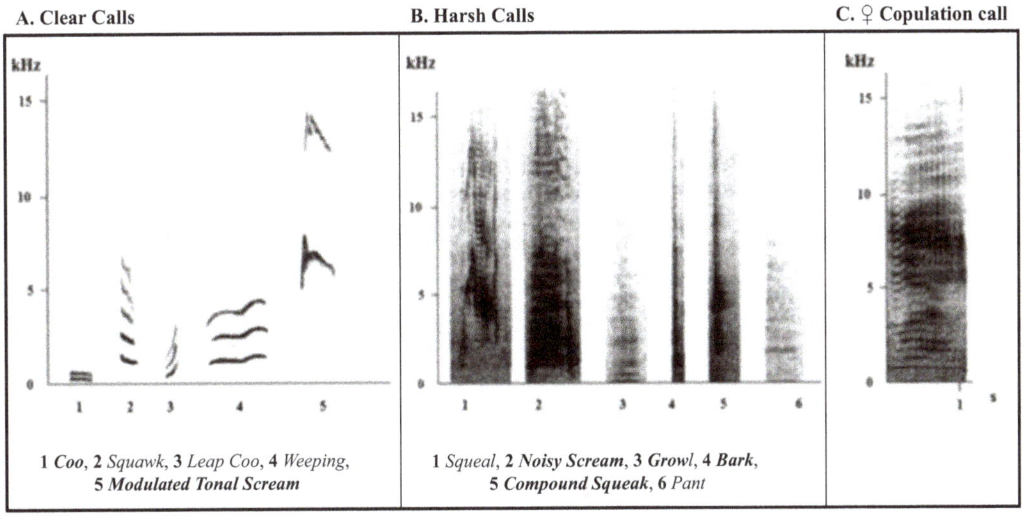

Fig. 7.2 Adapted from Bernstein et al. (2016). Calls in **bold** are homologous calls shared with the rest of the genus. (**a**) Representative spectrograms of the clear calls in the repertoire (**b**) Representative spectrograms of the harsh calls in the repertoire (**c**) Representative spectrogram of the female copulation call

Overall, the acoustic structure and context of emission was similar among the species investigated. A few main differences in context were a result of the group composition and various species-specific behaviors present in the congeneric species, but not in Tibetan macaques. For example, all of the infants died during my study on the vocal repertoire of Tibetan macaques, and therefore information about the calls associated with mother-infant interactions are limited (Bernstein et al. 2016). Also, an absence of calls from infants leaves out one vital age class that is necessary for a more complete comparison across species. Additionally, Tibetan macaques are one of two macaque species that show bridging behavior where two adults simultaneously lift an infant (Ogawa 1995, this volume). Bridging may be another context associated with infants where individuals emit calls that I was unable to observe.

Also, within the context of mating, female Tibetan macaques do not show any behaviors indicative of estrus (Li et al. 2005) and do not emit an estrous call. My study took place in the Valley of the Wild Monkeys, China, a site where the provisioned macaque groups that inhabit the mountainous sparsely covered terrain have been protected from hunting and trapping since the 1940s (Wada et al. 1987). During my observations, the Tibetan macaques did not emit an *alarm* or *food call.* Although Tibetan macaques were observed to emit *loud calls* that propagated over a large distance, these calls were clustered in the noisy scream category.

The female Tibetan macaque *copulation call* was the only call in my previous study that was a derived vocalization distinct from what has been reported in the rest of the genus (Bernstein et al. 2016). Usually, female macaques emit soft grunts during copulations. The Tibetan macaque females also emit quiet calls, but their acoustic structure does not involve biphasic "inhale-exhale" components, and they are instead shrill, undulating calls that are not always given in phrases. In

Table 7.2 Tibetan macaque vocal repertoire and congeneric comparisons

	Species	Call type	Context
Coo			*Foraging, provisioning, group movements, cohesion, dispersal*
	M. radiata	Contact whoo	Similar
	M. cyclopis	Contact coo	Similar
	M. sinica	Hum	Similar
	M. fuscata	Class II coo	Similar
	M. mulatta	Clear call	**Shut out**[a]
	M. fascicularis	Coo	Similar
	M. arctoides	Coo	Similar
	M. silenus	Whoo call	Similar
	M. sylvanus	Coo	**Limited use**[a]
Squawk			*♂ copulation call, dispersal, food begging, arrival of tourists*
	M. fuscata	Class IV high squawk	**Defensive submissive**[a]
	M. cyclopis	Squawk	**Defensive submissive**[a]
Squeal			*Submissive agonistic context, displacement by humans*
	M. radiata	Squeal	**Rejection of infants by mother, juvenile mobbing of adults**[a]
	M. cyclopis	Squeal	**Infant separation**[a]
	M. fuscata	Class VI squeal	**Estrus solicitation, infant separation**[a]
	M. silenus	Shriek	Submissive agonistic interactions
Noisy scream			*Agonistic, submissive intragroup/interspecies, provisioning, dispersal*
	M. cyclopis	Non-tonal scream	Similar
	M. radiata	Non-tonal scream	Intergroup interactions**, infant weaning**[a]
	M. mulatta	Non-tonal scream	Similar
	M. nemestrina	Agonistic scream	Similar
	M. cyclopis	Food yell	Similar
	M. sinica	Food yell	**Foraging discovery new food source**[a]
	M. fascicularis	Wraagh	Similar

(continued)

Table 7.2 (continued)

	Species	Call type	Context
Growl			*Agonistic, submissive intragroup/interspecies, provisioning, male-male mounts provisioning/ foraging, male-male mounts*
	M. cyclopis	Threat rattle	Similar
	M. radiata	Threat rattle	**Group movement**[a]
	M. silenus	Rattle	Similar
	M. fascicularis	Harr	**Juveniles threatened, during play**[a]
	M. fuscata	Class X gruff	**During threats and standoffs**[a]
Bark			*Intragroup/interspecies agonistic, provisioning, movements, weaning*
	M. radiata	Bark	Similar
	M. nemestrina	Bark	Similar
	M. tonkeana	Bark	Similar
	M. mulatta	Bark	Similar
	M. arctoides	Bark	Similar
	M. cyclopis	Bark	Similar
	M. fuscata	Bark	**Alarm response to dogs, snakes, raptors**[a]
Compound squeak			*Intragroup/interspecies agonistic context, weaning*
	M. cyclopis	Compound squeak	Similar
Leap Coo			Adults, *provisioning, foraging, group movements, arrival of tourists*; juveniles, *weaning*
	M. nemestrina	Leap coo	**During juvenile excitement**[a]
Weeping			*Rejection and weaning of juveniles*
	M. cyclopis	Weeping	Similar
	M. silenus	Weeping	Similar
	M. radiata	Pulse whoo	Similar
	M. sinica	Infant separation call	Similar
Modulated tonal scream			*Harassment of copulations, intragroup/interspecies agonistic context, provisioning, rejection and weaning of juveniles*
	M. sylvanus	Mod. tonal scream	**Forced separation from mother**[a]
	M. cyclopis	Tonal squeak	Similar
	M. radiata	Squeak	Similar
	M. fuscata	Squeak	**♀ homosexual solicitations**[a]

(continued)

Table 7.2 (continued)

	Species	Call type	Context
Pant			*Harassment of copulations, intragroup/interspecies agonistic context*
	M. mulatta	Pant threat	Similar

Context of emission for Tibetan macaques in *italics*
Comparisons made from acoustic spectra and context of emission provided in previous reports of macaque vocal repertoires (Tibetan macaques, *Macaca thibetana*, Bernstein et al. 2016; Formosan macaques, *Macaca cyclopis*, Hsu et al. 2005; Barbary macaques, *M. sylvanus*, Hammerschmidt and Fischer 1998; Tonkean macaques, *M. tonkeana*, Masataka and Thierry 1993; long-tailed macaques, *M. fascicularis*, Palombit 1992; bonnet macaques, *M. radiata*, Hohmann 1989; lion-tailed macaques, *M. silenus*, Hohmann and Herzog 1985; Sugiyama 1968; stump-tailed macaques, *M. arctoides*, Lillehei and Snowdon 1978; Japanese macaques, *M. fuscata*, Green 1975; Itani 1963; rhesus macaques, *M. mulatta*, Rowell 1962; Rowell and Hinde 1962)
[a]Tibetan macaques do not emit the homologous call in the **bold** context reported in the comparison species

comparison to the rest of the macaque genus, *female copulation call* rate was low and resembled that of *M. tonkeana* and *M. fuscata* (Masataka and Thierry 1993; Hohmann 1989; Oda and Masataka 1992; see Table 7.3). Similar to *M. radiata*, a closely related species in the same lineage, females do not emit an *estrous call* and *copulation calls* are the only auditory cue females emit in a copulation context. Males did emit *copulation calls* as well, but at a much higher frequency. Harassment of copulatory dyads is evident in Tibetan macaques, involves a wider range of age/sex classes than what is reported for most other members of the genus, and appears to be associated with the number of females in the audience.

7.5.3 Potential Effects of Phylogeny, Sociality, and Ecology

Phylogenetic Factors In the macaque species whose repertoires have been systematically analyzed and reported, there are two main types of acoustically structured *alarm calls*: *non-tonal barks* and *roars* (NT) or *tonal segmented* (TS) calls. Hohmann (1989) proposed that the type of *alarm call* (NT or TS) present in the repertoire of a macaque species is mostly conserved within phylogenetic groups (NT, *sylvanus* and *silenus*; TS, *sinica* and *fascicularis*). This pattern holds true for the *silenus*, *sinica*, and *fascicularis* groups (see Table 7.1). Indeed, species in the *silenus* group (*M. nemestrina* and *M. silenus*) have *non-tonal alarm calls* (Caldecott 1986; Hohmann and Herzog 1985). The only two terrestrial species of macaque with a purely graded signal system, *M. sylvanus* of the *sylvanus* lineage and *M. arctoides* of the *sinica* group, also have *non-tonal alarm calls* (Fischer and Hammerschmidt 2002; Chevalier-Skolnikoff 1974). The only exception in the *silenus* group, *M. tonkeana*, emits *tonal segmented alarm calls*. This pattern, where the acoustic structure of an *alarm call* is conserved within phylogenetic groups, is found in

Table 7.3 Copulation calls in *Macaca* and life history traits

Phylogenetic group	Species	♂ Copulation call	♀ Copulation call	Traits					
				Mount pattern	Seasonality	Number of males	♀♂ Ratio	Sexual dimorphism	Sex skin
Sylvanus	*M. sylvanus*	–	x	SME	2	4.7	1.1	0.37	Yes
Silenus	*M. silenus*	x	x	MME	1	1.6	9.9	0.38	Yes
	M. nemestrina	–	x	MME	0–1	2.4	6.3	0.54	Yes
	M. tonkeana	–	x	MME	0–1	2	3.4	0.5	Yes
Sinica	*M. radiata*	x	x	SME	2	7.3	1.7	0.54	No
	M thibetana	x	**Shrill**	SME	2	5.5	1.9	0.36	Little
	M. arctoides	**Grunt**	x	SME	1	?	1.32	0.37	No
Fascicularis	*M. fascicularis*	x	x	SME	1	4.7	4.8	0.41	Little
	M. fuscata	–	x, **Tonal estrus**	MME	2	3.4	1.3	0.32	Little
	M. mulatta	**Scream**	–	MME	2	4.5	2.9	0.22	Little
	M. cyclopis	x	x	MME	2	1.8	1	0.2	Little

Compilation from van Schaik et al. (1999), Maestripieri and Roney (2005), and Pradhan et al. (2006)
An x indicates the call is present in the repertoire of the species
Calls in **bold** are species-specific derived calls
Breeding seasonality: 0, 33% of births in 3-month period; 1, between 33% and 67%; 2, 67% in 3-month period (van Schaik et al. 1999)

sympatric species from different phylogenetic groups that may be prey to the same predators (*M. silenus* and *M. radiata*, Hohmann 1989).

Fooden (1976) sorted the macaque lineages based on the reproductive anatomy of males and females. Although his study relied only on behavioral and anatomical features, his classification is still mostly accurate even with more sophisticated genetic analyses. The evolution of different reproductive anatomy likely stemmed from each lineage going through a genetic bottleneck (Tosi et al. 2003). The three species that emit derived *copulation calls* (*M. thibetana*, *M. mulatta*, *M. arctoides*) are more closely related than they initially appear. Recent studies in the whole genome sequencing of Tibetan macaques have found that they are most closely related to Chinese rhesus macaques despite having been reported in different lineages (Fan et al. 2014). Analyses of mitochondrial loci in the macaque genus show stump-tailed macaques to potentially be a hybrid of proto-*M. thibetana* (Tosi et al. 2003).

Social Factors Most of the derived calls in the macaque genus are related to the context of mating, in particular, *female* and *male copulation calls*. Throughout the genus, males typically produce squeak-like *copulation calls*, while females emit grunts characterized by an "inhale-exhale element" (Pradhan et al. 2006). However, a few species stray from these genus characteristic calls and emit acoustically distinct calls associated with copulations. Two species emit acoustically distinct *male copulation calls*: *M. mulatta copulation screams* (Hauser 1993) and *M. arctoides copulation grunts* (Bauers 1993). Japanese macaques are the only macaque species that have studies reporting a distinct *estrous call* from the *female copulation call* (Oda and Masataka 1992), and my previous study reports an acoustically distinct shrill *female copulation call* in Tibetan macaques (Bernstein et al. 2016). All of these species that emit acoustically distinct calls in the context of copulations are seasonal breeders, have promiscuous mating systems, all exhibit the interruption of copulations by conspecifics, and have similar adult female-to-male ratios (1.3:2.9, Singh and Sinha 2004; Pradhan et al. 2006).

A large number of call types in the context of agonistic interactions is a characteristic of *M. tonkeana*. This species has high degrees of tolerance, small rank differences, a high frequency of conciliatory patterns, small interindividual distances, and a high rate of bidirectional agonistic interactions (Masataka and Thierry 1993). Context-dependent differences were also found in the derived *girney/greeting calls* of *M. cyclopis*. All of the *girney/greeting calls* were paired with particular contexts and age/sex classes, with acoustic structure changing according to the recipient of the *girney/greeting call* (Hsu et al. 2005).

Ecological Factors Overall, *loud calls* adhere to the bioacoustic requirements for long-distance propagation of sound (e.g., the repetition of units and phrases, no intergradations to other vocal patterns, a stereotyped structure, a large range of frequencies, and a concentration of energy on the lower frequencies; Hohmann and Herzog 1985). *Loud calls* are harsh calls where the fundamental frequency is in the lower frequency range. However, only some species show a large range of frequencies and utter *loud calls* that are composed of both tonal and non-tonal harsh units. Three macaque species utter *loud calls* that are acoustically distinct from the rest of the genus and are composed of tonal and harsh units in a wider range of frequencies (*M. silenus*, *loud call*, Hohmann and Herzog 1985; *M. fascicularis*, *krahoo*, Palombit 1992; *M. tonkeana*, *loud call*, Masataka and Thierry 1993). Also, for all three species, only males have been reported to emit these *loud calls*. The *loud calls* of *M. silenus* are mixed units with a noise-like inhalation phase followed by a tonal exhalation (Hohmann and Herzog 1985), the *krahoo* of *M. fascicularis* is distinct for its harsh "*kra*" syllable at the beginning of the call (Palombit 1992), and *M. tonkeana*'s *loud call* is characterized by a distinct flag and mast with a sharp upward and downward frequency modulation (Masataka and Thierry 1993). All three species live in rainforest habitats and are mostly arboreal.

Food calls are common in the genus, but so far only two species have been reported to give acoustically distinct calls strictly emitted in food-related contexts. *M. mulatta warbles*, *harmonic arches*, and *chirps* are acoustically distinct calls made exclusively in a food context (Hauser and Marler 1993). *M. cyclopis* emit *food yells* similar to the *M. sinica food call*, but are acoustically distinct (Hsu et al. 2005). Reports from these two species are from provisioned groups living in human habitations or captivity.

7.6 Discussion

Mammalian vocalizations consist mainly of grunts, harsh non-tonal threatening vocalizations, and the sometimes tonal alarm calls and screams (Seyfarth and Cheney 2012). Indeed, the macaque genus shared these basic vocalizations (e.g., grunts, *girneys*, *estrus calls*, *female copulation calls*; harsh non-tonal calls, *barks*, *growls*; *tonal alarm calls*; screams, *shrieks*, *tonal*, *non-tonal screams*) but also emitted additional tonal calls (e.g., *coos*, *squawks*, and *male copulation calls*). Overall, macaque repertoires can be described as a graded signal system with intergradations between calls, and the contexts of emission for shared calls were consistent.

Although primate vocalizations appear to be largely genetically predetermined (Newman and Symes 1982), there are differences in the flexibility and conservation of the call types investigated. The flexibility observed in the derived vocalizations seems to be a by-product of the characteristics of the contexts in which calls are emitted or species-specific features. The most salient difference found in my congeneric comparison were the calls emitted in the copulation context. Three species emit derived *copulation calls* (*M. thibetana female copulation call*, *M. mulatta male copulation scream*, *M. arctoides male copulation grunt*). These species that emit *copulation calls* uncharacteristic of the genus have species-specific features in their copulation styles.

7.6.1 The Tibetan Macaque Vocal Repertoire

In sum, the Tibetan macaque vocal repertoire was generally comparable to that of other macaque species. The context of emission was also similar across species, but with a few key differences. For example, my previous study could not investigate the vocalizations of infants because of a 100% mortality rate during the 2014 mating season (Bernstein et al. 2016). Future studies on the vocalizations of infants and infant-related contexts may find that Tibetan macaques do emit calls not previously described in the analysis of the repertoire. The main differences found between Tibetan macaques and the rest of the genus may not only be a result of a missing age class but also the method used to quantify the repertoire and species-specific aspects of their social behavior.

An absence of a *food call* or *alarm call* may be the result of site management. These call types may not be absent. It is possible that I simply did not observe them over the course of my study or that provisioning and predators being hunted out of the area have removed these contexts that would elicit calls they may have present in their repertoire. Although Tibetan macaques do emit calls that propagate over a large distance, they were not a distinct call type like the *loud calls* described for other species. Most of the species that emit *loud calls* live in rainforest habitats where certain acoustic requirements are necessary for sound propagation. Tibetan macaques inhabit mountainous terrain with sparse tree coverage, where the propagation of their calls is not limited by a densely covered forest (Xia et al. 2010). My quantitative method of relying solely on acoustic characters to define distinct call types may have clustered calls identified separately in other studies into more overarching categories. The same could be true for *greeting calls*. In some cases, variants of the *coo* call are considered as a separate category, but for my study, individual classification of *coo* call variants was not warranted.

The unique copulatory behavior of Tibetan macaques may be one reason an acoustically distinct female call was selected for. Also within the context of mating, female Tibetan macaques do not show any behaviors indicative of estrus (Li et al. 2005), and therefore, it is not surprising that their concealed ovulation would yield the absence of an *estrous call*. The similarity between *M. thibetana*, *M. tonkeana*, and *M. fuscata* in terms of a low *female copulation call* rate may imply a cost-benefit strategy that differs between the sexes. All of these species are seasonal breeders, living in multi-male, multi-female groups, and exhibit high rates of copulation interruptions (see Table 7.3). *Macaca tonkeana* and *M. fuscata copulation call* emission rate may also be low because females additionally produce *estrous calls* to solicit males. These species may have evolved an alternate strategy of mating promiscuously and a low call rate to reduce female competition. My preliminary investigation on the association between *female copulation call* rate and harassment of copulatory dyads does indicate that harassment increases with the number of females in the audience.

However, the presence of both *male* and *female copulation calls* might indicate that part of their function is to synchronize male and female mating behavior, especially in seasonally breeding species. Additionally, for males, it might not be so costly to call as it is for females. Adult females harass copulations, sneak copulate with lower-ranking males, and mate promiscuously with males outside of their consortship. In primates, there are very few studies that have investigated a cost-benefit analysis of signaling in the mating context (Hauser 1993). Future studies should investigate the costs and benefits of producing *copulation calls* along with emission rate and how intra- and intersexual selection plays a role in the mating strategies and auditory sexual signals of males and females.

7.6.2 Phylogenetic, Social, and Ecological Factors Influencing Macaque Vocal Repertoires

The type of *alarm call* was conserved within phylogenetic groups (*non-tonal*, *sylvanus* and *silenus*; *tonal segmented*, *sinica* and *fascicularis*) and followed the designation described by Hohmann (1989) with a few exceptions. *Macaca sylvanus* and *M. arctoides*, the exceptions in the *sinica* group, would be expected to have *non-tonal harsh alarm calls* since their repertoires consist of mostly harsh calls and their placement in the phylogeny of macaques is contested (*M. arctoides*, Li et al. 2009). *Macaca tonkeana*, the only exception in the *tonkeana* group, emit *tonal segmented alarm calls*, which may be the result of this species having a wide array of tonal calls in their repertoire (Masataka and Thierry 1993). In conclusion *alarm calls* are one salient example of a vocalization type that is mediated by phylogeny and a repertoire's acoustic structure, which is largely genetically predetermined (Newman and Symes 1982).

The species that emit derived species-specific *copulation calls* are all closely related. Phylogenetic and social factors associated with reproduction may be the selective factors under which derived calls have evolved in these three species. Future studies are needed to investigate the *copulation calls* of females in more detail, to understand their effect as an auditory sexual signal, and to make more in-depth comparisons of their acoustic structure with the rest of the genus.

Certain aspects of *M. tonkeana* social behavior may drive the need for a larger number of agonistic vocalizations (Masataka and Thierry 1993). The high degree of variability in the frequency of the acoustic structure of *girney/greeting calls* in *M. cyclopis* could also be a result of sociality since they were heavily context dependent. Some of these acoustically distinct *girney/greeting calls* are different from what has been reported in the rest of the genus and could be considered as derived calls.

Factors that influence call production and the extent to which a sound travels can sometimes be enhanced or restricted based on the type of environment that a species inhabits. The environment can affect the context in which a call is given, or change the motivational threshold to call (Green 1981). Macaques inhabit the widest range of habitats; therefore, environmental factors may drive the differences seen among the repertoires of various macaque species. The *loud calls* of *M. silenus* and *M. tonkeana*, and the *krahoo* of *M. fascicularis*, are acoustically distinct from the rest of the genus. Their *loud calls* are made up of tonal and harsh units in a wider range of frequencies, and their arboreal lifestyle and rainforest habitat may have selected for a call with an acoustic structure that enables the propagation of sound in a dense forest.

The acoustically distinct *warbles*, *harmonic arches*, and *chirps* of *M. mulatta* and the *food yells* of *M. cyclopis* and *M. sinica* are allospecific *food calls* emitted by species from provisioned groups living in human habitations or captivity. These allospecific calls are related to the quantity or quality of a food source (Gustison et al. 2012), and provisioning heightens these characteristics in a given environment. *Food calls* and *alarm calls* are special in that they both are emitted in response to non-conspecifics and they combine call elements to procure new meanings. Therefore, it is possible that other species that do not emit a food-specific call still have the basic acoustic requirements needed to flexibly alter the acoustic structure of a call to convey a food context-specific meaning.

7.7 Conclusions

The main differences in the genus are in the calls associated with copulation. Macaques are the most behaviorally diverse and widespread primate genus, yet reproductive features appear to be the most important discriminating factor among species. This may explain the flexibility of derived call types observed in the *copulation calls* of males and females. However, particularly for this genus, it is difficult to tease apart the effects of phylogeny and behavior on reproduction. Instead, a complex interplay of phylogenetic and social features of a species' reproduction drives the evolution of derived calls in the context of copulations. In conclusion, it is not likely that any particular factor is mutually exclusive. Instead, a complex interplay of phylogenetic, social, and ecological factors may shape the development of derived calls and the preservation of homologous calls across the macaque genus.

Acknowledgments I would like to thank the Forestry Bureau of Anhui Province, China and the Huangshan Garden Forest Bureau for permitting my research in the Valley of the Wild Monkeys, Mt. Huangshan, China. Thank you to the staff and Dr. Dongpo Xia and Dr. Binghua Sun's cooperation and help throughout the study. I am also grateful to Drs. Michael Huffman, David Hill, Fred Bercovitch, Peter Kappeler, Hideshi Ogawa, and Hiroki Koda for their extremely helpful comments in the preparation of this manuscript. I was funded by the Primatology and Wildlife Science Leading Graduate Program (PWS), and by funding allocated to Dr. Michael Huffman from PWS (U04-JSPS). Dr. Hiroki Koda's funding through the Japanese Society of the Promotion of Science (JSPS KAKENHI [Grant Number: 15K00203, 25285199 to HK as PI or co-PI]) made this study possible.

References

Bauers KA (1993) A functional analysis of staccato grunt vocalizations in the stumptailed macaque (*Macaca arctoides*). Ethology 94:147–161

Bernstein SK, Sheeran LK, Wagner RS, Li J, Koda H (2016) The vocal repertoire of Tibetan macaques (*Macaca thibetana*): a quantitative classification. Am J Primatol 78:937–949

Blumstein DT, Armitage KB (1997) Does sociality drive the evolution of communicative complexity? A comparative test with ground-dwelling sciurid alarm calls. Am Nat 150:179–200

Bouchet H, Blois-Heulin C, Lemasson A (2013) Social complexity parallels vocal complexity: a comparison of three non-human primate species. Front Psychol 4:390

Bruce KE, Estep DQ (1992) Interruption of and harassment during copulation by stumptail macaques, *Macaca arctoides*. Anim Behav 44:1029–1044

Caldecott JO (1986) Mating patterns, societies and the ecogeography of macaques. Anim Behav 34:208–220

Chevalier-Skolnikoff S (1974) Male-female, female-female, and male-male sexual behavior in the Stumptail Monkey, with special attention to the female orgasm. Arch Sex Behav 3:95–116

Fan Z, Zhao G, Li P et al (2014) Whole-genome sequencing of Tibetan macaque (*Macaca thibetana*) provides new insight into the macaque evolutionary history. Mol Biol Evol 31:1475–1489

Fischer J, Hammerschmidt K (2002) An overview of the Barbary macaque, *Macaca sylvanus*, vocal repertoire. Folia Primatol 73:32–45

Fitch WT, de Boer B, Mathur N, Ghazanfar AA (2016) Monkey vocal tracts are speech-ready. Sci Adv 2:1–7

Fooden J (1976) Provisional classification and key to living species of macaques (Primates: *Macaca*). Folia Primatol 25:225–236

Freeberg TM (2006) Social complexity can drive vocal complexity: group size influences vocal information in Carolina chickadees. Psychol Sci 17:557–561

Freeberg TM, Dunbar RIM, Ord TJ (2012) Social complexity as a proximate and ultimate factor in communicative complexity. Philos Trans R Soc Lond Ser B Biol Sci 367:1785–1801

Furrer RD, Manser MB (2009) The evolution of urgency-based and functionally referential alarm calls in ground-dwelling species. Am Nat 173:400–410

Green S (1975) Variation in vocal pattern with social situation in the Japanese monkey (*Macaca fuscata*): a field study. In: Rosenblum A (ed) Primate behavior: developments in field and laboratory research. Academic, New York, pp 1–102

Green S (1981) Sex differences and age graduations in vocalizations of Japanese and liontailed monkeys (*Macaca fuscata* and *Macaca silenus*). Am Zool 21:165–183

Gustison ML, le Roux A, Bergman TJ (2012) Derived vocalizations of geladas (*Theropithecus gelada*) and the evolution of vocal complexity in primates. Philos Trans R Soc Lond B Biol Sci 367:1847–1859

Hammerschmidt K, Fischer J (1998) The vocal repertoire of Barbary macaques: A quantitative analysis of a graded signal system. Ethology 104:203–216

Hauser MD (1993) Rhesus monkey copulation calls: honest signals for female choice? Proc Biol Sci 254(1340):93–96

Hauser MD, Marler P (1993) Food-associated calls in rhesus macaques (*Macaca mulatta*): I. Socioecological factors. Behav Ecol 4:194–205

Hohmann GM (1989) Vocal communication of wild bonnet macaques (*Macaca radiata*). Primates 30:325–345

Hohmann GM (1991) Comparative analyses of age-specific and sex-specific patterns of vocal behavior in 4 species of old-world monkeys. Folia Primatol 56:133–156

Hohmann GM, Herzog MO (1985) Vocal communication in lion-tailed macaques (*Macaca silenus*). Folia Primatol 45:148–178

Hsu MJ, Chen LM, Agoramoorthy G (2005) The vocal repertoire of Formosan macaques, *Macaca cyclopis*: acoustic structure and behavioral context. Zool Stud 44:275–294

Itani J (1963) Vocal communication of the wild Japanese monkey. Primates 4:11–67

Kudo H (1987) The study of vocal communication of wild mandrills in Cameroon in relation to their social structure. Primates 28:289–308

Lattenkamp EZ, Vernes SC (2018) Vocal learning: a language-relevant trait in need of a broad cross-species approach. Curr Opin Behav Sci 21:209–215

Li JH, Yin HB, Wang QS (2005) Seasonality of reproduction and sexual activity in female Tibetan macaques *Macaca thibetana* at Huangshan, China. Acta Zool Sin 51:365–375

Li J, Han K, Xing J et al (2009) Phylogeny of the macaques (Cercopithecidae: *Macaca*) based on *Alu* elements. Gene 448:242–249

Lillehei RA, Snowdon CT (1978) Individual and situational differences in the vocalizations of young stumptail macaques. Behaviour 65:270–281

Maestripieri DM, Roney JR (2005) Primate copulation calls and postcopulatory female choice. Behav Ecol 16:106–113

Masataka N, Thierry B (1993) Vocal communication of Tonkean macaques in confined environments. Primates 34:169–180

McComb K, Semple S (2005) Coevolution of vocal communication and sociality in primates. Biol Lett 1:381–385

Newman JD, Symes D (1982) Inheritance and experience in the acquisition of primate acoustic behavior. In: Snowdon CT, Brown CH, Petersen MR (eds) Primate communication. Cambridge University Press, New York, pp 59–70

Oda R, Masataka N (1992) Functional significance of female Japanese macaque copulatory calls. Folia Primatol 58:146–149

Ogawa H (1995) Bridging behavior and other affiliative interactions among male Tibetan macaques (*Macaca thibetana*). Int J Primatol 16:707–729

Palombit RA (1992) A preliminary study of vocal communication in wild long-tailed macaques (*Macaca fascicularis*). I. Vocal repertoire and call emission. Int J Primatol 13:143–182

Peckre L, Kappeler PM, Fichtel C (2019) Clarifying and expanding the social complexity hypothesis for communicative complexity. Behav Ecol Sociobiol 73(1):11

Pradhan GR, Engelhardt A, van Schaik CP, Maestripieri D (2006) The evolution of female copulation calls in primates: a review and a new model. Behav Ecol Sociobiol 59:333–343

Price JJ, Lanyon SM (2002) Patterns of song evolution and sexual selection in the oropendolas and caciques. Behav Ecol 15:485–497

Rowell TE (1962) Agonistic noises of the rhesus monkey (*Macaca mulatta*). Symp Zool Soc Lond 8:91–96

Rowell TE, Hinde RA (1962) Vocal communication by rhesus monkey (*Macaca mulatta*). Proc Zool Soc Lond 128:279–294

Seyfarth RM, Cheney DL (2010) Production, usage, and comprehension in animal vocalizations. Brain Lang 115:92–100

Seyfarth RM, Cheney DL (2012) Primate social cognition as a precursor to language. In: Gibson K, Tallerman M (eds) Oxford handbook of language evolution. Oxford University Press, Oxford, pp 59–70

Singh M, Sinha A (2004) Life history traits: ecological adaptations or phylogenetic relics? In: Thierry B, Singh M, Kaumanns W (eds) Macaque societies: a model for the study of social organization. Cambridge University Press, Cambridge, pp 80–83

Sugiyama Y (1968) The ecology of the liontailed macaque (*Macaca silenus* Linnaeus): a pilot study. J Bombay Nat Hist Soc 65:283–293

Thierry B, Heistermann M, Aujard F et al (1996) Long-term data on basic reproductive parameters and evaluation of endocrine, morphological, and behavioral measures of monitoring reproductive status in a group of semi-free ranging Tonkean macaques (*Macaca tonkeana*). Am J Primatol 39:47–62

Thierry B, Iwaniuk AN, Pellis SM (2000) The influence of phylogeny on the social behaviour of macaques (primates: Cercopithecidae, genus *Macaca*). Ethology 106:713–728

Tosi AJ, Morales JC, Melnick DJ (2003) Paternal, maternal, and biparental molecular markers provide unique windows onto the evolutionary history of macaque monkeys. Evolution 57:1419–1435

van Schaik CP, van Noordwijk M, Nunn CL (1999) Sex and social evolution in primates. In: Lee PC (ed) Comparative primate socioecology. Cambridge University Press, Cambridge, pp 204–240

Wada K, Xiong CP, Wang QS (1987) On the distribution of Tibetan and rhesus monkeys in Southern Anhui province, China. Acta Theriol Sin 7:148–176

Wilkinson GS (2003) Social and vocal complexity in bats. In: de Waal FBM, Tyack PL (eds) Animal social complexity: Intelligence, culture, and individualized societies. Harvard University Press, Cambridge, MA, pp 322–341

Xia DP, Li JH, Zhu Y, Sun BH, Sheeran LK, Matheson MD (2010) Seasonal variation and synchronization of sexual behaviors in free-ranging male Tibetan macaques (*Macaca thibetana*) at Huangshan, China. Zool Res 5:509–515

Chapter 8
Tibetan Macaque Social Style: Covariant and Quasi-independent Evolution

Krishna N. Balasubramaniam, Hideshi Ogawa, Jin-Hua Li, Consuel Ionica, and Carol M. Berman

8.1 Introduction: Primate Sociality and Social Structure

In animals, including primates, group living and the social tendencies that support it evolve when the benefits of living in groups (e.g., predator avoidance, cooperative resource sharing) outweigh the costs (e.g., competition for resources, disease risk) (Kappeler and Van Schaik 2002). Primate species share close evolutionary histories and are biologically, socially, and physiologically similar to humans (Cobb 1976; Kappeler et al. 2015; Suomi 2011). As such, they make excellent comparative models for understanding the evolution of aspects of human group living, an endeavor that is fundamentally important given that social connectedness is

K. N. Balasubramaniam (✉)
Department of Population Health and Reproduction, School of Veterinary Medicine, University of California at Davis, Davis, CA, USA

H. Ogawa
School of International Liberal Studies, Chukyo University, Toyota, Aichi, Japan
e-mail: hogawa@lets.chukyo-u.ac.jp

J.-H. Li
School of Resources and Environmental Engineering, Anhui University, Hefei, Anhui, China

International Collaborative Research Center for Huangshan Biodiversity and Tibetan Macaque Behavioral Ecology, Anhui, China

School of Life Sciences, Hefei Normal University, Hefei, Anhui, China
e-mail: jhli@ahu.edu.cn

C. Ionica
Biomedical Department, F. I. Rainer Anthropology Institute, Romanian Academy, Bucureşti, Romania

C. M. Berman
Department of Anthropology and Graduate Program in Evolution Ecology and Behavior, State University of New York at Buffalo, Buffalo, NY, USA
e-mail: cberman@buffalo.edu

J.-H. Li et al. (eds.), *The Behavioral Ecology of the Tibetan Macaque*, Fascinating Life Sciences, https://doi.org/10.1007/978-3-030-27920-2_8

associated with many benefits, including physical and emotional well-being, coping with environmental stressors, and enhancing survival, both in humans (e.g., Cohen et al. 2015; Holt-Lunstad et al. 2010; House et al. 1988) and in nonhuman primates (e.g., Ostner and Schülke 2014; Sapolsky 2005; Schülke et al. 2010; Silk et al. 2003, 2010; Young et al. 2014).

The complexity of primate social systems may be captured by studying three interrelated aspects of social life—social organization, mating systems, and social structure (Clutton-Brock and Harvey 1977; Hinde 1976; Kappeler and Van Schaik 2002; Koenig et al. 2013; Thierry et al. 2004). Social organization refers to the distribution and composition of individuals within social groups, specifically group sizes, sex ratios, and sex-typical philopatry versus dispersal strategies (Clutton-Brock and Harvey 1977). Social organization in turn gives rise to the other two aspects of social life. The first concerns mating systems which can be seen as the outcome of individual reproductive strategies adopted by resident and dispersing adults in a group (Greenwood 1980). The second is social structure, which arises from the patterning and distribution of agonistic, affiliative, cooperative, and/or conflict-managing interactions among group members (Hinde 1976; Kappeler and Van Schaik 2002; Sterck et al. 1997; van Schaik 1989, 1996). Social structure is of particular interest to behavioral ecologists because it provides both opportunities and constraints on individuals as they strive to maximize the benefits of group living while minimizing its costs. As such, social structure provides a rich window for testing hypotheses about adaptive and evolutionary strategies that individuals may use to gain fitness.

Among primates, social structure varies widely. Over the years, several conceptual frameworks have been proposed to explain how ecological factors might explain this variation. Wrangham (1980) was one of the earliest socioecologists to point out that, in several cases, (1) species with similar social organizations display sharply different social structures and to (2) argue that the distribution and abundance of resources needed by females to reproduce influenced their tendencies to form groups and to disperse or not. He also introduced the idea that these factors would in turn influence within- and between-group agonistic and affiliative social structure. This approach was extended by van Schaik (1989) and by Sterck et al. (1997) to include predation pressure and infanticidal risk as influential ecological factors. Sterck et al. (1997) described variation in the social structure of diurnal primates (1) emphasizing the influence of competitive regimes within groups on the nature and quality of female relationships (despotic vs. tolerant vs. egalitarian) and (2) pointing out that dominance structures were often associated with other aspects of social structure, e.g., kin bias and coalitionary support (Sterck et al. 1997). Since then, researchers have continued to emphasize power relationships as central to understanding variation in social structure.

In contrast to socioecological explanations, other schools of thought argue that aspects of social structure are more strongly influenced by inherent or intrinsic characteristics (Thierry 2004, 2007). Dominance hierarchies are once again central to this framework (de Waal and Luttrell 1989), but rather than adaptive responses to current ecological conditions, they are hypothesized to be either due to ecological adaptation in the distant past (Chan 1996; Kamilar and Cooper 2015; Matsumura 1999) or to emergent properties of self-organizational processes (Hemelrijk 1999,

2005; Puga and Sueur 2017). In either case, variation in social structure is posited to be strongly influenced by species' ancestral or phylogenetic histories (Balasubramaniam et al. 2012a; Kamilar and Cooper 2015; Thierry et al. 2008, 2000) due to tendencies for dominance structures to be structurally linked to entire suites of social traits and hence for these traits to covary with one another (Thierry et al. 2008; Thierry 2000). Such contrasting explanatory models have complicated our understanding of variation for dominance and other aspects of primate social structure.

In this chapter, we review our previous studies to illustrate the complexity of a few aspects of Tibetan macaque (*Macaca thibetana*) social structure, focusing on indicators of social style. The concept of social style posits that a suite of characteristics related to dominance and social tolerance covary (and may have coevolved) with one another and that species can be placed on a continuum ranging from extremely despotic to extremely tolerant (de Waal and Luttrell 1989; Thierry 2000, 2004; Thierry et al. 2008). Early studies on Tibetan macaques (Deng 1993; Ogawa 1995; Zhao 1996) detected a suite of moderately tolerant traits that led to their categorization as a "grade 3" species (Thierry 2000). Yet later work has revealed a mixture of characteristics that are associated with both despotism and tolerance, prompting a change in classification. Here we review this more recent line of evidence to describe how quasi-independent evolution, i.e., the apparent adaptive responses by animals to variation in current sociodemographic factors and ecological conditions, may have also influenced the evolution of such a mixture of social style traits in this species. We suggest that such effects may potentially mask underlying tendencies for the covariation of traits. Similarly, we describe comparative research across the macaque genus that suggests that aspects of social style may have been influenced by species' phylogenetic histories, by the covariation of traits to each other, and by adaptive responses to variation in current conditions. In doing so, we suggest that aspects of Tibetan social styles may represent flexible responses to different types of selection pressures that are nevertheless limited to a given species-typical range of possible responses, i.e., "social reaction norms" (Berman and Thierry 2010). We end by highlighting some avenues for future research that address some crucial gaps in our current understanding of the evolution of Tibetan macaque social styles and macaque social structure in general.

8.2 The Macaques and the Study of Variation in Social Structure

To date, comparative studies of variation in primate social structure have largely focused on members of the family Cercopithecinae, specifically baboons (genus: *Papio, Theropithecus*) and macaques (genus: *Macaca*) (Cords 2013). This is because these genera are represented by many species that are among the most geographically widespread and ecologically diverse of all primates. Moreover, they show marked variation in several aspects of social structure both within and across species (Cords 2013). Members of the genus *Macaca* range from North Africa in the West to Japan in the East and from China in the North to Sulawesi, Indonesia, in the South (Abegg and Thierry 2002; Thierry 2013; Thierry et al. 2004). All macaque species show a broadly similar social organization (Thierry 2013; Thierry et al. 2004). Barring a few exceptions (Sinha et al. 2005), they live in multi-male, multi-female social groups with varying sex ratios (although usually skewed toward females). Females are philopatric and males disperse from their natal groups when they reach maturity (Cords 2013; Paul and Kuester 1987). Yet macaque species vary broadly in several aspects of social structure related to aggression intensity, dominance relationships, maternal style, tendencies to reconcile following conflicts, and tendencies to show preferences for kin in affiliative and cooperative exchanges (Thierry 2000, 2007, 2013).

Early observations found these various tendencies appeared to be clustered, particularly among some better-studied species. This led to the theoretical concept of macaque "dominance styles" (de Waal and Luttrell 1989) and shortly thereafter renamed "social styles" (Thierry 2000), a concept with dominance relationships at its core. The social style concept posits that aspects of macaque social structure covary with one another and that species can be placed on a continuum (later simplified into a four-grade scale) that ranges from extremely despotic to extremely tolerant (Thierry 2000, 2007; Thierry et al. 2004). Specifically, species categorized as "grade 1" [extremely despotic, e.g., rhesus macaques (*Macaca mulatta*) and Japanese macaques (*M. fuscata*)] and "grade 2" [moderately despotic, e.g., long-tailed macaques (*M. fascicularis*) and pigtailed macaques (*M. nemestrina*)] may be expected to show intense aggression, steep hierarchies and highly asymmetric dominance relationships (Balasubramaniam et al. 2012a), low tendencies to reconcile after conflicts (Thierry et al. 2008), strong preferences toward kin (Thierry et al. 2008), protective maternal styles (Thierry et al. 2000), and community formation or substructuring in their affiliative relationships (Sueur et al. 2011). In contrast, "grade 3" [moderately tolerant, e.g., Barbary macaques (*M. sylvanus*) and bonnet macaques (*M. radiata*)] and "grade 4" [extremely tolerant, e.g., crested macaques (*M. nigra*) and Tonkean macaques (*M. tonkeana*)] species are hypothesized to display mild aggression, shallow hierarchies, frequent counter-aggression, strong tendencies to reconcile, l`ow degrees of kin bias, relaxed maternal styles, and dense well-connected grooming social networks.

Although the concept of social style is widely employed in the primate literature, certain issues remain. First, the extent to which one can speak of single species-typical social styles is not clear. In most Cercopithecine primates, female-female relationships define the core of the group's social structure (Cords 2013; Koenig et al. 2013; Sterck et al. 1997; van Schaik 1989). As such, most designations of social style among macaques are based on female-female relationships, particularly in captive groups where the number of males available to study is often limited (Balasubramaniam et al. 2012a, b; Thierry 2000; Thierry et al. 2004, 2008). However, a number of studies have found differences in social styles among males and females of the same species (Cooper and Bernstein 2008; Preuschoft et al. 1998; Richter et al. 2009; Tyrrell et al. 2018).

It is also unclear how consistent social style indicators are over time or among groups of the same species. Although several studies have found consistent styles among different captive groups (Demaria and Thierry 2001; Petit et al. 1997; Sueur et al. 2011; Thierry 1985), and among wild, free-ranging and captive groups of the same species (Balasubramaniam et al. 2012a, b), others have found variation in social style indicators (e.g., grooming kin bias, hierarchical steepness) as groups have grown and fissioned (Balasubramaniam et al. 2011; Berman and Thierry 2010; Zhang and Watanabe 2014). Such inconsistencies in the literature have been partly responsible for ongoing debates over the origins of social style and other aspects of social structure. Specifically, the extent to which variation in social structure may be linked to inherited characteristics [e.g., genetic polymorphisms, phylogenetic history (Blomberg et al. 2003; Thierry 2007, 2013; Thierry et al. 2000)] versus outcomes of adaptive responses of individuals to current socioecological factors (Clutton-Brock and Janson 2012; Isbell 2017; Koenig et al. 2013; Sterck et al. 1997) remains unclear and hotly debated.

Some proponents of the phylogenetic argument posit that aspects of macaque social style are structurally linked and hence covary with each other across species (Petit et al. 1997; Thierry 2000; Thierry et al. 2008). In an early extreme form, this idea, called the systematic variation hypothesis (Castles et al. 1996; Petit et al. 1997), posited that species that show a single indicator of a particular social style (e.g., steep dominance gradients in grade 1 species) may be expected to display all other aspects of that style. In other words, it predicts that all indicators for species showing extreme despotism or extreme tolerance should be in the extreme range and that all indicators for species showing intermediate styles should display intermediate indicators. However, there is now a consensus that this extent of covariation is not always detectable and may be different for different traits and species. For instance, Thierry et al. (2008) found that across the range of macaques with different social styles, post-conflict affiliation covaried systematically with explicit forms of contact, but not with grooming kin bias. Further, although most studies thus far have found the predicted clusters of social style indicators and marked differences between species at the extreme ends of the social style scale (e.g., Demaria and Thierry 2001; Sueur et al. 2011; Thierry 1985), the positions of species in intermediate grades remain unclear and do not always concur with the covariation hypothesis (summarized in Balasubramaniam et al. 2012b).

Many proponents of socioecological explanations for variation in social style hypothesize that current conditions such as the distribution and abundance of resources, predation pressure, and the risk of infanticide critically influence the number and spatial distribution of females in groups. In female philopatric groups, those factors in turn influence not only the number and distribution of males but also the social structure of the group. While social style indicators are expected to cluster to some extent, this framework (named the Ecological Model of Female Social Relationships or EMFSR: Koenig et al. 2013; Sterck et al. 1997) does not necessarily predict structural linkages or tight covariation among them.

In the next sections, we summarize our research findings on aspects of Tibetan macaque social structure for both males and females. We describe indicators associated with both despotic and tolerant social styles and evaluate the extent to which these different indicators seem to covary with one another, or are labile, independent responses by individuals to external factors, within this species. Then, we review our comparative studies that examine the extent to which social style indicators covary and display phylogenetic signals and/or are influenced by variation in current sociodemographic factors, across macaque groups and species belonging to all four social style grades.

8.3 Tibetan Macaques and the YA1 Group

Tibetan macaques are the largest macaque species (Fooden 1986). Preferring montane habitats, they are found in the subtropical and temperate forests that range from South Central to Western China as far as Tibet (Li 1999; Zhao 1996). Taxonomically, they are in the *sinica-arctoides* phylogenetic clade. They are most closely related to Assamese (*M. assamensis*) and Arunachal (*M. munzala*) macaques (Balasubramaniam et al. 2012a; Chakraborty et al. 2007; Purvis 1995; Tosi et al. 2003), although they morphologically resemble stump-tailed macaques (*M. arctoides*). Like all macaques, they live in multi-male, multi-female social groups that show female philopatry, male dispersal, and linear dominance hierarchies (de Vries 1998; Zhao 1996).

Work on the ecology and social structure of Tibetan macaques began with pioneering studies on Mt. Emei, Sichuan Province, China (Deng 1993; Li 1999; Zhao 1996; Zhao et al. 1991), and at Mt. Huangshan, Anhui Province, China (118.3°E, 30.2°N) (Li 1999; Wada et al. 1987). The study at Huangshan continued by focusing on the semi-provisioned Yulingkeng A1 or YA1 group (Li 1999). Huangshan is a popular tourist destination in East-Central China and home to several groups of Tibetan macaques with (apparently) nonoverlapping home ranges. The montane vegetation consists of mixed deciduous and evergreen forests at lower elevations and sparsely covered peaks at higher elevations (Li 1999; Zhao 1996; Zhao et al. 1991). The YA1 group has been monitored, and maternal kin relationships of adults have been recorded since 1986 (Li 1999). The data we review here were collected on the YA1 group over six observation periods spread across 12 years. During each of the six observation periods, the group's hierarchy was stable, and we used similar observation methods (focal animal sampling and all occurrences sampling). H. Ogawa collected data between 1991 and 1992, and C. Berman and colleagues collected data between 2000 and 2002. During this time, the group size varied from 39 to 52 individuals. Although wild, the macaques were managed for tourism for a period of time that overlapped with data collection—inconsistently during the first three periods, but consistently so for the second three periods. Park officials herded the animals from the forest into a viewing area where tourists could observe them from a pavilion, as the monkeys were provisioned with corn. Between feedings, the group spent most of their time in the surrounding forest.

8.4 Evidence of Female Despotism Contradicts Earlier Studies

When Thierry published his classic analysis of social style and proposed a four-grade social style scale (Thierry 2000; Thierry et al. 2000), comprehensive analyses of core social style indicators were available for a limited number of macaque species. Hence, the placement of many species, including Tibetan macaques, was tentative and based on their phylogenetic closeness to better-studied species (Thierry et al. 2000), or on indirect behavioral indicators. Thierry (2000) placed Tibetan macaques in grade 3 (moderately tolerant) based on their membership in the *sinica-arctoides* lineage (which contains other tolerant macaques) and on early studies that described evidence for multiple traits that were associated with a mildly tolerant social style: frequent ritualized affiliation among males, bidirectionality in silent bare-teeth displays, triadic infant handling interactions particularly among males, and tolerant responses to infant handling by mothers (Deng 1993; Ogawa 1995; Zhao 1996). It was not until 2004 that Berman and colleagues evaluated multiple other core indicators of social style (Berman et al. 2004). Using data collected on the YA1 group between 2000 and 2002, this study expected to confirm the tentative designation of Tibetan macaques as "tolerant" or "grade 3" species. They examined (1) the degree of aggressive asymmetry and counter-aggression in dominance relationships, (2) conciliatory tendencies, and (3) kin bias in a variety of affiliative and tolerant behaviors separately for males and females.

The results were surprising for both sexes: they supported a despotic rather than a tolerant social style. First, both males and females displayed high degrees of asymmetry in their dominance relationships. Three indices were used to evaluate the extent to which aggression was bidirectional, specifically the (1) directional inconsistency index (or DII), (2) the dyads-up index, and (3) percentage of counter-aggression, and all revealed extreme aggressive asymmetry typical of extremely despotic species (Table 8.1). Similarly, submissive signaling in the form of silent bare-teeth displays was almost always given in the same direction between dyads and in the same direction as other fearful and submissive interactions. Second, conciliatory tendencies were low. These were examined using the post-conflict matched-control pairs, or the PC-MC method (de Waal and Yoshihara 1983), which examined the timing of affiliative interaction during a 5-min window of time following an act of aggression (the PC or post-conflict sample), and compared it with the timing during a comparable period of time without aggression (the MC or matched sample). The method thus evaluated whether affiliation occurred earlier, later, or at the same time in the PC as the MC. The comparisons were used to calculate a corrected conciliatory tendency (CCT) that varies from 0 to 100% and

Table 8.1 Bidirectionality of aggression during each of three data analysis periods[a] for a single group of Tibetan macaques

	8/1/00–1/28/01	2/27/01–5/29/01	12/9/01–7/25/02
A. Directional inconsistency index			
Male-male dyads	3/583 (5.2%)	2/50 (4.0%)	2/108 (1.9%)
Female-female dyads	6/160 (3.8%)	3/166 (1.8%)	4/243 (1.6%)
B. Dyads-up index			
Male-male dyads	1/36 (2.8%)	2/36 (5.6%)	0/28 (0.0%)
Female-female dyads	3/78 (3.8%)	1/78 (1.3%)	0/45 (0.0%)
C. Percent counter-aggression			
Male-male dyads	3/37 (8.1%)	0/33 (0.0%)	0/73 (0.0%)
Female-female dyads	0/97 (0.0%)	4/114 (3.5%)	6/157 (3.8%)

All three measures of bidirectionality indicate a high degree of asymmetry in dominance relationships among both males and females, typical of a despotic species

A. The percentage of total aggressive interactions that were directed in the less frequent direction within dyads

B. The percentage of dyads for which the main direction of aggression was up the dominance hierarchy

C. The percentage of instances of aggression of any kind to which the target responded with aggression of any kind

[a]Adapted from Berman et al. (2004)

Table 8.2 Kin bias among Tibetan macaques, indicated by Partial Kr coefficients between interaction rates and degree of relatedness, controlling for rank distance[a]

Partner combination	% Time w/in 5 m	Grooming	Sit near	Approaches	Co-feed
♀-♀	0.16*	0.22**	0.25**	0.24*	0.18*
♂-♂	−0.14	0.13	−0.17(*)	−0.17	−0.14

The results suggest a significant degree of kin bias among females in affiliation and tolerance

$**p < 0.01$

$*p < 0.05$

$(*)p < 0.1$

[a]Adapted from Berman et al. (2004)

estimates the degree to which dyads are more likely to reconcile (i.e., engage in affiliation soon after a conflict than at other times) (Veenema et al. 1994). For female-female dyads, the conciliatory tendency was extremely low (4.2%), a figure that is low even among grade 1 macaque species. For males, it was a moderate 19.7% (see also Berman et al. 2006), but still within the range of despotic male macaques (unrelated male Japanese macaques: Petit et al. 1997).

Finally, females, but not natal males, consistently displayed high levels of affiliative kin bias and social tolerance. Mean rates of female affiliation (grooming, sitting near, co-feeding) with maternal kin were significantly greater than affiliation rates with nonkin (Table 8.2). Indeed, the intensity of grooming kin bias, measured as the observed-to-expected ratio of grooming bouts between kin, suggested that females groomed kin more than three times the value expected by chance. All the results were consistent across seasons and locations (provisioning area vs. forest), strongly supporting a despotic social style for Tibetan macaques. Although several core indicators were similar to those of grade 1 macaques, findings from this study led to a reclassification of Tibetan macaques as grade 2 rather than grade 1 or grade 3 species (Matsumura 1999; Thierry 2007) based on previous findings of ritualized affiliative behavior among males and of maternal tolerance for infant handling among females.

Although these findings greatly expand our understanding of social style among Tibetan macaques, they pose complications for possible explanations for the origins of such despotism. On the one hand, a phylogenetic explanation is difficult since Tibetan macaques are members of the *sinica-arctoides* lineage which is composed of predominantly tolerant macaque species. Yet, they display heightened levels of despotism that are characteristic of members of the *fascicularis* lineage (Matsumura 1999; Thierry et al. 2000) like pigtailed macaques (Castles et al. 1996) and Assamese macaques (Cooper and Bernstein 2002). Socioecological explanations based on the distribution of natural resources and predation pressure are also unlikely given that Tibetan macaques have a largely folivorous diet (Zhao et al. 1991) and experience (historically) moderate levels of predation pressure (Xiong 1984), factors that socioecological frameworks associate with tolerant or egalitarian societies (Sterck et al. 1997).

Berman and colleagues (Berman et al. 2004) speculated that indicators of a despotic social style in this group may have risen recently due to human activity; activities associated with tourism, particularly range restriction, herding, and food provisioning, likely elevated levels of intragroup aggression and/or competition (Berman and Li 2002; Hill 1999; Judge 2000; Marechal et al. 2011; Ram et al. 2003) and thereby generated atypical despotism in a species that would otherwise show greater social tolerance. Together, these results demonstrate that, rather than all aspects of female social style showing moderate despotism, some are in the range of extremely despotic species, and others are in the range of moderately tolerant species. Such a mixture of both despotic and tolerant social style traits in the same species suggests that structural linkage and potential covariation of traits may be offset, at least in part, by potentially adaptive individual responses to contemporary selection pressures posed by human management.

8.5 Males Exhibit Social Tolerance Despite Evidence for Despotism

In general, less research has focused on male-male relationships than female-female relationships among Cercopithecine primates and other primates that display female philopatry and male dispersal. When males disperse, they typically join groups that lack close kin, particularly in species that disperse repeatedly (as do Tibetan macaque males, whose average tenure in a group is about 5–6 years: Berman et al., unpublished data). As such, opportunities for long-term affiliative and cooperative relationships among males are limited. More fundamentally, affiliation and cooperation would not be expected among males, because males primarily compete for resources (i.e., conceptions with fertile females) that cannot be shared in the same way as food patches or other resources (Schülke and Ostner 2008, 2013; Sterck et al. 1997). As such, theoretical considerations do not generally predict high levels of affiliation and cooperation among males.

Socioecological models predict that the social strategies of primate males are driven by the availability and distribution of fertile females (Kappeler and Van Schaik 2002; Sterck et al. 1997; van Schaik 1989). When several females form a cohesive group, single males are not likely to be able to monopolize them, particularly when they breed seasonally. This leads to the formation of multi-male, multi-female societies in which males may coexist but attempt to outcompete one another through various forms of scramble or contest competition, particularly over access to females (Kappeler 2000; van Schaik 1996; Van Schaik et al. 2004). High levels of competition should theoretically preclude strong male-male affiliative relationships in these groups. However, some macaque males regularly engage in affiliative interactions that lead to strong social bonds and that enhance their fitness (Ostner and Schülke 2014; Schülke et al. 2010; Young et al. 2014).

Indeed, contrary to predictions, particularly high levels of male-male affiliation and tolerance have been reported among groups and species with relatively even sex ratios versus those with highly female-biased sex ratios. In these cases, male-male affiliation and tolerance have been hypothesized to function as coping mechanisms against intense conflict and competition both from other group males (e.g., access to females, revolutionary coalitions) and from external threats (e.g., predators, intergroup encounters, dispersing males: Cooper and Bernstein 2002; Ogawa 1995; Preuschoft and Paul 2000). The reasoning is that cooperation with some males against others may enhance a male's competitive abilities within his group and that group action may increase probabilities of success against external threats.

Previous research on Tibetan macaques suggests that males compete vigorously with one another for rank and access to fertile females, consistent with a despotic social style (Li and Wang 1996). Nevertheless, as mentioned earlier, they also show several forms of affiliative interaction with one another, consistent with their tendencies toward relatively even sex ratios (Li 1999). These include moderate conciliatory tendencies, ritualized greetings, and bridging behavior in which infants are used to facilitate friendly contact between males and to enhance the chances of forming affiliative social bonds (Ogawa 1995).

To further examine the ways in which male Tibetan macaques may engage in tolerant and cooperative interactions with kin and nonkin, despite strong indications of a despotic social style, Berman and colleagues examined cooperative strategies among males, focusing particularly on agonistic coalitions (Berman et al. 2007). They examined affiliation, tolerance, and agonistic support to test the hypothesis that increased tolerance in otherwise despotic males may occur when high-ranking males require support from other males to maintain their positions. In this group, conservative coalitions, in which two higher-ranking males aggressively target a lower-ranking male, are the most common, and these serve to reinforce the current hierarchy. Nevertheless, revolutionary coalitions in which two lower-ranking males challenge a higher-ranking male pose a threat particularly to alpha males. Although agonistic support is unrelated to kinship and rates of grooming, high-ranking males display tolerance in the form of co-feeding toward lower-ranking males that support them, and alpha males show the most cooperation with the males that targeted them in revolutionary coalitions. Berman and colleagues suggested that high-ranking males discourage revolutionary alliances by using two cooperative strategies. They primarily rely on conservative alliances, but also offer tolerance to potential rivals in cases in which conservative coalitions are less effective.

We suggest that, contrary to socioecological theory (van Hooff and van Schaik 1994; van Schaik 1996; Van Schaik et al. 2004), Tibetan macaque males may be "tolerant despots" (*cf* Kaburu and Newton-Fisher 2015). Just as Kaburu and Newton-Fisher use the term "egalitarian despots" to describe chimpanzees that display signs of social tolerance which emerge strategically from a more despotic dominance style, we suggest that social tolerance in male Tibetan macaques is not shaped by external factors like increased between-group competition or predation pressure which were low or absent in this population. Rather, we suggest that social tolerance may be an adaptive outgrowth of within-group despotism, in that it allows despotic males to enhance their competitive abilities by selectively displaying tolerance to other males. Like the findings for female social styles, findings for male-male relationships also suggest a mixture of both despotic and tolerant social style indicators. Such evidence for unlinked traits that may be independently influenced by different, sometimes opposing, types of selection pressures may mask or offset potential underlying evidence for the structural linkage or covariation of traits.

8.6 Comparative Studies Provide Evidence for Both Covariation and Quasi-independent Evolution

Comparative research among primate species provides powerful opportunities to discern the extent to which variation in a characteristic is likely to be related to variation in ecological factors and/or evolutionary trends (Nunn 2011). This is particularly the case for comparative studies of macaque social structure, given their similar social organizations but varying social structures (Thierry 2007). To date, comparative studies of macaque social structure have largely focused on evolutionary explanations, specifically on the influence of phylogenetic relatedness. However, they have found somewhat mixed evidence for the phylogenetic model. An early study found that 7 out of a set of 22 macaque social behavioral traits analyzed appeared to be strongly influenced by phylogenetic relatedness (Thierry et al. 2000). Later work found that aspects of macaque social structure related to post-conflict affiliation and explicit contact, but not grooming kin bias, were more similar among more closely related species and that they appeared to covary continuously across species in predicted directions after accounting for phylogeny (Thierry et al. 2008). Most of these studies focused on captive groups in which it was relatively easy to restrict the analysis to behavior outside of the feeding context and hence to avoid the possibility that behavior related to social style might be masked by direct feeding competition.

Building on these pioneering studies, Balasubramaniam et al. (2012a, b) conducted three comparative studies of variation in other aspects of macaque social structure, specifically dominance hierarchies and grooming social network structure, which until then had not been tested in a comparative framework that involved multiple macaque groups and species. These studies included both captive and free-living groups of macaques, including Tibetan macaques at Huangshan, and specifically examined the potential influence of living condition on social style traits. We review the major findings from these three studies below, focusing on their implications for both the characterization of Tibetan macaque social style and for the evolutionary origins of macaque social structure in general.

In the first of two comparative studies of macaque dominance social structure that tested the predictions of phylogenetic models (Balasubramaniam et al. 2012a), Balasubramaniam and colleagues examined whether the steepness of hierarchies (de Vries et al. 2006) and degrees of dominance asymmetry or counter-aggression showed strong phylogenetic signals, i.e., whether they were more similar among more closely related species (Blomberg et al. 2003; Kamilar and Cooper 2015; Matsumura 1999; Thierry et al. 2008). They assembled a behavioral dataset on dyadic aggressive interactions and submissive displacements among adult females, for each of 14 groups of macaques representing 9 species. For each group, they calculated the steepness of the dominance hierarchy as the absolute slopes of plots between normalized David's scores (NDS: de Vries et al. 2006) and ordinal ranks of individuals attributed by David's scores. David's scores are calculated for each individual as the differences between its wins and losses in dyadic agonistic interactions, which are weighted by the relative wins and losses of all of the other individuals in the group (Gammell et al. 2003; de Vries et al. 2006). To date, hierarchical steepness represents the most comprehensive measure of the "dominance gradient" and hence the degree of despotism in the dominance hierarchy of a social group. In addition to steepness, they also calculated the degree of "dominance asymmetry," by estimating the proportion of aggressive interactions that involved counter-aggression from the initial recipient.

To test predictions of phylogenetic signals, they used multiple macaque Bayesian and maximum likelihood phylogenetic trees, both self-reconstructed and extracted from the online database for primate phylogenies *10kTrees* (Arnold et al. 2010). The results were consistent with the phylogenetic model in that both hierarchical steepness and counter-aggression showed strong, significant phylogenetic signals. Moreover, phylogenetic signals were consistent and strong when they examined a subset dataset of just free-living groups representing seven species of macaques.

In a second comparative study that used the same dataset, Balasubramaniam et al. (2012b) tested the covariation hypothesis and found mixed support for it. They asked whether hierarchical steepness and counter-aggression both covaried with the hypothesized placement of species on Thierry's four-grade social style scale and whether evolutionary shifts in social style grade corresponded consistently to covariant changes in hierarchical steepness and counter-aggression. As predicted by the covariation hypothesis, steepness and counter-aggression were strongly correlated with the placement of species on macaque social style scales in the predicted directions. Yet the nature of these relationships differed for each trait. On the one hand, hierarchical steepness appeared to covary continuously with the scale. Macaques in grades 1 and 2, including Tibetan macaques, showed the steepest hierarchies, grade 3 species showed moderately steep hierarchies, and grade 4 species showed the shallowest hierarchies. On the other hand, counter-aggression appeared highly dichotomous. Species from grades 1–3 all showed similar and markedly low levels of counter-aggression (range, 0.6–6.4% of all aggressive interactions) in comparison with the two grade 4 species (crested macaques, 50.8% of aggressive interactions; Tonkean macaques, 60.4%). This suggested that the covariation between dominance traits and social scale were driven primarily by species at the extreme ends of the scale (grade 1 vs. grade 4 macaques).

In comparison, the characteristics of species in intermediate grades 2 and 3 were inconsistent. For instance, grade 1 rhesus and Japanese macaques showed both steep dominance hierarchies (range, 0.68–0.99) and low levels of counter-aggression (range, 0–7.1% of all aggressive interactions), while grade 4 Sulawesi crested and Tonkean macaques showed the opposite characteristics (steepness range, 0.27–0.53; counter-aggression range, 50–61% of all aggressive interactions). However, grade 2 species, rather than showing more shallow hierarchies than grade 1, fell within the range for grade 1 species (steepness range, 0.85–0.98). In fact, two grade 2 species—Tibetan and long-tailed macaques—showed the steepest hierarchies of all species examined in the study, i.e., exceeding the values for grade 1 Japanese and rhesus macaques. Further, Tibetan macaques showed the lowest levels of counter-aggression (mean = 0.64%; range, 0.53–0.76% of all aggressive interactions), even lower than grade 1 macaques.

The case of grade 3 macaques was even less consistent. Although hierarchical steepness of Barbary macaques was intermediate to those for grades 2 and 4 as predicted for a moderately tolerant social style, low levels of counter-aggression seen in this same group were more indicative of a more despotic style (but see Thierry and Aureli 2006), indicating possible evidence for quasi-independent evolution. Moreover, grade 3 bonnet macaques showed both steeper hierarchies and low levels of counter-aggression, within the range of despotic species.

To test the covariation hypothesis statistically, Balasubramaniam et al. (2012b) used phylogenetic independent contrast approaches of ancestral trait reconstruction to determine whether major evolutionary shifts from despotic to tolerant social style grades corresponded to shifts in hierarchical steepness and counter-aggression in the evolutionary history of macaques. The findings were telling. For both steepness and counter-aggression, the magnitudes of independent contrasts were greatest at the nodes in the phylogenetic tree that split grade 4 macaques from all other grades. This was consistent with the finding that there were indeed marked differences between macaques at the extreme ends of the scale. However, contrary to predictions, evolutionary shifts from higher to lower social style scales, or from tolerant to despotic societies, did not result in consistent increases in steepness contrasts or decreases in counter-aggression contrasts. Moreover, the magnitude of steepness and counter-aggression contrasts were no greater at nodes in the phylogeny where there were evolutionary shifts in social style, compared to "neutral" nodes in which there were no shifts in style. In summary, the authors found strong evidence for phylogenetic signals in dominance traits and mixed evidence for the covariation hypothesis. Specifically, covariation seemed to more strongly influence the underlying differences in dominance hierarchies between some species and phyletic lineages (macaques across grades 1 and 4), but less so others (species in grades 2 and 3).

In a third study, Balasubramaniam and colleagues (2018a) examined interspecific covariation in higher-order aspects of both dominance hierarchies and social grooming relationships across an even wider range of macaque populations. To capture higher-order aspects of social structure, they used social network analysis (SNA: Farine and Whitehead 2015; Lusseau 2003). Beyond using only direct interactions among dyads, network approaches incorporate both the direct social connections and indirect pathways of interactions among individuals (Brent et al. 2011; Farine and Whitehead 2015; Kasper and Voelkl 2009; Sueur et al. 2011). Such approaches may identify consistent patterns or "motifs" of interactions that in turn describe group social structure. The focus on social grooming, a core aspect of affiliative and cooperative social structure, expanded on previous studies that had predominantly focused on aspects of post-conflict affiliation, affiliative kin bias, and dominance hierarchies (but see the inclusion of macaque grooming traits by Schino and Aureli (2008) in their interspecies analyses).

In comparison to these latter traits which may be strongly influenced by intrinsic factors such as personality (Capitanio 1999; Krause et al. 2010), matrilineal inheritance (Cords 2013; Kapsalis 2004; Sade 1972), and phylogenetic histories (Thierry 2007; Thierry et al. 2008, 2000), the patterning and distribution of grooming relationships has been assumed to be more responsive to environmental variability, such as group sizes (Berman and Thierry 2010; Kasper and Voelkl 2009; Majolo et al. 2009), the presence of infants (Fruteau et al. 2011; Gumert 2007; Henzi and Barrett 2002; Tiddi et al. 2010), and the distribution of resources (Balasubramaniam and Berman 2017; Carne et al. 2011; Ventura et al. 2006). Nevertheless, across the macaque genus, the structure of grooming social networks is also likely to be strongly influenced by varying tendencies for species to affiliate with kin, i.e., grooming kin bias (Berman 2011; Berman and Kapsalis 2009; Berman and Thierry 2010; Kapsalis 2004; Thierry et al. 2008).

Under the covariation framework, more despotic, nepotistic species that show steeper dominance hierarchies, and in which individuals prefer to interact more with close kin compared to distant kin, may be expected to show more modular or substructured grooming networks (Sueur et al. 2011). On the other hand, more tolerant species in which hierarchies are shallower, and in which individuals form affiliative relationships with a wider range of conspecifics, may be expected to show more dense, well-connected social networks that are less modular (Sueur et al. 2011). In support of this covariation framework, a study by Sueur and colleagues found that more intolerant or nepotistic grade 1 species like rhesus and Japanese macaques showed less dense but more centralized, modular, or substructured social networks than more socially tolerant, less nepotistic grade 4 species like Tonkean and Sulawesi crested macaques (Sueur et al. 2011). Balasubramaniam and colleagues used a wider range of macaques to ask whether (1) aspects of dominance and grooming social networks showed strong phylogenetic signals and (2) they covaried with each other independently of the effects of phylogeny and/or extrinsic sociodemographic factors like group size, sex ratio, and living condition.

The authors assembled a broad comparative dataset—38 dominance matrices and 34 datasets of social grooming—from captive, free-ranging, and wild groups representing 10 species of macaques. From these, they estimated two dominance social network traits, specifically triangle transitivity (Shizuka and Mcdonald 2012) and group-wide dominance certainty (Fujii et al. 2013). From the grooming datasets, they estimated three aspects of grooming social network structure: the (1) centrality coefficient, i.e., the extent to which dominant individuals are also highly central in their grooming network (Handcock et al. 2006); (2) Newman's modularity, i.e., the degree to which networks were substructured into communities of closely interacting individuals (Newman and Girvan 2004); and (3) clustering coefficient (Csardi and Nepusz 2006), which is an indicator of the localized density of a group's social connections. As in previous efforts, they replicated their comparative analyses across multiple phylogenetic trees to account for phylogenetic branch length uncertainty.

As in the previous comparative studies, the results showed mixed evidence for the phylogenetic model. Consistent with findings from the previous study on dominance steepness and counter-aggression (Balasubramaniam et al. 2012a, b), dominance transitivity and certainty both showed strong phylogenetic signals (Fig. 8.1). On the other hand, aspects of grooming social networks showed only moderate or weak phylogenetic signals (Fig. 8.1). Moreover, contrary to the covariation hypothesis, grooming network measures did not covary with dominance traits across species (Fig. 8.2). Instead, two aspects of grooming social networks, modularity and clustering coefficient, were more strongly predicted by group size, a sociodemographic factor, than by species' social styles or dominance traits. Specifically, larger groups of macaques showed more modular, less densely clustered grooming networks than smaller groups, with the results being independent of groups' living condition and variation in sampling effort.

Taken together, findings from these interspecies comparative studies suggest that different aspects of social structure may have been subject to different evolutionary and ecological selection pressures. Specifically, aspects of dominance social structure seem to be strongly influenced by phylogenetic relationships and to show tendencies to covary across some (but not all) species and lineages. In comparison to dominance, grooming social networks seem weakly influenced by phylogeny. Rather than covarying with dominance traits across groups and species, grooming networks appear to be more responsive to variation in current sociodemographic factors like group sizes and living condition. The detection of more clear-cut differences in dominance traits between the grade 4 Sulawesi macaques and members of grades 1 and 2 in the *fascicularis* lineage suggests that the influence of

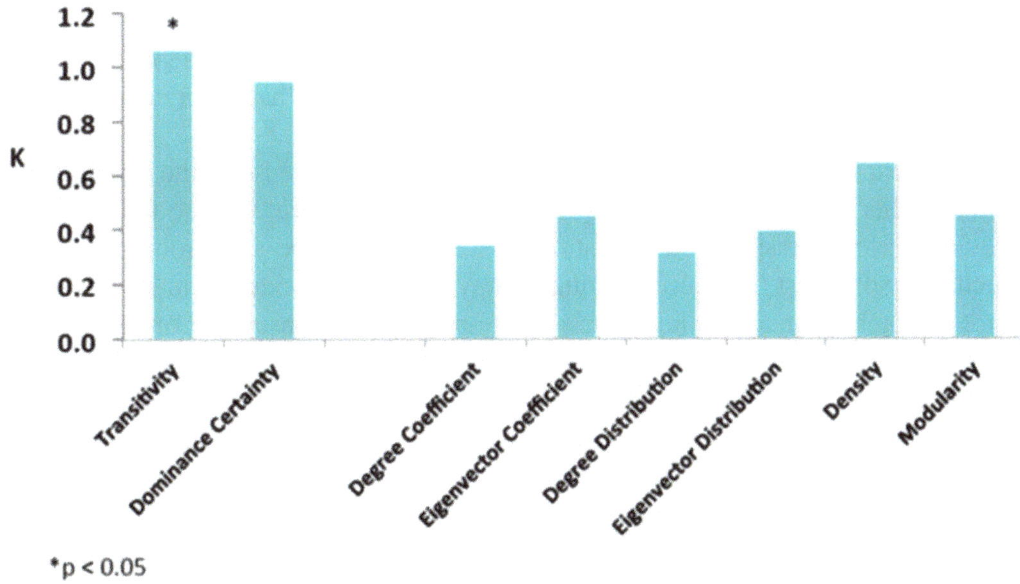

Fig. 8.1 Bar plot of the *K*-statistic calculated to determine the strength of phylogenetic signals among aspects of dominance and grooming social network structure for ten species of macaques. On the one hand, values of *K* approaching or greater than 1 indicate strong phylogenetic signals in the dominance traits. On the other hand, values of $K < 1$ indicate moderate-to-weak signals in the grooming traits

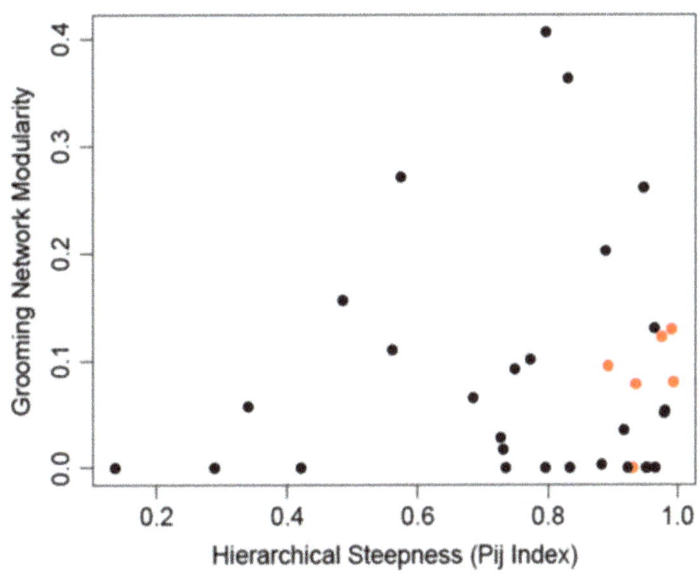

Fig. 8.2 Scatterplot showing the lack of a relationship between directional consistency index (DCI) of dominance hierarchies and Newman's community modularity of grooming social networks for 34 groups of macaques across 10 species. Red dots represent scores for the YA1 Tibetan macaque group at each of six study periods

phylogeny and the structural linkage of traits may be visible across higher organizational scales (e.g., across some species and lineages), but perhaps less so at lower organizational scales (e.g., across species within a lineage, across groups). In other words, they suggest that the structural covariation of aspects of macaque social structure has occurred over longer evolutionary timescales that separate entire phylogenetic lineages, rather than across shorter timescales that separate species or populations within lineages.

Consistent with findings at the interspecific level, and indeed with the conclusions from the other within-species studies we have reviewed above, we also found a lack of evidence for intraspecific covariation between dominance and grooming social structure among just Tibetan macaques. They displayed moderate-to-steep dominance hierarchies and very low levels of counter-aggression on the one hand and moderate-to-low levels of grooming modularity on the other hand. Moreover, aspects of dominance hierarchies in Tibetan macaques were not correlated with aspects of grooming social network across six study periods of hierarchical stability (e.g., hierarchical steepness vs. grooming modularity: $n = 6$, $r = 0.26$, $p = 0.62$). Thus, findings from our comparative studies are consistent with those from our previous studies on just Tibetan macaques. Both suggest that aspects of Tibetan macaque social style, while having potentially been shaped by phylogenetic relatedness and structural linkage in the evolutionary past, have been subject to somewhat unlinked, independent trajectories influenced by current socioecological and demographic conditions more recently in their evolutionary history.

8.7 Discussion and Future Directions

Aspects of social structure represent higher-order phenomena, i.e., phenotypically visible, group level outcomes of individuals striving to access the benefits of group living while minimizing its costs. Given the importance of social structure for individual fitness, primatologists are interested in the evolutionary underpinnings of variation in social structure (Clutton-Brock and Janson 2012; Koenig et al. 2013; Thierry 2004). Theoretical models based on species' phylogenetic history (Matsumura 1999; Thierry 2007; Thierry et al. 2000), inherent self-organization by individuals (Hemelrijk 1999, 2005; Puga and Sueur 2017), and adaptations to variation in current ecological conditions (Kappeler and Van Schaik 2002; Sterck et al. 1997; van Schaik 1989) have all been proposed, but currently there is no consensus regarding the relative extents to which these factors influence social structure.

In this chapter, we demonstrate how research on the genus *Macaca*, and in particular on Tibetan macaques, has provided useful empirical tests of some of these conflicting frameworks. Among macaque researchers, some proponents of phylogenetic arguments argued that aspects of social structure, specifically those that constitute social style, are structurally linked and covariant with each other (Petit et al. 1997; Thierry 2000; Thierry et al. 2008). Citing the detection of strong phylogenetic signals in some core aspects of macaque social style, they predicted that evolutionary change in one trait would co-occur with evolutionary change in a suite of other traits, with such covariation being more perceptible between species rather than within groups of the same species (Petit et al. 1997; Thierry 2000; Thierry et al. 2008).

Our findings on Tibetan macaques, and from comparative studies across a range of macaques, reveal incomplete evidence for such covariation across species and social style traits. We suggest that the pattern of covariation (and the lack thereof) that we found is consistent with the notion suggesting that covariation was likely present in the evolutionary past of the genus but that it has been subsequently partially offset by more contemporary, quasi-independent responses by macaques to variation in current conditions.

First, we found a mixture of both despotic and tolerant social style traits in Tibetan macaques. While aspects of the dominance structure in Tibetan macaques are within the range of despotic macaques, the structure of affiliative grooming social networks, male-male ritualized affiliation, and cooperative exchange are all characteristic of more tolerant societies. Covariation by itself, i.e., that has not been influenced by adaptive responses, would have been suggested if all these characteristics had showed moderate despotism or moderate tolerance. Second, our comparative studies revealed evidence for covariation between dominance traits and social style grade across some macaques in different phylogenetic lineages, but not across species within the same lineage. Third, the finding that grooming social network structure was more strongly influenced by group size than by phylogeny and dominance traits suggests that covariation was apparently offset by adaptive responses by macaques more recently in their evolutionary history. In other words, different aspects of social style and social structure in general likely evolved along more independent evolutionary trajectories, with some traits more strongly influenced by past ancestry and structural linkage than others, and others apparently more labile to current conditions.

This view is consistent with the concept of "social reaction norms," or the idea that species may respond similarly to major ecological changes, but may have inherently different ranges of responses to the same conditions, i.e., ranges that are related at least in part to their ancestry (Berman and Thierry 2010). The concept of social reaction norms can be seen as a group-level analog of the concept of interindividual variation in "behavioral syndromes," in which suites of correlated behaviors reflect individual variation in behavioral plasticity and in individual's capacities to show adaptive responses across multiple environmental situations and contexts (Sih et al. 2004, 2012).

Having said that, there remain many gaps in our understanding of social style and social structure in general in Tibetan macaques and in the macaque genus as a whole. Social style indicators have yet to be examined in many species, and a number of additional hypotheses about their origins remain to be explored in detail. We end this chapter by proposing and elaborating on five potentially important avenues of future research on social styles.

Paternal Kin Relatedness In animal social groups, the relatedness of individuals is strongly associated with the form, frequency, and distribution of cooperative and competitive social interactions (Hamilton 1964). In most Cercopithecine primates, females are philopatric, and there is ample evidence that matrilineal kin bias structures many aspects of social behavior within groups, more so in despotic species than tolerant species (Berman 2011; Berman and Thierry 2010; Thierry 2007; Thierry et al. 2008). As such, maternal kin bias has been considered one of the core indicators of a group's social style (Thierry 2000). In 1979, Altmann suggested that in addition to matrilineal ties, the structure of within-group relatedness may also be influenced by mating patterns, specifically by the degree to which individual males are able to monopolize fertile females (Altmann 1979). Specifically, the frequent replacement of highly successful males (i.e., males with high reproductive skew) may lead to age-structured paternal sibships within groups. If this occurs, and if individuals are able to recognize their paternal kin, inherent tendencies to favor kin could influence the patterning and distribution of cooperative and competitive interactions in a group and hence aspects of social style (Widdig 2013).

Recently, Schülke and Ostner hypothesized that the evolution of social tolerance in some macaque species may in fact be explained by high male reproductive skew (Schülke and Ostner 2008, 2013). High mean levels of paternal relatedness might explain why females in tolerant species distribute their affiliative interactions across a wide range of partners. As evidence for this hypothesis, they found that species of macaques classified as less nepotistic and more tolerant also showed higher male reproductive skew (Schülke and Ostner 2008). However, their analyses were limited by a small sample size and to comparisons of intolerant but seasonally breeding macaques with tolerant but nonseasonal macaques and hence were unable to control for the effects of seasonality on reproductive skew. In a second study, they found that paternal kin bias accounted for a small but significant proportion of the variation in the strength of grooming social bonds in a captive group of rhesus macaques (Schülke et al. 2013). Whether the reproductive skew argument can help explain some or all of the indicators of tolerance in Tibetan macaques is uncertain. Although male mating success is highly skewed toward dominants and alpha males have tenures of only a year or two, whether or not this translates into reproductive skew awaits genetic confirmation of paternity. Also needed are broader comparative studies that assess possible links between male reproductive skew, paternal relatedness, and multiple aspects of female social structure.

Disease Risk as a Selection Pressure Traditional socioecological models have speculated that multiple socioecological factors—natural resource abundance, clumped, human provisioned food, and changes in predation pressure over evolutionary time—may have all impacted variation in primate social structure (Kappeler and Van Schaik 2002; Sterck et al. 1997; van Schaik 1989). More recently, the influence of another ecological factor—disease risk—has also been proposed (Kappeler et al. 2015; Nunn 2012), but it has remained relatively understudied. Disease risk may influence primate group living and social life in at least two ways. First, strong parasite prevalence may encourage more spatially widespread social groups and subgrouping because it minimizes parasite transmission through social contact (Kappeler et al. 2015; Nunn 2012; Nunn et al. 2011). Indeed, comparative studies across a range of primates have found strong, positive correlations between parasite prevalence or diversity and community substructuring or modularity of affiliative social networks (Griffin and Nunn 2012; Nunn et al. 2015). This relationship is consistent with the hypothesis that increased parasite risk may have led to the evolution of increased levels of despotism in some primate groups and species, especially given that modular social networks have already been shown to be a characteristic of such societies (Sueur et al. 2011).

On the other hand, increased parasite risk may also cause animals to seek out more social partners, because strong and diverse social ties may enhance their social capital and lower stress levels associated with decreased disease resistance (Cohen et al. 2007; Kaplan et al. 1991; Sapolsky 2005). In accordance with this "social buffering hypothesis," increased disease risk could lead to the evolution of greater social tolerance (Silk et al. 2003, 2010; Young et al. 2014).

Evidence from macaque species, particularly wild Japanese macaques (MacIntosh et al. 2012; Romano et al. 2016) and captive rhesus macaques (Balasubramaniam et al. 2016, 2018b), suggests the transmission of gastrointestinal nematode parasites and gut pathogenic bacteria, respectively, maybe strongly influenced by the structure of affiliative grooming or huddling social networks. However, at this time, no comparable data are available for Tibetan macaques. Nevertheless, given the now well-established impact that parasite risk has on aspects of primate sociality and health (Kappeler et al. 2015), future research should focus on further unraveling such links.

Intraspecific Variation In general, comparative studies of variation in primate social organization and social structure have tended to focus more on interspecific than intraspecific variation, with the latter often overlooked or not appropriately accounted for (Clutton-Brock and Janson 2012; Nunn 2011). Given this, a long-term goal of Tibetan macaque research should be to establish the ecological and evolutionary underpinnings of intraspecific variation in social structure. To date, much of our understanding of intraspecific variation in Tibetan macaque social structure stems from longitudinal data collected on a single group at Mt. Huangshan. In one study, Berman and colleagues found that the extent of grooming kin bias among females in YA1 varied across seven different data collection periods between December 1991 and November 2004 (Berman et al. 2008). This variation was positively correlated to group size, but unaffected by other socioecological factors (e.g., within-group competition, human presence), leading to speculation that increased time constraints faced by the macaques when the group size is larger may influence their social styles (Berman et al. 2008). Later, Balasubramaniam and colleagues revealed that the steepness of the dominance hierarchy of YA1 females also varied significantly across six of these periods (Balasubramaniam et al. 2011). Yet contrary to expectation, steepness was negatively correlated to group size, leading to the speculation that the macaques responded to human presence by forming within-group alliances that may in turn have impacted their dominance style. So whether and how socioecological factors may impact one or more aspects of Tibetan macaque social structure remains unclear.

Given that the social groups studied at both Mt. Huangshan and Mt. Emei were subjected to varying degrees of tourism, provisioning, and range restriction (Berman and Li 2002), comparisons of their social styles with each other and with those of other wild populations are crucial. Evidence for intergroup differences in aspects of social structure is available for other free-living macaque species, but somewhat mixed. For instance, Majolo et al. (2009) found stark intergroup differences in the distribution of grooming and post-conflict reconciliation between two wild Japanese macaque groups of different sizes within the same population. Further, Duboscq et al. (2013) found between-group differences in rates of approaches, silent bare-teeth displays, and percentages of counter-aggression by females in two groups of wild crested macaques at Tangkoko Nature Reserve. On the other hand, a study on free-ranging rhesus macaques at Cayo Santiago revealed no differences in female dominance hierarchies, aggression intensities, and post-conflict affiliation rates across three groups of different sizes (Balasubramaniam et al. 2014). Such findings warrant establishing potential links (or the lack thereof) between socioecological factors and intraspecific variation in social structure across groups and populations of Tibetan macaques as well as other species.

Comparative Studies of Male Social Styles Most of what we know about the concept of macaque social styles has emerged from studies of female-female social relationships in captive groups. In comparison, the social relationships of males have been less intensely studied, with studies on free-ranging and wild populations of some species having just emerged during the last decade. In the light of this, we suggest that the time is right for conducting comparative studies of the evolution of male social styles.

Male relationships among the female philopatric Cercopithecine primates are expected to be predominantly intolerant or antagonistic. Coalitions are expected to be primarily (1) conservative to maintain dominance rank and (2) based on partner availability rather than the establishment of long-term affiliative social bonds among unrelated males (Cords 2013; Kappeler 2000; Preuschoft and Paul 2000). Yet, there is also evidence for overt affiliation, social bond investment, and male-male social tolerance in some species (Ostner and Schülke 2014; Schülke et al. 2010), including Tibetan macaques (see above). Evolutionary explanations for such evidence remain somewhat unclear. One possible explanation, proposed above for Tibetan macaques, involves the emergence of male "tolerant despots," i.e., the idea that male social tolerance may be an emergent outcome of a group- or species-typical tendency to exhibit a generally despotic social style. This may occur when selective tolerance enhances males' abilities to compete in a despotic society. Such indicators of social tolerance maybe a consequence of males' counter-strategies to deal with generally heightened levels of within-group resource competition (Ostner and Schülke 2014; Schülke et al. 2010).

Findings from Assamese macaques also seem to support this explanation. For instance, both male and female Assamese macaques show steep, linear dominance hierarchies that are indicative of high within-group competition and a despotic social style (Balasubramaniam et al. 2012a, b; Cooper 1999; Cooper and Bernstein 2002, 2008). Yet, males also show moderate levels of post-conflict affiliation (Cooper 1999; Cooper and Bernstein 2002, 2008) and form long-term social bonds based on partner preference rather than partner availability (Kalbitz et al. 2016; Ostner and Schülke 2014; Schülke et al. 2010). Findings from other species suggest that male social tolerance, rather than being emergent outcomes of within-group despotism, may have socioecological underpinnings. For instance, grooming and alliance formation among male Japanese macaques appear to be more strongly influenced by between-group mating competition than by within-group competition (Horiuchi 2007). In bonnet macaques, the distribution and abundance of natural resources may strongly influence male-to-female sex ratios and, thereby, male social strategies (Ram et al. 2003; Sinha et al. 2005).

When abundant and widespread resources give rise to large, multi-male-multi-female groups with near-even adult sex ratios, reduced within-group competition may give rise to unusual levels of social tolerance, ritualized greetings, huddling, and social grooming among males (Adiseshan et al. 2011; Silk 1994, 1999). When resources are clumped or seasonal, intense female-female competition may lead to group-fissioning, resulting in the emergence of atypical unimale troops in several species (Ram et al. 2003; Sinha et al. 2005). In these contexts, males actively herd females, defend them against immigrating males by engaging in intergroup encounters, and remain intolerant of resident juvenile and subadult males (Ram et al. 2003; Sinha et al. 2005). Such stark differences and explanations for male social strategies warrant more comparative studies across macaque species that examine the relative effect(s) of intrinsic characteristics (e.g., female social styles, male despotism) and extrinsic factors (e.g., resource abundance, intergroup competition, predation) on male social styles.

Phylogeographic Approaches Finally, comparative studies have now established that several (but not all) aspects of macaque social structure (Balasubramaniam et al. 2012a, 2018a; Thierry et al. 2008), and more broadly primate sociality (Kamilar and Cooper 2015), are associated at least in part with species' ancestral relationships. What is unclear is whether such phylogenetic signals are reflections of true genetic linkage or coevolution versus artifacts of adaptive responses by ancestral species to major environmental changes (Balasubramaniam et al. 2018a). To address this gap, we recommend that future research should focus on establishing links between ancestral state reconstructions of primate social structure (Pagel et al. 2004; Revell 2012) and potential changes in historic phylogeographic ranges (Abegg and Thierry 2002; Lemey et al. 2009; Ree and Smith 2008). Such approaches would help establish whether, for instance, some aspects of social structure coevolved with major dispersal events or climatic changes toward more resource abundant (or sparser) environments in the evolutionary past.

Acknowledgments We thank the Huangshan Monkey Management Center and the Huangshan Garden Forest Bureau for granting permission to collect field data on the YA1 Tibetan macaque group. We thank the following organizations: the Leakey Foundation, the Wenner-Gren Foundation, and the National Geographic Society (C.M.B.); the National Natural Science Foundation of China, Key Teacher Program of the Ministry of Education of China, and the Excellent Youth Foundation of Anhui (J.L.) for providing funding to carry out fieldwork at Huangshan. We are grateful to May Lee Gong, Krista Jones, Ming Li, Stephan Menu, Stephanie Pieddesaux, Justin Sloan, and Lei Zhang for their role as field assistants at Huangshan. We give special thanks to Xinming Chen and his family for housing and accommodation in China. For the comparative studies, we thank several past and current colleagues in primatology and their research teams, specifically Marina Butovskaya, Matthew Cooper, Arianna De Marco, Julie Duboscq, Sabina

Koirala, Bonaventura Majolo, Andrew MacIntosh, Richard MacFarland, Sandra Molesti, Odile Petit, Gabriele Schino, Sebastian Sosa, Cedric Sueur, Bernard Thierry, and Frans de Waal for contributing their datasets and playing a role in manuscript publication. Further, we also thank Katharina Dittmar, Brenda McCowan, and Brianne Beisner for providing much-needed input in phylogenetic and social network analyses, respectively. We thank the National Science Foundation (NSF), USA, for partially funding the comparative studies. Finally, we are grateful to Bernard Thierry and Bonaventura Majolo for their reviews and feedback while constructing this chapter.

References

Abegg C, Thierry B (2002) Macaque evolution and dispersal in insular south-east Asia. Biol J Linn Soc 75(4):555–576

Adiseshan A, Adiseshan T, Isbell LA (2011) Affiliative relationships and reciprocity among adult male bonnet macaques (*Macaca radiata*) at Arunachala Hill, India. Am J Primatol 73:1107–1113

Altmann J (1979) Age cohorts as paternal sibships. Behav Ecol Sociobiol 6:161–164

Arnold C, Matthews LJ, Nunn CL (2010) The *10kTrees* website: a new online resource for primate phylogeny. Evol Anthropol 19:114–118

Balasubramaniam KN, Berman CM (2017) Grooming exchange for resource tolerance: biological markets principles in a group of free-ranging rhesus macaques. Behaviour 154(11):1145–1176

Balasubramaniam KN, Berman CM, Ogawa H, Li J (2011) Using biological markets principles to examine patterns of grooming exchange in *Macaca thibetana*. Am J Primatol 73 (12):1269–1279

Balasubramaniam KN, Dittmar K, Berman CM, Butovskaya M, Cooper MA, Majolo B, Ogawa H, Schino G, Thierry B, de Waal FBM (2012a) Hierarchical steepness and phylogenetic models: phylogenetic signals in *Macaca*. Anim Behav 83:1207–1218

Balasubramaniam KN, Dittmar K, Berman CM, Butovskaya M, Cooper MA, Majolo B, Ogawa H, Schino G, Thierry B, de Waal FBM (2012b) Hierarchical steepness, counter-aggression, and macaque social style scale. Am J Primatol 74:915–925

Balasubramaniam KN, Dunayer ES, Gilhooly LJ, Rosenfield KA, Berman CM (2014) Group size, contest competition, and social structure in Cayo Santiago rhesus macaques. Behaviour 151:1759–1798

Balasubramaniam KN, Beisner BA, Vandeleest J, Atwill ER, McCowan B (2016) Social buffering and contact transmission: network connections have beneficial and detrimental effects on Shigella infection risk among captive rhesus macaques. PeerJ 4:e2630

Balasubramaniam KN, Beisner BA, Berman CM, De Marco A, Duboscq J, Koirala S, Majolo B, MacIntosh AJJ, McFarland R, Molesti S, Ogawa H, Petit O, Schino G, Sosa S, Sueur C, Thierry B, de Waal FBM, McCowan BJ (2018a) The influence of phylogeny, social style, and sociodemographic factors on macaque social network structure. Am J Primatol 80(1): e227227

Balasubramaniam KN, Beisner BA, Guan J, Vandeleest J, Fushing H, Atwill ER, McCowan B (2018b) Social network community structure is associated with the sharing of commensal *E. coli* among captive rhesus macaques (*Macaca mulatta*). PeerJ 6:e4271

Berman CM (2011) Kinship: family ties and social behavior. In: Campbell CJ, Fuentes A, MacKinnon KC, Panger M, Bearder SK (eds) Primates in perspective, 2nd edn. Oxford University Press, New York, pp 576–587

Berman CM, Kapsalis E (2009) Variation over time in grooming kin bias among female rhesus macaques on Cayo Santiago supports the time constraints hypothesis. Am J Phys Anthropol 48:89–90

Berman CM, Li J (2002) Impact of translocation, provisioning and range restriction on a group of *Macaca thibetana*. Int J Primatol 23:287–293

Berman CM, Thierry B (2010) Variation in kin bias: species differences and time constraints in macaques. Behaviour 147(13):1863–1887

Berman CM, Ionica CS, Li J (2004) Dominance style among *Macaca thibetana* on Mt. Huangshan, China. Int J Primatol 25:1283–1312

Berman CM, Ionica CS, Dorner M, Li JH (2006) Post-conflict affiliation between former opponents in *Macaca thibetana*: for males only? Int J Primatol 27:827–854

Berman CM, Ionica CS, Li J (2007) Supportive and tolerant relationships among male Tibetan macaques at Huangshan, China. Behaviour 144:631–661

Berman CM, Ogawa H, Ionica CS, Yin H, Li J (2008) Variation in kin bias over time in a group of Tibetan macaques at Huangshan: contest competition, time constraints or stress response? Behaviour 145:863–896

Blomberg SP, Garland T, Ives A (2003) Testing for phylogenetic signal in comparative data: behavioral traits are more labile. Evolution 57:717–745

Brent LJN, Lehmann J, Ramos-Fernandez G (2011) Social network analysis in the study of nonhuman primates: a historical perspective. Am J Primatol 73(8):720–730

Capitanio JP (1999) Personality dimensions in adult male rhesus macaques: prediction of behaviors across time and situation. Am J Primatol 47:299–320

Carne C, Viper S, Semple S (2011) Reciprocation and interchange of grooming, agonistic support, feeding tolerance, and aggression in semi-free-ranging Barbary macaques. Am J Primatol 73:1127–1133

Castles DL, Aureli F, de Waal FBM (1996) Variation in conciliatory tendency and relationship quality across groups of pigtail macaques. Anim Behav 52:389–403

Chakraborty D, Ramakrishnan U, Panor J, Mishra C, Sinha A (2007) Phylogenetic relationships and morphometric affinities of the Arunachal macaque *Macaca munzala*, a newly described primate from Arunachal Pradesh, Northeastern India. Mol Phylogenetics Evol 44:838–849

Chan LKW (1996) Phylogenetic interpretations of primate socioecology: with special reference to social and ecological diversity in *Macaca*. In: Martins EP (ed) Phylogenies and the comparative method in animal behaviour. Oxford University Press, Oxford, pp 324–360

Clutton-Brock TH, Harvey P (1977) Primate ecology and social organization. J Zool 183:1–39

Clutton-Brock TH, Janson CH (2012) Primate socioecology at the crossroads: past, present, and future. Evol Anthropol 21:136–150

Cobb S (1976) Social support as a moderator of life stress. Psychosom Med 38:300–314

Cohen S, Janicki-Deverts D, Miller GE (2007) Psychological stress and disease. JAMA 298:1685–1687

Cohen S, Janicki-Deverts D, Turner RB, Doyle WJ (2015) Does hugging provide stress-buffering social support? A study of susceptibility to upper respiratory infection and illness. Psychol Sci 26:135–147

Cooper MA (1999) Social tolerance in Assamese macaques (*Macaca assamensis*). Ph.D. Dissertation. University of Georgia, Athens, GA

Cooper MA, Bernstein IS (2002) Counter aggression and reconciliation in Assamese macaques (*Macaca assamensis*). Am J Primatol 56:215–230

Cooper MA, Bernstein IS (2008) Evaluating dominance styles in Assamese and rhesus macaques. Int J Primatol 29:225–243

Cords M (2013) The behavior, ecology, and social evolution of Cercopithecine monkeys. In: Mitani JC, Call J, Kappeler PM, Palombit RA, Silk JB (eds) The evolution of primate societies. University of Chicago Press, Chicago, pp 91–112

Csardi G, Nepusz T (2006) The igraph software package for complex network research. InterJ. Complex Syst. 1695

de Vries H (1998) Finding a dominance order most consistent with a linear hierarchy: a new procedure and review. Anim Behav 55:827–843 de Vries H, Stevens JMG, Vervaecke H (2006) Measuring and testing the steepness of dominance hierarchies. Anim Behav 71:585–592

de Waal FBM, Luttrell LM (1989) Towards a comparative ecology of the genus *Macaca*: different dominance styles in rhesus and stumptailed macaques. Am J Primatol 19:83–109

de Waal FBM, Yoshihara D (1983) Reconciliation and redirected affection in rhesus monkeys. Behaviour 85:224–241

Demaria C, Thierry B (2001) A comparative study of reconciliation in rhesus and Tonkean macaques. Behaviour 138:397–410

Deng ZY (1993) Social development of infants of *Macaca thibetana* at Mt. Emei, China. Folia Primatol 60:28–35

Duboscq J, Micheletta J, Agil M, Hodges K, Thierry B, Engelhardt A (2013) Social tolerance in wild female crested macaques (*Macaca nigra*) in Tangkoko-Batuangus Nature Reserve, Sulawesi, Indonesia. Am J Primatol 75:361–375

Farine DR, Whitehead H (2015) Constructing, conducting and interpreting animal social network analysis. J Anim Ecol 84:1144–1163

Fooden J (1986) Taxonomy and evolution of the *sinica* group of macaques: 5. Overview of natural history. Fieldiana Zool 29:1–22

Fruteau C, van de Waal E, Van Damme E, Noe R (2011) Infant access and handling in sooty mangabeys and vervet monkeys. Anim Behav 81:153–161

Fujii K, Fushing H, Beisner BA, McCowan B (2013) Computing power structures in directed biosocial networks: flow percolation and imputed conductance. Technical Report, Department of Statistics, UC Davis

Gammell MP, de Vries H, Jennings DJ, Carlin CM, Hayden TJ (2003) David's score: a more appropriate dominance ranking method than Clutton-Brock et al.'s index. Anim Behav 66:601–605

Greenwood PJ (1980) Mating systems, philopatry and dispersal in birds and mammals. Anim Behav 28:1140–1162

Griffin RH, Nunn CL (2012) Community structure and the spread of infectious disease in primate social networks. Evol Ecol 26(4):779–800

Gumert MD (2007) Grooming and infant handling interchange in *Macaca fascicularis*: the relationship between infant supply and grooming payment. Int J Primatol 28:1059–1074

Hamilton WD (1964) The genetical evolution of social behaviour I/II. J Theor Biol 7:1–52

Handcock M, Hunter D, Butts C, Goodreau S, Morris M (2006) Statnet: an R package for the statistical analysis and simulation of social networks. University of Washington. http://www.csde.washington.edu/statnet

Hemelrijk C (1999) An individual-orientated model of the emergence of despotic and egalitarian societies. Proc R Soc Lond B 266:361–369

Hemelrijk C (2005) Self-organisation and evolution of biological and social systems. Cambridge University Press, Cambridge

Henzi SP, Barrett L (2002) Infants as a commodity in a baboon market. Anim Behav 63:915–921

Hill DA (1999) Effects of provisioning on the social behaviour of Japanese and rhesus macaques: implications for socioecology. Primates 40:187–198

Hinde RA (1976) Interactions, relationships and social structure. Man 11:1–17

Holt-Lunstad J, Smith TB, Layton JR (2010) Social relationships and mortality risk: a meta-analytic review. PLoS Med 7:e1000316

Horiuchi S (2007) Social relationships of male Japanese macaques (*Macaca fuscata*) in different habitats: a comparison between Yakushima island and Shimokita peninsula populations. Anthropol Sci 115:63–65

House JS, Landis KR, Umberson D (1988) Social relationships and health. Science 241:50–54

Isbell LA (2017) Socioecological model. In: Bezanson M, MacKinnon KC, Riley E, Campbell CJ, Nekaris KAI, Estrada A, Di Fiore AF, Ross S, Jones-Engel LE, Thierry B, Sussman RW, Sanz C, Loudon J, Elton S, Fuentes (eds) The international encyclopedia of primatology. Wiley Online Library

Judge P (2000) Coping with crowded conditions. In: Aureli F, de Waal FBM (eds) Natural conflict resolution. University of California Press, Berkeley, pp 129–154

Kaburu SSK, Newton-Fisher NE (2015) Egalitarian despots: hierarchy steepness, reciprocity and the grooming-trade model in wild chimpanzees, *Pan troglodytes*. Anim Behav 99:61–71

Kalbitz J, Ostner J, Schulke O (2016) Strong, equitable and long-term social bonds in the dispersing sex in Assamese macaques. Anim Behav 113:13–22

Kamilar J, Cooper N (2015) Phylogenetic signal in primate behaviour, ecology and life history. Philos Trans R Soc B 368:20120341

Kaplan JR, Heise ER, Manuck SB, Shively CA, Cohen S, Rabin BS, Kasprowicz AL (1991) The relationship of agonistic and affiliative behavior patterns to cellular immune function among cynomolgus monkeys Macaca- fascicularis living in unstable social groups. Am J Primatol 25 (3):157–174

Kappeler PM (2000) Primate males: causes and consequences of variation in group composition. Cambridge University Press, Cambridge

Kappeler PM, Van Schaik CP (2002) Evolution of primate social systems. Int J Primatol 23:707–740

Kappeler PM, Cremer S, Nunn CL (2015) Sociality and health: impacts of sociality on disease susceptibility and transmission in animal and human societies. Philos Trans R Soc B 370:20140116

Kapsalis E (2004) Matrilineal kinship and primate behavior. In: Chapais B, Berman CM (eds) Kinship and behavior in primates. Oxford University Press, New York, pp 153–176

Kasper C, Voelkl B (2009) A social network analysis of primate groups. Primates 50:343–256

Koenig A, Scarry CJ, Wheeler BC, Borries C (2013) Variation in grouping patterns, mating systems and social structure: what socio-ecological models attempt to explain. Philos Trans R Soc B 368:20120348

Krause J, James R, Croft DP (2010) Personality in the context of social networks. Philos Trans R Soc B 365:4099–4106

Lemey P, Rambaut A, Drummond AJ, Suchard MA (2009) Bayesian phylogeography finds its roots. PLoS Comput Biol 5:e1000520

Li J (1999) The Tibetan macaque society: a field study. Anhui University Press, Hefei

Li J, Wang Q (1996) Dominance hierarchy and its chronic changes in adult male Tibetan macaques (*Macaca thibetana*). Acta Zool Sinica 42:330–334

Lusseau D (2003) The emergent properties of a dolphin social network. Proc R Soc Lond B 270: S186–S188

MacIntosh AJJ, Jacobs A, Garcia C, Shimizu K, Mouri K, Huffman MA, Hernandez AD (2012) Monkeys in the middle: parasite transmission through a social network of a wild primate. PLoS One 7:e51144

Majolo B, Ventura R, Koyama NF, Hardie SM, Jones BM, Knapp LA, Schino G (2009) Analyzing the effects of group size and food competition on Japanese macaque social relationships. Behaviour 146:113–137

Marechal L, Semple S, Majolo B, Qarro M, Heistermann M, MacLarnon A (2011) Impacts of tourism on anxiety and physiological stress levels in wild male Barbary macaques. Biol Conserv 144(9):2188–2193

Matsumura S (1999) The evolution of "egalitarian" and "despotic" social systems among macaques. Primates 40:23–31

Newman MEJ, Girvan M (2004) Finding and evaluating community structure in networks. Phys Rev E 69:1–15

Nunn CL (2011) The comparative approach in evolutionary anthropology and biology. University of Chicago Press, Chicago

Nunn CL (2012) Primate disease ecology in comparative and theoretical perspective. Am J Primatol 74(6):497–509

Nunn CL, Thrall PH, Leendertz FH, Boesch C (2011) The spread of fecally transmitted parasites in socially structured populations. PLoS One 6:e21677

Nunn CL, Jordan F, McCabe CM, Verdolin JL, Fewell JH (2015) Infectious disease and group size: more than just a numbers game. Philos Trans R Soc B 370:20140111

Ogawa H (1995) Bridging behavior and other affiliative interactions among male Tibetan macaques (*Macaca thibetana*). Int J Primatol 16:707–729

Ostner J, Schülke O (2014) The evolution of social bonds in primate males. Behaviour 151:871–906

Pagel M, Meade A, Barker D (2004) Bayesian estimation of ancestral character states on phylogenies. Syst Biol 53(5):673–684

Paul A, Kuester J (1987) Dominance, kinship and reproductive value in female Barbary macaques (*Macaca sylvanus*) at Affenberg Salem. Behav Ecol Sociobiol 21:323–331

Petit O, Abegg C, Thierry B (1997) A comparative study of aggression and conciliation in three cercopithecine monkeys (*Macaca fuscata*, *Macaca nigra* and *Papio papio*). Behaviour 134:415–432

Preuschoft S, Paul A (2000) Dominance, egalitarianism, and stalemate: an experimental approach to male-male competition in Barbary macaques. In: Kappeler PM (ed) Primate males: causes and consequences of variation in group composition. Cambridge University Press, Cambridge, pp 205–216

Preuschoft S, Paul A, Kuester J (1998) Dominance styles of female and male Barbary macaques (*Macaca sylvanus*). Behaviour 135:731–755

Puga I, Sueur C (2017) Emergence of complex social networks from spatial structure and rules of thumb: a modelling approach. Ecol Complex 31:189–200

Purvis A (1995) A composite estimate of primate phylogeny. Philos Trans R Soc B 348:405–421

Ram S, Venkatachalam S, Sinha A (2003) Changing social strategies of wild female bonnet macaques during natural foraging and on provisioning. Curr Sci 84:780–790

Ree RH, Smith SA (2008) Maximum likelihood inference of geographic range evolution by dispersal, local extinction, and cladogenesis. Syst Biol 51:4–14

Revell LJ (2012) phytools: an R package for phylogenetic comparative biology (and other things). Methods Ecol Evol 3:217–233

Richter C, Mevis L, Malaivijitnond S, Schülke O, Ostner J (2009) Social relationships in free-ranging male *Macaca arctoides*. Int J Primatol 30(4):625–642

Romano V, Duboscq J, Sarabian C, Thomas E, Sueur C, MacIntosh AJJ (2016) Modeling infection transmission in primate networks to predict centrality-based risk. Am J Primatol 78:767–779

Sade DS (1972) A longitudinal study of social behavior of rhesus monkeys. In: Tuttle R (ed) The functional and evolutionary biology of primates. Aldine-Atherton, Chicago, pp 378–398

Sapolsky RM (2005) The influence of social hierarchy on primate health. Science 308:648–652

Schino G, Aureli F (2008) Tradeoffs in primate grooming reciprocation: testing behavioral flexibility and correlated evolution. Biol J Linn Soc 95:439–446

Schülke O, Ostner J (2008) Male reproductive skew, paternal relatedness, and female social relationships. Am J Primatol 70:695–698

Schülke O, Ostner J (2013) Ecological and social influences on sociality. In: Mitani JC, Call J, Kappeler PM, Palombit RA, Silk JB (eds) Evolution of primate societies. University of Chicago Press, Chicago, pp 195–219

Schülke O, Bhagavatula J, Vigilant L, Ostner J (2010) Social bonds enhance reproductive success in male macaques. Curr Biol 20:2207–2210

Schülke O, Wenzel S, Ostner J (2013) Paternal relatedness predicts the strength of social bonds among female rhesus macaques. PLoS One 8:e59789

Shizuka D, Mcdonald DB (2012) A social network perspective on measurements of dominance hierarchies. Anim Behav 83:925–934

Sih A, Bell A, Chadwick Johnson J (2004) Behavioral syndromes: an ecological and evolutionary overview. Trends Ecol Evol 19:372–378

Sih A, Cote J, Evans M, Fogarty S, Pruitt J (2012) Ecological implications of behavioural syndromes. Ecol Lett 15:278–289

Silk JB (1994) Social relationships of male bonnet macaques: male bonding in a matrilineal society. Behaviour 130:271–291

Silk JB (1999) Male bonnet macaques use information about third-party rank relationships to recruit allies. Anim Behav 58:45–51

Silk JB, Alberts SC, Altmann J (2003) Social bonds of female baboons enhance infant survival. Science 302:1231–1234

Silk JB, Beehner JC, Bergman C, Crockford AL, Engh LR, Moscovice RM, Wittig RM, Seyfarth RM, Cheney DL (2010) Strong and consistent social bonds enhance the longevity of female baboons. Curr Biol 20:1359–1361

Sinha A, Mukhopadhyay K, Datta-Roy A, Ram S (2005) Ecology proposes, behaviour disposes: ecological variability in social organization and male behavioural strategies among wild bonnet macaques. Curr Sci 89(7):1166–1179

Sterck EHM, Watts DP, van Schaik CP (1997) The evolution of female social relationships in nonhuman primates. Behav Ecol Sociobiol 41(5):291–309

Sueur C, Petit O, De Marco A, Jacobs AT, Watanabe K, Thierry B (2011) A comparative network analysis of social style in macaques. Anim Behav 82(4):845–852

Suomi SJ (2011) Risk, resilience, and gene-environment interplay in primates. J Can Acad Child Adolesc Psychiatry 20:289–297

Thierry B (1985) Patterns of agonistic interactions in three species of macaque (*Macaca mulatta*, *M. fascicularis*, *M. tonkeana*). Aggress Behav 11:223–233

Thierry B (2000) Covariation of conflict management patterns across macaque species. In: Aureli F, de Waal FBM (eds) Natural conflict resolution. University of California Press, Berkley, CA, pp 106–128

Thierry B (2004) Social epigenesis. In: Thierry B, Singh M, Kaumanns W (eds) Macaque societies: a model for the study of social organization. Cambridge University Press, Cambridge, pp 267–290

Thierry B (2007) Unity in diversity: lessons from macaque societies. Evol Anthropol 16:224–238

Thierry B (2013) The macaques: a double-layered social organization. In: Campbell CJ, Fuentes A, MacKinnon KC, Bearder SK, Stumpf RM (eds) Primates in perspective. Oxford University Press, Oxford, pp 229–240

Thierry B, Aureli F (2006) Barbary but not barbarian: social relations in a tolerant macaque. In: Hodges JK (ed) Biology and behaviour of Barbary macaques. Nottingham University Press, Nottingham, pp 1–18

Thierry B, Iwaniuk AN, Pellis SM (2000) The influence of phylogeny on the social behaviour of macaques. Ethology 106:713–728

Thierry B, Singh M, Kaumanns W (2004) Macaque societies: a model for the study of social organization. Cambridge University Press, Cambridge

Thierry B, Aureli F, Nunn CL, Petit O, Abegg C, de Waal FBM (2008) A Comparative study of conflict resolution in macaques: insights into the nature of covariation. Anim Behav 75:847–860

Tiddi B, Aureli F, Schino G (2010) Grooming for infant handling in tufted capuchin monkeys: a reappraisal of the primate infant market. Anim Behav 79:1115–1123

Tosi AJ, Morales JC, Melnick DJ (2003) Paternal, maternal, and biparental molecular markers provide unique windows onto the evolutionary history of macaque monkeys. Evolution 57:1419–1435

Tyrrell M, Berman CM, Agil M, Sutrisno T, Engelhardt A (2018) Social style among wild male crested macaques (*Macaca nigra*) in Tangkoko Nature Reserve, Sulawesi, Indonesia. In Proceedings of the American Society of Primatologists, San Antonio, TX

van Hooff JARAM, van Schaik C (1994) Male bonds: affiliative relationships among nonhuman primates. Behaviour 130:309–337

van Schaik CP (1989) The ecology of social relationships amongst female primates. In: Standen V, Foley RA (eds) Comparative socio-ecology: the behavioral ecology of humans and other animals. Blackwell, Oxford, pp 195–218 van Schaik CP (1996) Social evolution in primates: the role of ecological factors and male behaviour. Proc Br Acad 88:9–31

van Schaik CP, Pandit AP, Vogel ER (2004) A model for within-group coalitionary aggression among males. Behav Ecol Sociobiol 57:101–109

Veenema HC, Das M, Aureli F (1994) Methodological improvements for the study of reconciliation. Behav Process 31:29–38

Ventura R, Majolo B, Koyama NF, Hardie SM, Schino G (2006) Reciprocation and interchange in wild Japanese macaques: grooming, cofeeding and agonistic support. Am J Primatol 68:1138–1149

Wada KC, Xiong C, Wang Q (1987) On the distribution of Tibetan and rhesus monkeys in Southern Anhui, China. Acta Theriol Sinica 7:148–176

Widdig A (2013) The impact of male reproductive skew on kin structure and sociality in multi-male groups. Evol Anthropol 22:239–250

Wrangham RW (1980) An ecological model of female-bonded primate groups. Behaviour 75:262–300

Xiong CP (1984) Ecological studies of the stump-tailed macaque. Acta Theriol Sinica 4:1–9

Young C, Majolo B, Heistermann M, Schulke O, Ostner J (2014) Responses to social and environmental stress are attenuated by strong male bonds in wild macaques. PNAS 111:18195–18200

Zhang P, Watanabe K (2014) Intraspecies variation in dominance style of *Macaca fuscata*. Primates 55:69–79

Zhao QK (1996) Etho-ecology of Tibetan macaques at Mount Emei, China. In: Fa JE, Lindburg DG (eds) Evolution and ecology of macaque societies. Cambridge University Press, Cambridge, pp 263–289

Zhao QK, Deng ZY, Xu J (1991) Natural foods and their ecological implications for *Macaca thibetana* at Mount Emei, China. Folia Primatol 57:1–15

Part III
Evolution of Rituals: Insights from Bridging Behavior

Chapter 9
Preliminary Observations of Female-Female Bridging Behavior in Tibetan Macaques (*Macaca thibetana*) at Mt. Huangshan, China

Grant J. Clifton, Lori K. Sheeran, R. Steven Wagner, Jake A. Funkhouser, and Jin-Hua Li

9.1 Introduction

Several species of the genus *Macaca* engage in an affiliative behavior commonly referred to as "bridging." Bridging is a triadic behavior in which two older individuals lift and hold an infant or juvenile between them while teeth chattering and/or licking the infant/juvenile's genitals (Ogawa 1995a). Bridging and similar affiliative

G. J. Clifton (✉)
Primate Behavior and Ecology Program, Central Washington University, Ellensburg, WA, USA

L. K. Sheeran
Primate Behavior and Ecology Program, Central Washington University, Ellensburg, WA, USA

Department of Anthropology and Museum Studies, Central Washington University, Ellensburg, WA, USA
e-mail: SheeranL@cwu.edu

R. S. Wagner
Department of Biological Sciences, Central Washington University, Ellensburg, WA, USA
e-mail: WagnerS@cwu.edu

J. A. Funkhouser
Primate Behavior and Ecology Program, Central Washington University, Ellensburg, WA, USA

Department of Anthropology, Washington University in St. Louis, St. Louis, MO, USA
e-mail: jakefunkhouser@wustl.edu

J.-H. Li
School of Resources and Environmental Engineering, Anhui University, Hefei, Anhui, China

International Collaborative Research Center for Huangshan Biodiversity and Tibetan Macaque Behavioral Ecology, Anhui, China

School of Life Sciences, Hefei Normal University, Hefei, Anhui, China
e-mail: jhli@ahu.edu.cn

J.-H. Li et al. (eds.), *The Behavioral Ecology of the Tibetan Macaque*, Fascinating Life Sciences, https://doi.org/10.1007/978-3-030-27920-2_9

triadic behaviors have been recorded in Barbary (*M. sylvanus*; Deag and Crook 1971; Taub 1984; Paul et al. 1996; Kubenova et al. 2017), stump-tailed (*M. arctoides*; Estrada and Sandoval 1977; Estrada and Estrada 1984), Assamese (*M. assamensis*; Kubenova et al. 2017), and Tibetan macaques (*M. thibetana*; Ogawa 1995a, b, c; Zhao 1996; Bauer et al. 2013). Adult males' bridging tends to occur in non-agonistic contexts and is often followed by other affiliative behaviors such as grooming (Ogawa 1995a, b; Deag and Crook 1971). Adult males of all species use younger male infants as the bridge more often than female infants and juveniles of both sexes (Ogawa 1995a; Deag 1980). In *M. sylvanus*, subordinate, lower-ranked males often bring an infant to a dominant individual to engage in affiliative behaviors (Deag and Crook 1971; Deag 1980). Deag and Crook (1971) hypothesized that subordinate males use infants as an "agonistic buffer" when initiating interactions with dominant males in order to reduce the probability of aggression. Alternatively, Taub (1984) proposed that these interactions were related to paternal caretaking and instead suggested the "enforced babysitting" hypothesis. Rather than using the infant as a buffer, Taub suggested that natal males in *M. sylvanus* use infants that are matrilineally related to them in order to develop bonds with infants and to inform others of their relatedness to the individual (1984).

Paul, Kuester, and Arnemann (1996) tested both the agonistic buffer and enforced babysitting hypotheses in dyadic male-infant interactions and triadic male-infant interactions (i.e., bridging) in *M. sylvanus*. DNA evidence conflicted with the enforced babysitting hypothesis by demonstrating that males did not interact with infants related to them and did not gain any additional reproductive benefits from interacting with particular females' infants. Conversely, frequency of male-infant interactions increased during intervals of high male-male tension, which supported the agonistic buffering hypothesis in *M. sylvanus*.

Ogawa (1995a, b, c) investigated male-male bridging behavior in Tibetan macaques (*M. thibetana*) and tested the agonistic buffering and enforced babysitting hypotheses. Ogawa found that bridging almost always occurred in non-agonistic contexts and that the frequency of bridging between males positively correlated with rates of grooming (Ogawa 1995a). Subordinate males initiated bridges with dominant males more often than vice versa, and initiating males were more likely to use the infants that were preferred in social interactions by dominant males. Aggression never occurred in male-male dyadic interactions in which a bridge took place but occasionally happened in non-bridging dyadic interactions. Ogawa (1995a) suggested that in macaque species with a higher than average socionomic sex ratio, such as Barbary and Tibetan macaques, males may use bridging to reduce social tensions caused by increased male-male competition and to reduce the possibility of aggression in future interactions. While Ogawa's (1995a) findings did not support the enforced babysitting hypothesis, natal group males did not prefer to bridge with infants related to them, suggesting that male-male bridging is not related to kinship. Adult males preferred to use male infants rather than female infants, and individual males had particular immature males that they used in bridging interactions more often than they used others. However, there was no evidence to suggest that infants gained any direct benefits from bridging: male and female infants experienced equal mortality rates despite the sex bias toward male infants in male-male bridging.

Although all age and both sex classes engage in bridging, there has been little mention of bridging interactions involving females. Ogawa (1995c) studied bridging interactions between adult males and adult females. In contrast to the trend in male-male bridging, higher-ranked males were more likely to initiate male-female bridging interactions than were lower-ranked males. Additionally, males bridged more females they were in consort with. Ogawa (1995c) concluded that male-female bridging likely facilitates mating bonds.

Female-female bridging has been reported in *M. thibetana* (Ogawa 2006; Bauer et al. 2013) and *M. arctoides* (Estrada and Estrada 1984), and similar triadic female-female-infant interactions have been described in *M. sylvanus* (Deag and Crook 1971), but no studies have investigated female-female bridging in detail. When considering the existence of maternal relatedness and strong matrilines among females as well as the differing forms of within-sex competition between males and females, it is possible that female-female bridging does not serve the same purpose as male-male bridging. Moreover, maternity and females' matrilineal relationships may influence how they bridge with one another.

This study describes and analyzes female-female bridging behavior in a group of habituated Tibetan macaques and compares female-female bridging behavior to previous studies on male-male bridging behavior. We predicted female-female bridging would differ from male-male bridging and that these differences might aid in development of further hypotheses to explore female-female bridging in this species.

9.2 Methods

We conducted observations from 01 August to 21 September 2014 at the Vallcy of Wild Monkeys tourist site near Mt. Huangshan, Anhui Province, China (30°29′N, 118°11′W). One group of Tibetan macaques known as Yulingkeng A1 (YA1) ranges in the forest area surrounding platforms built to facilitate tourism. YA1 monkeys are free-ranging but are habituated to human presence and are provisioned with corn several times a day by park guards. Researchers have tracked and recorded matrilineal kinship, births, immigrations, and deaths since 1985 (Wada and Xiong 1996). At the study's start, the group consisted of 41 individuals: 9 adult males, 8 adult females, 4 subadult males, 4 subadult females, 6 juvenile males, 7 juvenile females, 2 male infants, and 1 female infant; however, the female infant died on 15 August 2014 (Table 9.1). We classified females as adult if they had given birth to least one offspring and were ≥ 6 years old and as subadults if nulliparous and 4–5 years old. We classified infants as under 1 year of age, with the two male infants being roughly 2 months of age and the sole female infant being roughly 5 months of age at the start of the study. To establish a dominance hierarchy for use in this investigation, Lori K. Sheeran (LKS) collected dominance interaction behavioral data from 14 July to 27 August 2014 (36 days) from 7:00 to 12:00 and 14:00 to 17:00 daily. LKS utilized all-occurrence sampling to collect dominance data (Altmann 1974). Agonistic data

Table 9.1 Interaction matrix of bridging events by initiator and receiver

Bridge initiator	Bridge receiver												
ID	YH	YM	YCY	THY	YXX	HH	TH	TXX	TR	TRY	TT	YZ	Total
YH	–	0	0	0	2	1	11	1	0	0	0	1	16
YM	2	–	0	0	0	0	0	0	0	0	0	0	2
YCY	6	0	–	0	3	0	24	6	4	0	0	1	44
THY	1	0	0	–	1	0	1	0	0	0	0	1	4
YXX	20	0	2	1	–	1	1	4	0	0	0	0	29
HH	0	0	0	0	1	–	0	0	0	0	0	0	1
TH	9	0	1	0	0	0	–	0	0	0	0	0	10
TXX	6	0	3	0	0	0	2	–	0	0	0	0	11
TR	0	0	0	0	0	0	0	1	–	0	0	0	1
TRY	0	0	0	0	0	0	0	1	0	–	0	0	1
TT	0	0	0	0	0	0	0	0	0	0	–	0	0
YZ	0	0	0	0	0	0	0	0	0	0	0	–	0
Total	44	0	6	1	7	2	39	13	4	0	0	3	

Full names of individuals are displayed in Table 9.2

Table 9.2 Select results and demographics per individual

Female individual	Date of birth	Age group	Elo score[a]	Focal time (hour)	Bridge initiations/h	Bridge receptions/h
YeXiaXue (YXX)	2010–05	Subadult	1316	2.83	2.826	0.354
HuaHong (HH)	2003–??	Adult	1173	2.00	0.498	0
YeChunYu (YCY)	2009–03	Subadult	1172	3.13	2.874	0
YeHong (YH)	2003–??	Adult	1136	3.53	1.134	1.416
YeMai (YM)	1990–04	Adult	1096	2.23	0.45	0
TouRui (TR)	2004–??	Adult	1057	2.53	0	0.396
TouHong (TH)	2003–??	Adult	987	2.72	1.464	2.196
TouXiaXue (TXX)	2008–03	Adult	885	3.13	0.318	0.96
TouRongYu (TRY)	2009–03	Subadult	838	1.00	0	0
TouTai (TT)	1991–04	Adult	766	2.23	0	0
YeZhen (YZ)	1992–01	Adult	729	2.97	0	0.336
TouHuaYu (THY)	2009–04	Subadult	656	2.67	0.378	0
Mean		–	–	2.58	0.828	0.474
Median		–	–	2.70	0.414	0.168

[a]We used Elo-rating scores as a metric of dominance status. Larger Elo scores indicate higher dominance status

consisted of fear-grin, scream, flee, displace, threat, lunge, chase, grab, slap, and bite as defined by Berman et al. (2004). Winners of these interactions were defined by either the actor in directional agonism or the recipient of a submissive behavior. Losers of these interactions were defined by those receiving directed agonism or the actors of submission. To derive a ranking order and dominance score for each individual, we analyzed these data using Elo-rating procedures (Neumann et al. 2011) in R (R Core Team 2016).

We established interobserver agreement of 100% for all adult male and female, subadult female, and infant identities on 08 August 2014. We identified juveniles by the presence of their mother or their age and sex rather than individual identities due to the difficulty of learning individual juvenile identities within the limited timeframe of our study. We collected data from 8:00 to 12:00 and 13:00 to 18:00. To record bridging events, Grant J. Clifton (GJC) used 2-min focal samples (Altmann 1974) of adult and subadult females. We selected focal order randomly each day using a random number generator. Within focal samples, we recorded behaviors from an ethogram (Ogawa 1995a). Proximity refers touching or within arm's length (Sheeran et al. 2010). GJC also used all-occurrence sampling (Altmann 1974) to record bridges occurring outside of focal samples. If a bridge occurred or was suspected to occur between non-focal individuals, then the current 2-min focal

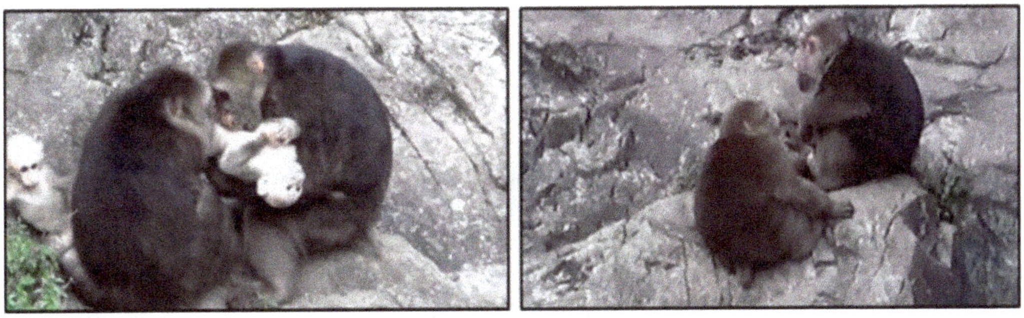

Fig. 9.1 Successful (left) and failed (right) bridge initiations (photo credit: Anne Salow)

was suspended, and the interacting dyad was observed for 2 min or until one individual left proximity. If a bridge occurred during a focal, we extended the focal for two additional minutes after the bridge to record context, but did not include the additional focal time in the focal data.

We used an ethogram modified from Ogawa (1995a) to record affiliative behaviors that often occurred before and/or during bridging. A bridge was defined as two individuals holding an infant or juvenile between one another with at least one individual teeth chattering or genital licking the infant. The bridge initiator was considered the individual who first teeth chattered or genital licked while lifting the infant or juvenile. A bridge began when both the initiator and receiver held up an infant. The bridge ended either when one of the individuals put down the infant or when no affiliative behaviors continued (other than grooming). Bridges that occurred within short succession of one another were considered separate events as long as at least one individual stopped holding the infant in between events.

We considered a bridge successful when the receiver held the infant or juvenile in the bridging position after receiving an initiation (Fig. 9.1). If a receiver teeth chattered or genital licked but turned away while the initiator was attempting a bridge, then it was considered failed (Fig. 9.1). Following Ogawa (1995a), we classified as Type I bridges in which the initiator brings an infant to a receiver or as Type II if the initiator approaches a receiver who is holding an infant (Fig. 9.2). Ogawa also observed bridges in which both males simultaneously approached and grabbed a single infant. In his study, he classified them as Type III bridges; we, on the other hand, did not observe Type III occur among females, so they were left out of our analysis.

We analyzed bridging behavior by initiator, receiver, and dyad. For individual and group comparisons, we used bridges recorded from focal animal sampling and converted bridges to rates to account for the uneven distribution of focal times. To calculate the rate of bridge initiations and bridge receptions per individual, we divided the number of bridge initiations and receptions from within the individual's focal samples by the total focal time for that individual. To calculate the rate of bridging in dyadic interactions, we calculated the total number of initiations by each individual and divided by the sum of the total focal follows for both individuals. Since focal and all-occurrence bridging dyads were significantly correlated (Kendall's tau; $\tau = 0.65$, $n = 22$, $p < 0.01$) and since the proportions of Type I

Fig. 9.2 Type I (left) and Type II (right) bridge initiations

and Type II bridges were not significantly different between focal and all-occurrence sampling (Fisher's exact test; $p > 0.05$, $df = 2$), we pooled both sample sets for categorical analyses of bridge initiation type. However, we only included bridges recorded via focal animal sampling in our analysis of success rates. Because of our limited size and aim (to describe and categorize female-female bridging), we mainly used descriptive and simple parametric statistics. We used Pearson's correlation coefficient to analyze the relationship between dominance status (Elo scores) and bridge initiation/reception rates, paired-sample t tests to examine differences in initiations rate by subordinate or dominant individual within each dyad, chi-square tests of independence, Fisher's exact, and binomial tests to investigate differences in the distribution of nonparametric frequencies of bridging event types.

9.3 Results

We recorded a total of 31 h of focal data and 119 bridging events (27 Type I; 92 Type II; 76 successful and 43 unsuccessful; Table 9.1). All four subadults and seven of eight adults were observed involved in at least one bridge (Table 9.1). Overall, we recorded 119 bridging events, 46 (38.7%) of which were recorded via focal sampling.

Of all observed bridging events, we recorded 27 (22.7%) that occurred via Type I initiation and 90 occurred via Type II initiation, of which 76 (63.9%) were successful bridges and 43 (36.1%) were failed bridges. In focal animal sampling, we recorded 7 (13.2%) via Type I initiations and 39 (84.8%) via Type II initiations, of which 21 (45.7%) were successful bridges and 25 (54.3%) were failed bridges. Table 9.2 displays all bridging initiations and receptions for each possible dyad and focal times and bridging rates for adult and subadult females.

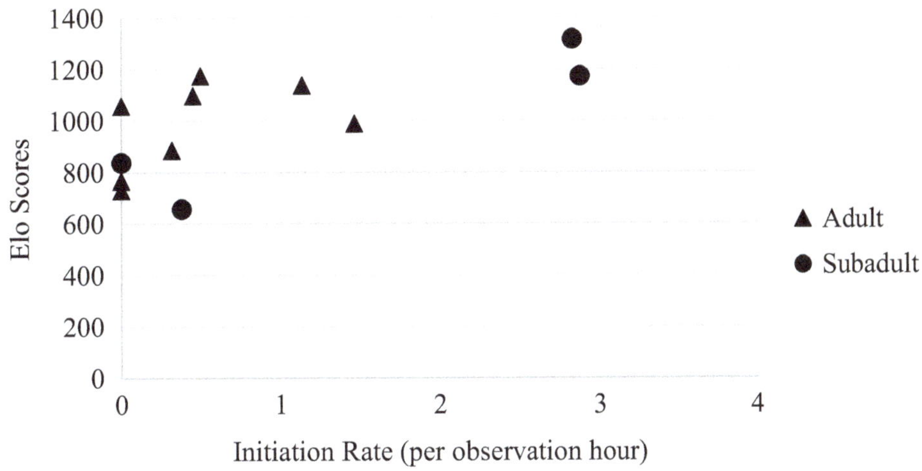

Fig. 9.3 Scatterplot of dominance status (Elo score) vs. bridge initiations (per hour of observation)

Female-female bridges were often subtle and did not involve audible vocalizations. Like male-male bridges, at least one female would lick the infant's genitals during bridges and sometimes touched the infant's genitals. Females often approached mothers who were holding infants and would peer at the infant before initiating the bridge by grabbing it. Successful bridges were never observed to be followed by aggression during our study, but receivers were occasionally observed committing aggression toward failed initiators.

The median bridge initiation rate for adults was 0.378 initiations per hour, whereas the median bridge initiation rate for subadults was 1.596 initiations per hour, but the sample size of the subadult group was too small to accurately test for significance. However, two subadult females, YCY and YXX, initiated far more often than any other adult or subadult (Table 9.1). The median bridge reception rate for adults was 0.366 receptions per hour, whereas no subadults were observed to be recipients of any bridging events. Like initiations, there were large differences in the number of bridges received by each individual within groups, with the two mothers of infants, YH and TH, receiving a large majority of bridges (Table 9.1). Individuals who initiated bridges were not more likely to have their bridges reciprocated by those with whom they initiated ($r_s = -0.45$, $n = 17$, $p > 0.05$).

We analyzed the relationship between dominance status and bridge initiations using a Pearson's rank correlation. Social rank was positively correlated with bridge initiations ($r(12) = 0.686$, $p = 0.014$; Fig. 9.3), but social rank and bridge receptions were not significantly correlated ($r(12) = 0.06$, $p = 0.85$; Fig. 9.4).

To test for significance differences between the mean rates of bridge initiations by either dominant or subordinate individuals within each dyad, we used a paired-samples t test. In contrast to what has been reported for males, within each dyad, subordinate individuals did not initiate bridges significantly more than that dominant individuals. We found no significant difference between the mean rate of bridge initiation between dominant or subordinate individuals ($t(65) = 1.398$, $p = 0.167$, Fig. 9.5). While this difference is not significant, descriptively, we found that

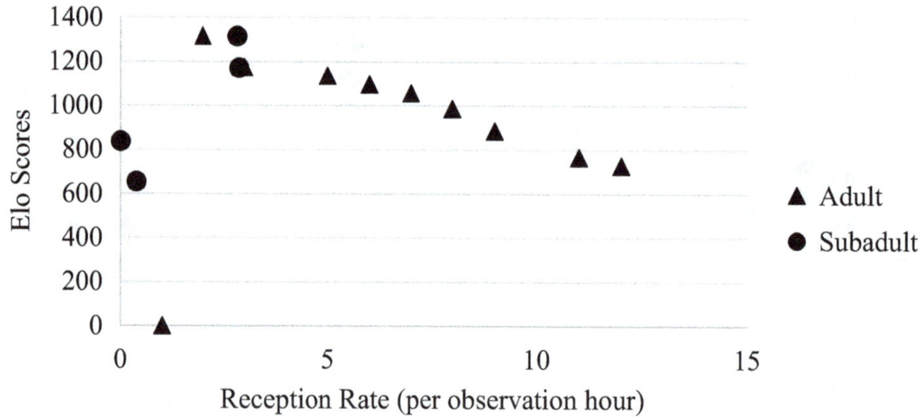

Fig. 9.4 Scatterplot of dominance status (Elo score) vs. bridge receptions (per hour of observation)

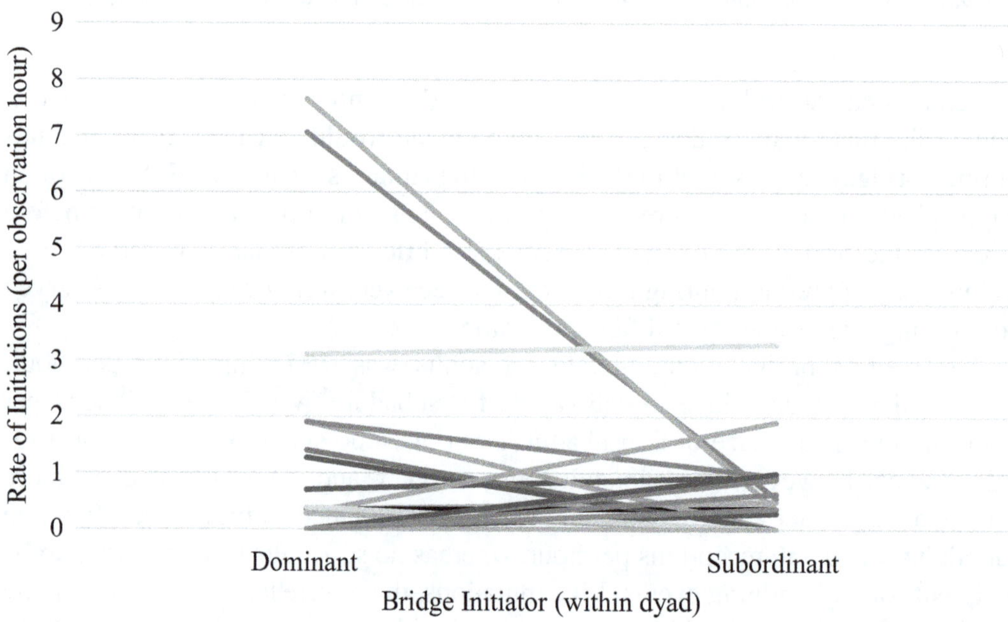

Fig. 9.5 Bridge initiation rates for the dominant vs. subordinate individual within each dyad. Each line represents a single dyad for a total of 22 dyads

dominant individuals were found to initiate bridges more (0.415 ± 1.35 initiations per observation hour) than subordinate individuals (0.195 ± 0.52 initiations per observation hour).

A majority of bridges occurred via Type II interactions. In bridges observed via focal sampling, all Type I bridges occurred by adults. Subadult females were observed initiating Type I interactions via all-occurrence sampling, but Type I interactions were still more likely to be initiated by an adult than a subadult (Fisher's exact test; $p < 0.05$, $df = 1$; Fig. 9.6). Of all but two Type II bridges, the receiver of the bridge was the mother of the infant or juvenile being used, and in the two exceptions, the same subadult female, YXX, was the receiver. Similarly, in all but two Type I interactions initiated by adults, the initiator was holding her own

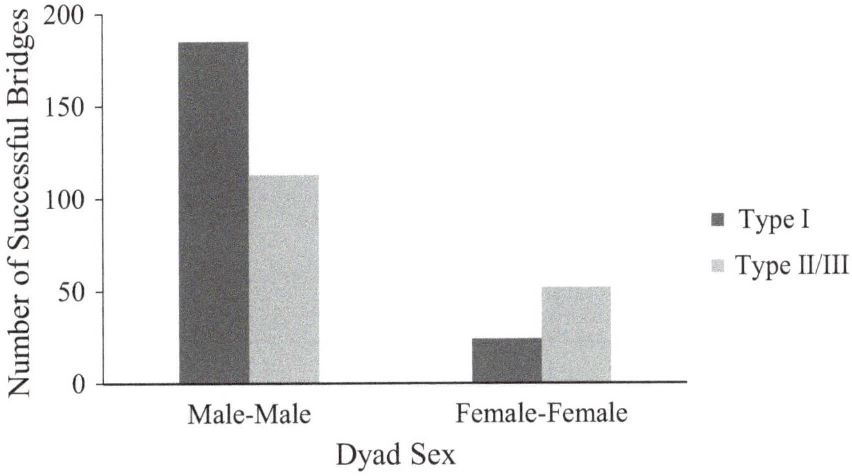

Fig. 9.6 Bar graph of successful Type I and Type II/III initiations between male and male

offspring. These two exceptions were initiated by the same individual, TXX, using TH's infant in both cases.

In Ogawa's (1995a) study of male-male bridging, 185 (62.1%) bridges occurred via Type I initiations, 70 (23.5%) occurred with Type II initiations, and 43 (14.4%) occurred via Type III interactions. We did not observe female-female bridging via Type III initiations, so to compare males and females, we pooled Type II and Type III male bridges together for a total of 185 (62.1%) Type I and 113 (37.9%) Type II/III interactions. Since Ogawa (1995a) did not mention the occurrence of failed male-male bridge initiations, we only used successful female-female bridge initiations to compare differences between male-male and female-female bridges to ensure validity between our dataset and Ogawa's. Of successful female-female bridges, 24 (31.6%) occurred via Type I initiations, and 52 (68.4%) occurred via Type II initiations. Even after pooling the male Type II and Type III initiations together, which creates a more conservative estimate than if Type III initiations were excluded, females were still significantly more likely than males to use Type II bridge initiations over Type I initiations (chi square test for independence; $\chi^2 = 28.85$, $df = 1$, $p < 0.01$; Fig. 9.6).

Bridges initiated by adults were significantly more likely to be successful than bridges initiated by subadults ($\varphi = 0.46$, $df = 1$, $p < 0.01$). In focal samples, all Type I bridges were successful, but we recorded two failed Type I bridge initiations via all-occurrence sampling. However, these two failed initiations were from a subadult. Receivers showed no significant preference in accepting bridges that utilized either infants or juveniles, and we found no significant difference in success rates of bridges that were preceded by grooming vs. those that were not preceded by grooming, although successful bridges were significantly more likely than unsuccessful bridges to be followed by grooming ($\varphi = 0.45$, $df = 1$, $p < 0.01$).

The two surviving infants in the group were used as a bridge in 88 out of 103 bridging cases, which is significantly more often than expected based on the proportions of infants and juveniles in the group (exact binomial test, $df = 1$,

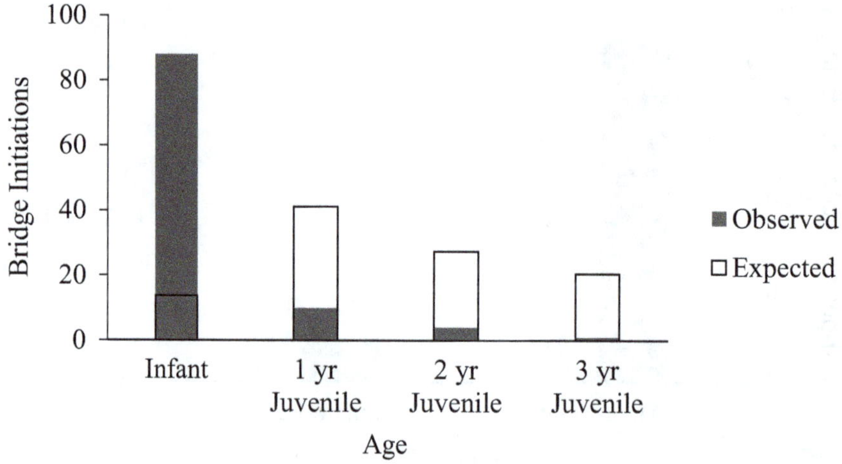

Fig. 9.7 Observed vs. expected use of infants and juveniles in bridge initiations after the death of TR's infant ($n = 103$); expected values are calculated based on the proportion of individuals in each age group

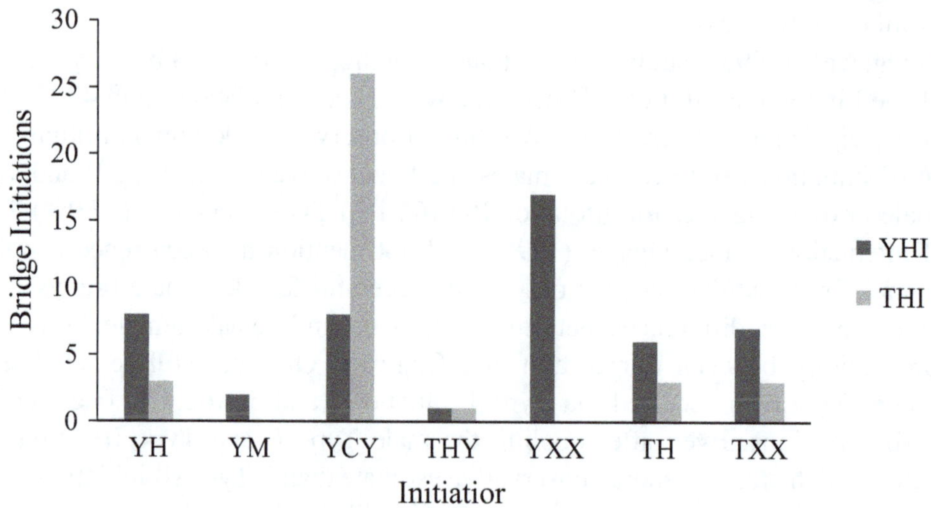

Fig. 9.8 Bar graph showing the number of bridge initiations using TH's infant (THI) and YH's infant (YHI) for each individual observed bridging with an infant at least once

$p < 0.01$, Fig. 9.7). Of individuals who used infants to bridge, some individuals demonstrated a clear preference toward a specific infant over the other (Fisher's exact test; $df = 6$, $p < 0.01$, Fig. 9.8), but our small sample size did not allow us to test the significance of the pairwise comparisons.

9.4 Discussion

The agonistic buffering hypothesis was not supported among female-female bridges within our dataset for this group of Tibetan macaques. In contrast to what has been reported in males (Ogawa 1995a; Zhao 1996; Bauer et al. 2013), social rank was positively correlated with bridge initiations. Initiators were also not more likely to be subordinate to those who received their bridge attempts, which is evidence in contrast to the idea that subordinate females initiate bridges with dominant individuals to reduce the probability of future aggression (Ogawa 1995a). Further departing from the trends seen in males, females were not more likely to initiate bridges with those from which they received bridges. These data suggest the females in this group do not bridge in order to avoid conflicts, as has been observed for males.

Although there were no significant differences in bridging initiations between adult and subadult groups in this population, two of the subadult individuals, YCY and YXX, initiated far more bridges than any other individual. In contrast, the other two subadults, THY and TRY, rarely bridged. Interestingly, YCY and YXX were both much larger and more physically mature than either THY or TRY, who were both more juvenile-like in appearance and behavior. The mother that bridged the most out of the mothers without infant offspring, TXX, was relatively young for a mother and would normally be considered a subadult if not for the fact that she previously gave birth. This individual was also the only mother to initiate Type I bridges utilizing another mother's offspring and to initiate a Type II bridge toward a receiver holding her offspring. These bridges occurred toward subadult YCY.

As a group, adults received most of the recorded bridges, though statistically there was no difference between the rate at which adults and subadults were recipients. In particular, the infant mothers TH and YH each received far more bridges than any other individual. Out of bridges that involved adults, all except for two involved the mother of the infant. In addition, female bridges were more likely to occur by Type II initiation than by Type I. Ogawa (1995a) reported far more Type I initiations between males than Type II initiations. The comparison of female-female bridges and male-male bridges from Ogawa (1995a) demonstrates that this difference in Type I vs. Type II initiations between intra-sexed bridging events is significant, meaning that females are more likely to approach and initiate a bridge with a receiver who is holding an infant, whereas males are more likely to initiate while holding the infant. Of female-female Type II bridges, a large majority involved the mother as the receiver, and all of those that involved the mother as the receiver used her infant. Type I bridges were also more likely to be successful than Type II bridges, which may suggest an importance of the infant in the bridge rather than the relationship between the other individuals. In addition, individuals were not more likely to reciprocate bridges, which further suggests that the relationship between the older members of the bridging dyad is not as important as the infant. These trends support motherhood and interest in infants as factors in how female-female bridging is both initiated and received.

Female-female bridging in our dataset did show some consistencies with male-male bridging, however. Similar to males, successful bridges never occurred in aggressive contexts and were not followed by aggression. Females also showed a bias toward initiating bridges with their own infants over other juveniles. Although initiators preferred to use infants, receivers did not appear to show a preference, as bridges initiated with juveniles were equally likely to be successful, though we should caution that the few observed juvenile bridges limits the power of our analysis. With that being said, it is nonetheless consistent with bridging trends between males observed in this same study group (Bauer et al. 2013). Moreover, among individuals that often bridged with infants, most showed a preference for one infant over the other. Interestingly, mothers of multiple offspring only initiated and received bridges utilizing their youngest offspring, further supporting the hypothesis that initiators are biased toward using younger individuals.

Unfortunately, the death of TR's infant, the only female infant, prevented us from investigating whether there are sex biases in the infants used in female-female bridges. YCY initiated four bridges using TR's infant, demonstrating that female-female bridges do occur with female infants, but we did not observe enough occurrences during this time for any sex biases to be tested. Of juvenile 1-year-olds, only females were used; however, there were five female 1-year-old juveniles and only one male 1-year-old juvenile. Moreover, TH was the mother of the single male 1-year-old juvenile, and she had a younger infant that she most often used in bridges. Without a more equally distributed sex ratio of infants and juveniles, it is impossible to state with these preliminary data whether or not females prefer to bridge with infants of a specific sex.

The preliminary results we present here can be used to generate future hypotheses regarding the function of bridging in females. In our dataset, the few individuals in the group make cross-comparisons between age classes difficult and limit a majority of the statistical analysis to less powerful, nonparametric, and descriptive procedures. Our study spanned a short time period, which limited the amount of focal data for each individual. This makes it unclear as to whether certain individuals never bridge or just do so less frequently than others. However, our study was conducted on the same group under similar conditions as past studies on bridging in male Tibetan macaques (Ogawa 1995a, b, c; Bauer et al. 2013; see also Zhao 1996 studying this species at Mt. Emei), which enabled us to directly compare male-male and female-female bridging behavior in Tibetan macaques within this particular group.

Despite the limitations, this study is the first to investigate bridging behavior between females in any species of macaque. Our results suggest that female bridging behavior may not follow the same pattern as male bridging and indicate that this behavior is related to female interest in infants than it is to the agonistic buffering hypothesis. Younger, higher-ranking female individuals may bridge more often because it allows them to gain alloparental experience from interacting with infants. Alternatively, female bridging may be a by-product of interest in mothers and infants, similar to what has been supported in female-infant interactions in bonnet macaques, *M. radiata* (Silk 1999). Future studies should follow the format of Ogawa

(1995a) by observing bridges in relation to all female-female and female-infant interactions over the course of the entire year to enable a more definitive test of the agonistic buffering hypothesis.

Acknowledgments We thank Dao Zhang and the Chen family for logistical support while we were in China. We thank the members of the Huangshan Scenic Bureau for allowing us to conduct research at the site. We thank Anne Salow, Gregory Fratellone, and Sofi Bernstein for help with the data collection in the field and Dr. Dominic Klyve for assistance with statistical analysis. Grant J. Clifton particularly thanks Dr. Audrey Huerta and Dr. Linda Raubeson of CWU's Science Honors Research Program for their support. This research was funded by CWU's Science Honors Program and the National Science Foundation (OISE-1065589). Our research was reviewed and approved CWU's Institutional Animal Care and Use Committee (A011401).

References

Altmann J (1974) Observational study of behavior: sampling methods. Behaviour 49:277–266

Bauer B, Sheeran LK, Matheson MD, Li J, Wagner RS (2013) Male Tibetan macaques' (*Macaca thibetana*) choice of infant bridging partners. Zool Res 35(3):222–230

Berman CM, Ionica CS, Li JH (2004) Dominance style among *Macaca thibetana* on Mt. Huangshan, China. Int J Primatol 25(6):1283–1312

Deag JM (1980) Interactions between males and unweaned Barbary macaques: testing the agonistic buffering hypothesis. Behaviour 75:54–81

Deag JM, Crook JH (1971) Social behavior and "agonistic buffering" in the wild Barbary macaque *Macaca sylvana* L. Folia Primatol 15:183–200

Estrada A, Estrada R (1984) Female-infant interactions among free-ranging stumptail macaques (*Macaca arctoides*). Primates 25(1):48–61

Estrada A, Sandoval JM (1977) Social relations in a free-ranging troop of stumptail macaques (*Macaca arctoides*): male-care behavior I. Primates 18(4):793–813

Kubenova B, Konecna M, Majolo B, Smilauer P, Ostner J, Schulke O (2017) Triadic awareness predicts partner choice in male–infant–male interactions in Barbary macaques. Anim Cogn 20:221–232. https://doi.org/10.1007/s10071-016-1041-y

Neumann C, Duboscq J, Dubuc C, Ginting A, Irwan AM, Agil M, Widdig A, Engelhardt A (2011) Assessing dominance hierarchies: validation and advantages of progressive evaluation with Elo-rating. Anim Behav 82(4):911–921. https://doi.org/10.1016/j.anbehav.2011.07.016

Ogawa H (1995a) Bridging behavior and other affiliative interactions among male Tibetan macaques. Int J Primatol 16(5):707–729

Ogawa H (1995b) Recognition of social relationships in bridging behavior among Tibetan macaques (*Macaca thibetana*). Am J Primatol 35:305–310

Ogawa H (1995c) Triadic male-female-infant relationships and bridging behavior among Tibetan macaques (*Macaca thibetana*). Folia Primatol 64:153–157

Ogawa H (2006) Wiley monkeys: social intelligence of Tibetan macaques. Kyoto University Press, Kyoto

Paul A, Kuester J, Arnemann J (1996) The sociobiology of male-infant interactions in Barbary macaques, *Macaca sylvanus*. Anim Behav 51:155–170

R Core Team (2016) R: a language and environment for statistical computing. R Foundation for Statistical Computing, Vienna

Sheeran LK, Matheson MD, Li JH, Wagner RS (2010) A preliminary analysis of aging and potential social partners in Tibetan macaques (*Macaca thibetana*). In: Banak SD (ed) Collected readings in biological anthropology in honor of Professor LS Penrose and Dr. Sahrah B. Holt. Unas Letras Industria Editorial, Merida, pp 349–358

Silk J (1999) Why are infants so attractive to others? The form and function of infant handling in bonnet macaques. Anim Behav 57:1021–1032

Taub DM (1984) Male caretaking behavior among wild Barbary macaques (*Macaca sylvanus*). In: Taub DM (ed) Primate paternalism. Van Nostrand Reinhold, New York, pp 20–55

Wada K, Xiong C (1996) Population changes of Tibetan monkeys with special regard to birth interval. In: Shotake T, Wada K (eds) Variations in the Asian macaques. Tokai University Press, Tokyo, pp 133–145

Zhao QK (1996) Male-infant-male interactions in Tibetan macaques. Primates 37(2):135–143

Chapter 10
Bridging Behavior and Male-Infant Interactions in *Macaca thibetana* and *M. assamensis*: Insight into the Evolution of Social Behavior in the *sinica* Species-Group of Macaques

Hideshi Ogawa

10.1 Introduction

Species in the genus *Macaca* have many common features: forming multi-male multi-female social groups with female philopatry, male dispersal, and a linear dominance hierarchy (Thierry et al. 2004). However, there are variations in their "dominance style" and the degree of kin-biased social relationships between females (Berman and Thierry 2010; de Waal and Luttrell 1989). Recently, the variations in social relationships between females have been systematically studied (Balasubramaniam et al. 2012). In addition to social relationships between females, there are many inter-species differences in social relationships between males. Compared to females, however, there have been fewer comparative studies on social interactions between males. For example, frequent triadic male-infant interactions were reported in some macaque species, but not reported in most of other macaque species (Deag and Crook 1971; Kalbitz et al. 2017; Ogawa 1995, 2006). Therefore, in this paper, I focus on social interactions between adult males and infant handling by adult males in macaques.

Macaques are divided into several species-groups. Though there are several categorizations, one is the *sinica* species-group comprised of toque macaques (*Macaca sinica*) in Sri Lanka, Bonnet macaques (*M. radiata*) in India, Assamese macaques (*M. assamensis*) from India to Southeast Asia, and Tibetan macaques (*M. thibetana*) in China. Delson (1980) included stump-tailed macaques (*M. arctoides*), which are distributed in India eastward to Southeast Asia, in the *sinica* species-group. Arunachal macaques (*M. munzala*) in the Arunachal area of India and white-cheeked macaques (*M. leucogenys*) in Tibet are recently listed as new species in the *sinica* species-group (Li et al. 2015; Sinha et al. 2005).

H. Ogawa (✉)
School of International Liberal Studies, Chukyo University, Toyota, Aichi, Japan
e-mail: hogawa@lets.chukyo-u.ac.jp

J.-H. Li et al. (eds.), *The Behavioral Ecology of the Tibetan Macaque*, Fascinating Life Sciences, https://doi.org/10.1007/978-3-030-27920-2_10

Assamese macaques have been traditionally divided into two subspecies, based on morphological traits (Fooden 1982). Eastern Assamese macaques (*M. assamensis assamensis*), which inhabit areas east of the Brahmaputra River, have shorter tails than western Assamese macaques (*M. assamensis pelops*), which inhabit areas west of the Brahmaputra River. Though Assamese macaques are widely distributed in Asia, direct observations of wild Assamese macaques are difficult, because they are patchy distributed in mountainous areas (Chalise et al. 2013; Wada 2005). There are only a few direct observations of wild Assamese macaques (Kalbitz et al. 2017). However, free-ranging Assamese macaques in several social groups are provisioned and habituated to humans at several locations. I observed free-ranging provisioned Assamese macaques in Thailand, India, and Nepal and compared these populations' social behaviors to those of Tibetan macaques in China.

As may occur with regard to genetic and morphological traits, inter-species and inter-subspecies differences in social behaviors could be affected by evolutionary divergence among populations. My aim in this study is to reconstruct the evolutionary processes of macaques' social behaviors by comparing the male-male and male-infant interactions of Tibetan, eastern Assamese, and western Assamese macaques at various sites.

10.2 Methods

10.2.1 Study Sites and Study Periods

Tibetan macaques are distributed in China, and Assamese macaques are distributed in Nepal, India, Bhutan, Bangladesh, China, Myanmar, Laos, Thailand, and Vietnam (Groves 2001; Fooden 1982; Wada 2005). Among these countries, I conducted field researches at several sites in Nepal, India, China, and Thailand (Fig. 10.1).

10.2.1.1 Western Assamese Macaque (*M. a. pelops*)

Site 1 Ramdi Village (27°54′N, 83°38′E, 437 m). I stayed at Ramdi Village, Pyuthan District, Nepal, from 20 to 23 March 2016 with Pavan Paudel. There were two provisioned groups of Assamese macaques around the village.

Site 2 Nagarjun (27°44′N, 85°17′E, 1350–2093 m). Shivapuri-Nagarjun National Park, Kathmandu District, Nepal, has two separate areas, the Shivapuri area and the Nagarjun area. I stayed at Nagarjun for a total of 113 days between 2011 and 2017. Wild Assamese macaques in several social groups inhabit Nagarjun, as well as wild rhesus macaques (*M. mulatta*) (Chalise et al. 2013; Wada 2005). There is a royal palace and an active army camp in the national park, and soldiers live in the army camp. The social group of Assamese macaque was unintentionally provisioned by food waste of the army camp that was available for monkeys. Though I did not

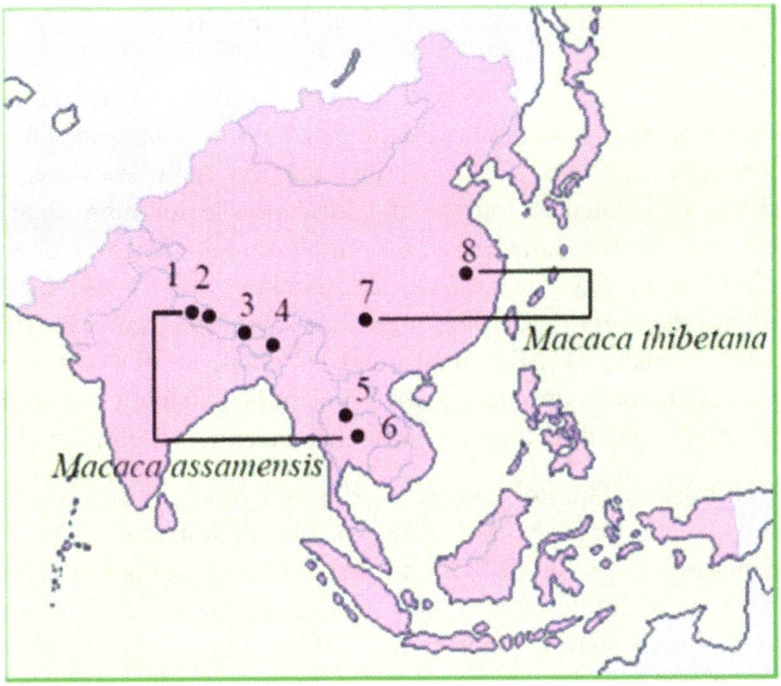

Fig. 10.1 Location of the study sites. 1, Ramdi Village (27°54′N, 83°38′E); 2, Nagarjun (27°44′N, 85°17′E); 3, Siliguri (26°54′N, 88°28′E); 4, Tukeswari Temple (26°03′N, 90°38′E); 5, Wat Tham Pla (20°19′N, 99°51′); 6, Wat Tham Pha Tha Pol (16°31′N, 100°40′E); 7, Mt. Emei (29°33′N, 103°20′E); 8, Huangshan (30°29′N, 118°11′E)

recommend it, when soldiers finished lunch and dinner, they usually disposed of their leftovers at a dumping site in the army camp. Monkeys consumed the garbage and became habituated to humans (Koirala et al. 2017). I identified all adult males and females in the AA group, based on their physical characteristics, and observed those monkeys with Sabina Koirala. The AA group was composed of 56 monkeys including 7 adult males, 13 adult females, 4 male infants (<1 year old), and 1 female infant in January 2015.

Site 3 Siliguri (26°54′N, 88°28′E, 199 m). I visited Coronation Bridge at Siliguri, West Bengal State, India, on 2 and 5 March 2009. Assamese macaques were provisioned and stayed along the road near Coronation Bridge.

10.2.1.2 Assamese macaque (*M. a. assamensis* or *M. a. pelops*)

Site 4 Tukreswari Temple (26°03′N, 90°38′E, 47 m). I stayed near Tukreswari (Tukeswari) Temple, Assam State, India, from 22 to 25 February 2009 with Mayur Bawri. Assamese macaques in two social groups were provisioned and habituated to humans on a hill surrounded by cultivated fields (Cooper and Bernstein 2008).

10.2.1.3 Eastern Assamese Macaque (*M. a. assamensis*)

Site 5 Wat Tham Pla (20°19′N, 99°51′E, 843 m). I stayed at Wat Tham Pla (Tham Pla Temple), Pong Ngam, Mae Sai, Chiang Rai Province, Thailand, for a total of 110 days between 2008 and 2012. At this temple, Assamese macaques were provisioned and habituated to tourists and local people for more than 20 years. The monkeys slept on steep hills behind the temple at night, and they often stayed in the temple area in the daytime, because the monkeys were given foods such as bananas and peanuts every day. In this area, there were 193 monkeys with 26 adult males, 54 adult females, 24 male infants, and 16 female infants in 4 social groups (the A, B, C, and D groups) in January 2012. I identified all adult males and females in the social groups, based on their physical characteristics (Ogawa et al. 2009).

Site 6 Wat Tham Pha Tha Pol (16°31′N, 100°40′E, 54 m). I stayed at Wat Tham Pha Tha Pol (Tham Pha Tha Pol Temple), non-hunting area, Amphoe Noen Maprang, Phitsanulok Province, Thailand, from 9 to 11 January 2009 with Eishi Hirasaki. This non-hunting area is surrounded by steep hills, but Assamese macaques were provisioned by tourists and local people at the temple outside of the non-hunting area.

10.2.1.4 Tibetan Macaque (*M. thibetana*)

Site 7 Mt. Emei (29°33′N, 103°20′E, 1260–2100 m). I visited Mt. Emei (Emeishan), Sichuan Province, China, from 28 May to 1 June 1990 with Ming Li. Tibetan macaques were provisioned and habituated to tourists in this area (Zhao and Deng 1988). We made a round trip along the walking routes and observed monkeys in six social groups.

Site 8 Huangshan (30°29′N, 118°11′E, 700–800 m). I stayed at Huangshan (Mt. Huang), Anhui Province, China, for a total of 382 days between 1989 and 1993. Wild Tibetan macaques in several social groups inhabited this area, as well as wild rhesus macaques (Wada et al. 1987). Since 1986, one social group, the Yulingkeng Group, has been provisioned for observations. Unlike now, no tourists visited there. The Yulingkeng Group had 42 monkeys with 7 adult males, 9 adult females, 3 male infants, and 3 female infants in November 1992. I identified all monkeys, based on their physical characteristics, and observed them sometimes with Kazuo Wada, Chenpei Xiong, Jinhua Li, Ming Li, and Qishan Wang (Ogawa 2006).

10.2.2 Sampling Methods

I used all occurrence behavior sampling in all of the study sites. In addition, I used focal animal sampling on adult males and adult females for 10 h each in both the mating and birth seasons, at Nagarjun in Nepal, Wat Tham Pla in Thailand, and Huangshan in China. For the focal sampling, I observed Assamese macaques at Nagarjun in Shivapuri-Nagarjun National Park from 30 July to 11 September 2014 in the birth season. S. Koirala and I observed them from 13 November 2014 to 15 January 2015 in the mating season and combined the data. I observed Assamese macaques at Wat Tham Pla in Thailand from 25 July to 7 September 2008 in the birth season and from 23 December 2010 to 8 January 2011 and from 31 December 2011 to 13 January 2012 in the mating season. I observed Tibetan macaques at Huangshan from 16 February to 26 April 1992 in the birth season and from 16 September to 4 November 1992 in the mating season.

The numbers of focal animals were 7 males and 7 females in the birth season and 7 males and 7 females in the mating season at Nagarjun; 10 males (5 in the A Group, 2 in the C Group, and 3 in the D Group) and 5 females (2 in the A Group, 1 in the C Group, and 2 in the D Group) in the birth season and 10 males (4 in the A Group, 2 in the C Group, and 4 in the D Group) and 5 females (3 in the A Group, 1 in the C Group, and 1 in the D Group) in the mating season at Wat Tham Pla; and 5 males and 9 females in the birth season and 6 males and 9 females in the mating season at Huangshan.

I calculated the frequency of behaviors, based on focal animal sampling. I combined the data by all occurrence behavior sampling and focal animal sampling to examine the presence/absence of behaviors and the rate of behaviors in which male and female infants were used in each study site.

10.2.3 Definition of Behavior

Bridging behavior is defined as "two individuals simultaneously lift up an infant" (Ogawa 1995, 2006, p. 52). If the two individuals are male, it is also called a triadic male-infant interaction. The typical interaction has the following sequence: (1) One adult holds a male or female infant ventrally. (2) The adult monkey carries the infant to another individual and presents it to the recipient. Another individual sometimes approaches the monkey who is holding the infant. (3) The two adults sat facing each other, one adult pulls up the infant's shoulder, the other pulls up its hip, the infant lays on its back, and the two adults lift up the infant together. As a result, the infant forms a bridge between them. While lifting up the infant, one or both adults often suck and/or touch the infant's genitalia with the expression of teeth chattering. Infants are handled gently and rarely show resistance or give signs of distress. After bridging, the two adults frequently stay in close proximity and groom each other, so this behavior may reduce social tension between the individuals and promote and maintain an affiliative relationship between males or between males and females.

Table 10.1 Presence/absence of social behavior in each site

	Assamese macaque						Tibetan macaque	
Species	Western Assamese macaque			?	Eastern Assamese macaque			
Country	Nepal		India		Thailand		China	
Site	Ramdi	Nagarjun	Siliguri	Tukreswari	Tham Pla	Tham Pha Tha Pol	Emei	Huangsham
Presence of behavior								
Bridging between males	No	No	No#	Yes	Yes	Yes	Yes	Yes
Sucking of an infant genitalia by males	No	No	No#	No#	Yes	Yes	Yes	Yes
Penis sucking between adult males	No	No	No#	No#	No	No#	Yes	Yes

Yes: the behavior was recorded. No: the behavior was not recorded. No#: Behavior was not recorded, possibly due to the short observational period at the study site. The study sites are arranged from west (left) to east (right). ?: It was not sure whether macaques here were Western Assamese macaques or Eastern Assamese macaques

10.3 Results

10.3.1 Bridging Behavior

Bridging behavior was recorded in Tibetan macaques at Huangshan and Mt. Emei in China and in Assamese macaques at Wat Tham Pha Tha Pol and Wat Tham Pla in Thailand and Tukreswari Temple in India (Table 10.1). However, no bridging behavior was recorded among any adults in Assamese macaques at Siliguri in India, Nagarjun in Nepal, and Ramdi Village in Nepal (Table 10.1). Although male Assamese macaques in these sites sometimes held an infant in front of another male, they never lifted the infant together with another male.

Figure 10.2 shows the frequency of bridging behavior at Nagarjun in Nepal, Wat Tham Pla in Thailand, and Huangshan in China, in which I observed 5–10 adult males and 5–9 adult females for 10 h each in the birth and mating seasons, respectively. Adult males of Tibetan macaque performed bridging behavior at the

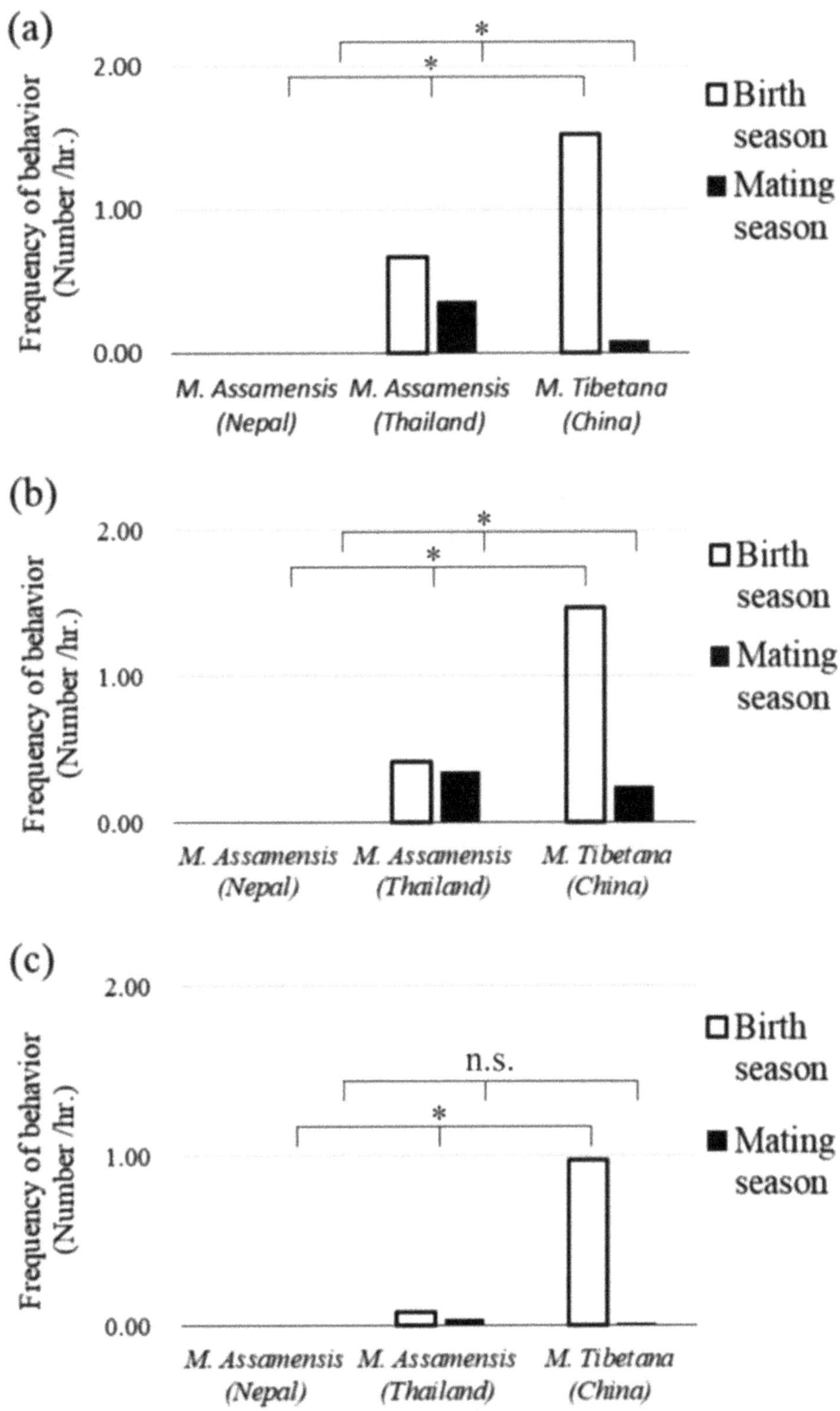

Fig. 10.2 Frequency of bridging behavior in the three study sites. Nepal: at Nagarjun, Shivapuri-Nagarjun National Park, Kathmandu District; Thailand: at Wat Tham Pla, Chiang Rai Province; China: at Huangshan, Anhui Province. $^{*}p < 0.05$. *n.s.* not significant. (**a**) Bridging between males, (**b**) bridging between females, (**c**) bridging between a male and a female

frequency of 1.52 (average number of behaviors of each individual per hour) in the birth season and 0.08 in the mating season. Adult males of eastern Assamese macaque in Thailand performed bridging behavior at the frequency of 0.67 in the birth season and 0.36 in the mating season, whereas adult males of western Assamese macaques in Nepal did not perform bridging behavior at all. The frequencies of bridging behavior between adult males were significantly higher in Tibetan and eastern Assamese macaques than that in western Assamese macaques who did not perform bridging behavior (Kruskal-Wallis test, $n_1 = 5$, $n_2 = 10$, $n_3 = 7$, $H = 12.964$, $p < 0.05$ in the birth season; $n_1 = 6$, $n_2 = 5$, $n_3 = 7$, $H = 13.112$, $p < 0.05$ in the mating season). Tibetan and eastern Assamese macaques used a male infant in bridging more frequently than expected, if the number of male and female infants in the study group were considered (chi-squared test, $df = 1$, $\chi^2 = 70.2$, $p < 0.05$ for Tibetan macaques; $df = 1$, $\chi^2 = 19.5$, $p < 0.05$ for eastern Assamese macaques).

Adult females of Tibetan macaque performed bridging behavior at the frequency of 1.47 in the birth season and 0.41 in the mating season. Adult females of eastern Assamese macaque performed bridging behavior at the frequency of 0.24 in the birth season and 0.34 in the mating season, whereas adult females of western Assamese macaque did not perform bridging behavior at all. The frequencies of bridging behavior between adult females were also significantly higher in Tibetan and eastern Assamese macaques than that in western Assamese macaques who did not perform bridging behavior (Kruskal-Wallis test, $n_1 = 9$, $n_2 = 10$, $n_3 = 7$, $H = 12.341$, $p < 0.05$ in the birth season; $n_1 = 9$, $n_2 = 5$, $n_3 = 7$, $H = 10.189$, $p < 0.05$ in the mating season).

Bridging behavior between adult males and adult females occurred in Tibetan macaques at the frequency of 0.98 in the birth season and 0.01 in the mating season and in eastern Assamese macaques at the frequency of 0.09 in the birth season and 0.04 in the mating season, whereas western Assamese macaques did not perform bridging behavior between males and females. The frequencies of bridging behavior were significantly higher in Tibetan and eastern Assamese macaques in the birth season than that in western Assamese macaques who did not perform bridging behavior, though it was not significant in the mating season (Kruskal-Wallis test, $n_1 = 14$, $n_2 = 20$, $n_3 = 14$, $H = 12.741$, $p < 0.05$ in the birth season; $n_1 = 15$, $n_2 = 10$, $n_3 = 14$, $H = 2.785$, n.s. in the mating season).

10.3.2 Dyadic Male-Infant Interactions

Adult males of Tibetan and Assamese macaques groomed and held an infant in their social group. Figure 10.3 shows the frequency of social grooming and holding an infant by adult males at Huangshan in China, Wat Tham Pla in Thailand, and Nagarjun in Nepal. There was no significant difference in the frequency of social grooming among the three sites (Kruskal-Wallis test, $n_1 = 5$, $n_2 = 10$, $n_3 = 7$, $H = 0.863$, n.s. in the birth season; $n_1 = 6$, $n_2 = 5$, $n_3 = 7$, $H = 1.380$, n.s. in the

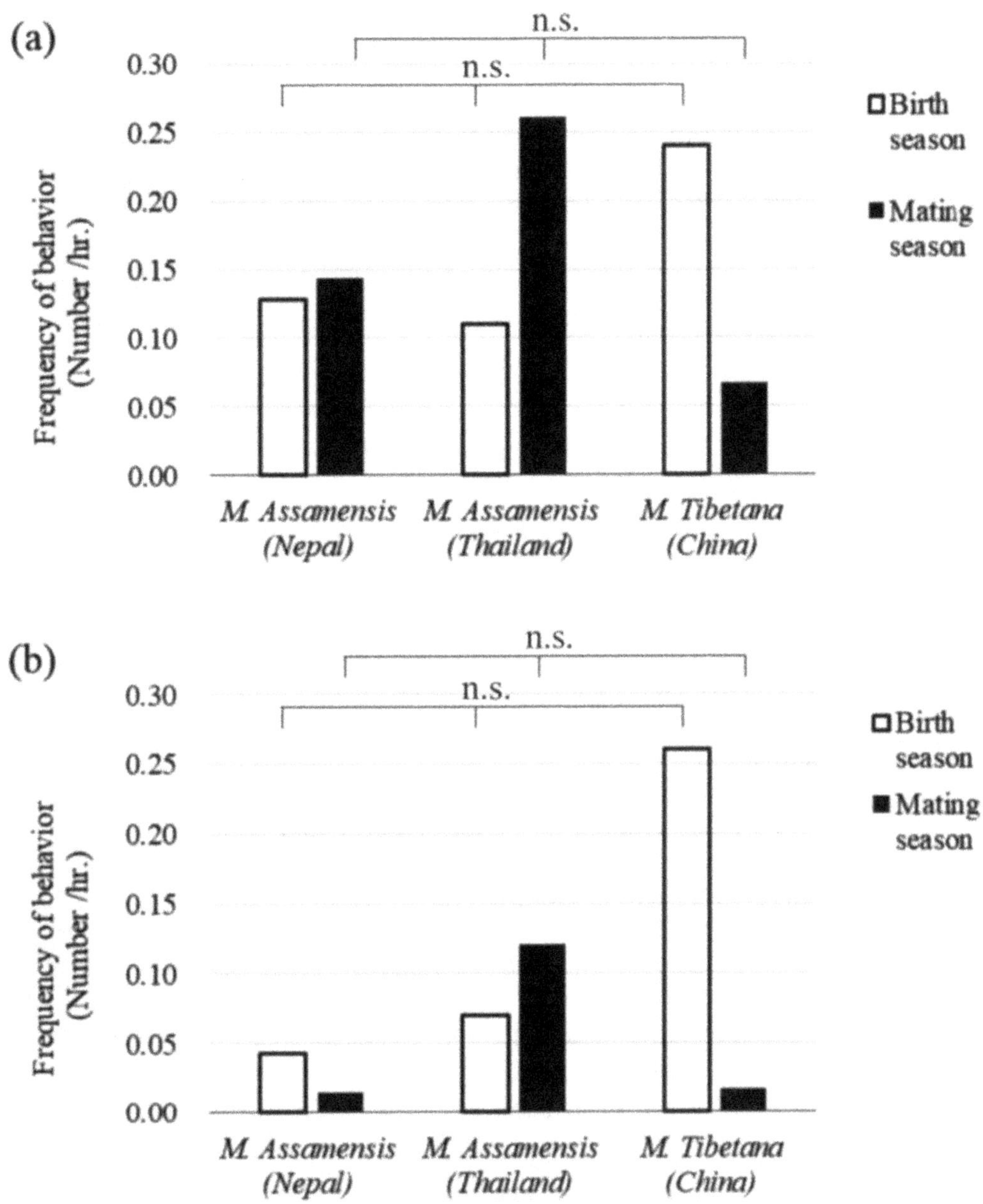

Fig. 10.3 Frequency of male-infant interactions in the three study sites. Nepal: at Nagarjun, Shivapuri-Nagarjun National Park, Kathmandu District; Thailand: at Wat Tham Pla, Chiang Rai Province; China: at Huangshan, Anhui Province. *n.s.* not significant. (**a**) Grooming an infant by adult males, (**b**) holding an infant by adult males

mating season). There was no significant difference in the frequency of holding among the three sites (Kruskal-Wallis test, $n_1 = 5$, $n_2 = 10$, $n_3 = 7$, $H = 1.022$, n.s. in the birth season; $n_1 = 6$, $n_2 = 5$, $n_3 = 7$, $H = 1.504$, n.s. in the mating season).

Adult males of Tibetan and Assamese macaques touched the genitalia of an infant, though this behavior was not confirmed at the sites in which the observational periods were short.

In addition to genital manipulation, Tibetan macaques in China and Assamese macaques at Wat Tham Pha Tha Pol and Wat Tham Pla in Thailand and Tukreswari Temple in India sucked an infant's genitalia during bridging behavior and during dyadic interactions between adult males and infants (Table 10.1). However, Assamese macaques at Siliguri in India, Nagarjun in Nepal, and Ramdi in Nepal did not suck the infant's genitalia (Table 10.1). During penis sucking behavior by an adult male on an infant, the adult male usually lifted up the infant and sucked his penis, sometimes turning the infant upside down to do so. Male infants sometimes jumped onto an adult male's face, in order to have his penis sucked.

10.3.3 Penis Sucking Between Adult Males

Only Tibetan macaques performed penis sucking behavior between adult males (Table 10.1). Assamese macaques in any study sites never performed penis sucking behavior between adult males (Table 10.1).

10.4 Discussion

The current comparative study on social behavior revealed that Tibetan macaques and eastern Assamese macaques in Thailand performed bridging behavior between males, between females, and between males and females. Males of these macaques also sucked the genitalia of an infant during bridging behavior and dyadic interactions with an infant. On the contrary, western Assamese macaques in Nepal did not perform either bridging or genital sucking behavior. Assamese macaques at Tukreswari Temple also performed bridging behavior. The macaque species there can be regarded as eastern Assamese macaques, because this site is east (on the south bank) of the Brahmaputra River. Although they have been introduced as western Assamese macaque (Biswas et al. 2011; Cooper and Bernstein 2008), their phylogenetic status remains in question. Therefore, the facts that Tibetan and eastern Assamese macaques have bridging and genital sucking behaviors are generally consistent with the genetic and morphological evidence that Tibetan and eastern Assamese macaques shared a close phylogenetic relationship (Biswas et al. 2011; Chakraborty 2007; Sukmak et al. 2014). The fact that western Assamese macaques in Nepal did not show bridging behavior indicates that they are different from Assamese macaques in other populations. Though Assamese macaques in Nepal have been regarded as the same species with Assamese macaques in other populations, the classification in the *sinica* species-group is still controversial (Khanal et al. 2018). Further studies are needed to understand phylogenetic relationships among the species in the *sinica* species-group including Arunachal and white-cheeked macaques, which were recently recorded as new species (Li et al. 2015; Sinha et al. 2005; but see Biswas et al. 2011).

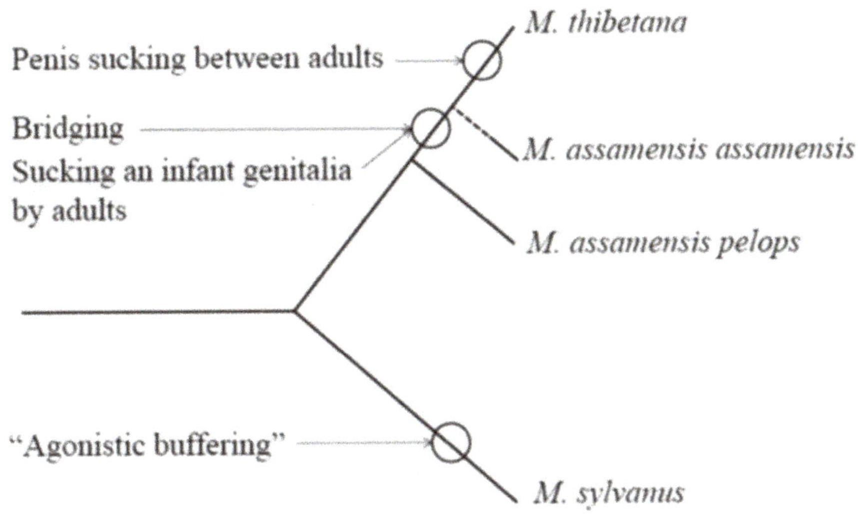

Fig. 10.4 Evolutionary scheme of bridging and penis sucking behaviors based on this comparative study, which shows that bridging and genital sucking behaviors evolved in the clade of *Macaca thibetana* and *M. assamensis assamensis*. In this figure, "agonistic buffering" in *M. sylvanus* evolved due to convergent evolution

Thus, the information on genetic and behavioral data is not complete. However, here I propose the hypothesis on the evolution of bridging and penis sucking behavior in macaques, based on the results of this study. Although it is not clear whether these social behaviors are genetically programmed or socially learned by cultural transmission, the difference of social behaviors by both of the two mechanisms is affected by the divergent evolutionary processes occurring in species, subspecies, and populations. I hypothesize that bridging and genital sucking behaviors evolved in the clade of Tibetan and eastern Assamese macaques in the *sinica* species-group, assuming that the Tibetan macaque was split from the eastern Assamese macaque (Fig. 10.4). As well as bridging behavior, many interactions between males and infants in eastern Assamese macaques in Thailand were similar to those in Tibetan macaques, though there might be some differences in frequencies, social contexts, functions, and the body form in the behavior (Ogawa, personal observation). In addition, eastern Assamese macaques in Thailand had various affiliative behaviors with body contact and sexual behaviors, most of which were similar to those of Tibetan macaques: social mount, hugging (embracing), genital showing, genital inspection, mating, and sexual harassment (Ogawa, personal observation).

It is also possible that the western Assamese macaques in Nepal lost the bridging behavior at some point in its evolutionary history. For example, similar to Tibetan and eastern Assamese macaques, stump-tailed macaques also frequently handled a newborn baby and performed bridging behavior between females one of which was the infant's mother (Estrada 1984; Maruhashi et al. 2018). Bridging behavior might occur between a mother and another female in the common ancestor of the *sinica* species-group including stump-tailed macaque, with some species engaging in frequent bridging between males and some species losing the bridging behavior.

However, it is more parsimonious to argue that bridging behavior evolved in the clade of Tibetan and eastern Assamese macaque, rather than it was lost in western Assamese macaques, because there is no report that toque and bonnet macaques have bridging behavior, except for one study reporting triadic male-infant interactions in bonnet macaques (Silk and Samuels 1984). I predict that Arunachal and white-cheeked macaques will not have bridging behavior, either, if they are phylogenetically closer to western Assamese macaques.

If bridging behavior evolved in the *sinica* species-group, especially in the clade of Tibetan and eastern Assamese macaques, this explains why similar behavior has not been reported in other species-groups of macaques, though there is one report of triadic male-infant interactions in long-tailed macaque (*M. fascicularis*) (de Waal et al. 1976). The only exception is that Barbary macaques (*M. sylvabus*) have a behavior called "agonistic buffering" (Deag and Crook 1971). The behavioral sequence, social context, and functions of "agonistic buffering" have some similarities with bridging behavior in Tibetan and eastern Assamese macaques (Deag and Crook 1971; Paul 1999; Taub 1980). Indeed, two males of Barbary macaques handle an infant together and manipulate and/or suck the genitalia of the infant; however, they seem not to lift up the infant together in the same way that Tibetan and eastern Assamese macaques do. The "agonistic buffering" in Barbary macaques and bridging behavior in the *sinica* species-group might be evolutionarily different, if the origin of these behaviors was not very old in the *sylvanus* and *sinica* species-groups.

Besides, several species in the genus *Papio* also show triadic male-infant interactions (Packer 1980; Strum 1984). However, these interactions of baboons are quite different from bridging and "agonistic buffering" in macaques, because, unlike macaques, one of the two male baboons handles an infant in agonistic interactions. Bridging behavior in the *sinica* species-group, "agonistic buffering" in Barbary macaques, and triadic male-infant interactions in baboons might occur due to convergent evolution.

Among macaques, only male Tibetan macaques directly sucked the penis of another adult male, though bridging behavior was frequently accompanied with sucking the genitalia of an infant across the taxa I studied. Penis sucking behavior between adult males evolved only in the clade to which Tibetan macaques belong. Generally, monkeys form a social relationship with another individual by dyadic interactions, and they sometimes use a third individual to regulate the social relationship. Ogawa (2006, p. 66) discussed that "during bridging behavior, a monkey may be using the penis of an infant as a substitute for his own penis because the use of the infant's penis is more effective." However, male Tibetan and eastern Assamese macaques performed bridging behavior and genital sucking of an infant. Tibetan macaques might expand penis sucking behavior between males and infants toward penis sucking behavior between adult males. The socionomic sex ratios (number of adult males/number of adult females) in the study groups were 0.78 in Tibetan macaques, 0.54 in western Assamese macaques, and 0.48 in eastern Assamese macaques during the study periods. The high socionomic sex ratio in Tibetan macaques might promote penis sucking and other affiliative behaviors between adult males in this species (Ogawa 1995).

In order to examine these hypotheses and predictions and to understand the evolution of the *sinica* species-group and the genus *Macaca*, further behavioral, genetic, and morphological studies are needed in Tibetan macaques, western and eastern Assamese macaques, and other macaque species in various areas.

10.5 Summary

1. Tibetan macaques and eastern Assamese macaques in Thailand performed bridging behavior, but western Assamese macaques in Nepal did not, though male western Assamese macaques in Nepal held an infant in front of another male.
2. Tibetan macaques and eastern Assamese macaques in Thailand sucked the genitalia of an infant, but western Assamese macaques in Nepal did not.
3. Only in Tibetan macaques did adult males perform penis sucking behavior with another adult male.

Acknowledgments I am grateful to the following people and organizations: in China, Dr. K. Wada at Kyoto University; Prof. Q. Wang, Mr. C. Xiong, Dr. J. Li, and Dr. M. Li at Anhui University; Mr. S. Zheng at Huangshan; Takashima Fund of Primate Society of Japan; JICA (Japan International Cooperation Agency); and JSPS KAKENHI Grant Numbers J1740480, 13740498; in Thailand, Dr. S. Malaivijitnond and Prof. S. Hannongbua at Chulalongkorn University; Dr. Y. Hamada at Kyoto University; Mr. L. Por, Mr. S. Win, Mr. S. Tirapanyo, and Mr. I. Ito, chief monks of Wat Tham Pla; National Research Council of Thailand (NRCT); and JSPS KAKENHI Grant Number 20255006; in Nepal, Dr. M. Chalise, Ms. S. Koirala, and Mr. B. Pandey at Tribhuvan University; Dr. Y. Kawamoto at Kyoto University; park rangers and staff of Shivapuri-Nagarjun National Park; Nepal Biodiversity Research Society; and JSPS KAKENHI Grant Numbers JP25440253 and JP16K07539; and in India, Mr. M. Bawri at Gauhati University; Dr. A. Sinha and Mr. N. Sharma at National Institute of Advanced Studies, Bangalore; Dr. M. Cooper at the University of Tennessee; Mr. G. S. S. Sarma at Tukreswari; and HOPE Project.

References

Balasubramaniam KN, Dittmar K, Berman CM, Butovskaya M, Cooper MA, Majolo B, Ogawa H, Schino G, Thierry B, de Waal FBM (2012) Hierarchical steepness and phylogenetic models: phylogenetic signals in *Macaca*. Anim Behav 83:1207–1218

Berman CM, Thierry B (2010) Variation in kin bias: species differences and time constraints in macaques. Behaviour 147:1863–1887

Biswas J, Borah DK, Das A, Das J, Bhattacharjee PC, Mohnot SM, Horwich RH (2011) The enigmatic Arunachal macaque: its biogeography, biology and taxonomy in northeastern India. Am J Primatol 73:1–16

Chakraborty D, Ramakrishnan U, Panor J, Mishra C, Sinha A (2007) Phylogenetic relationships and morphometric affinities of the Arunachal macaque, *Macaca munzala*, a newly described primate from Arunachal Pradesh, northeastern India. Mol Phylogenet Evol 44:838–849

Chalise MK, Ogawa H, Pandey B (2013) Assamese monkeys in Nagarjun forest of Shivapuri Nagarjun National Park, Nepal. Tribhuvan Univ J 18(1–2):181–190

Cooper MA, Bernstein IS (2008) Evaluating dominance styles in Assamese and rhesus macaques. Int J Primatol 29:225–243

De Waal FBM, Luttrell LM (1989) Toward a comparative socioecology of the genus *Macaca*: different dominance styles in rhesus and stumptail monkeys. Am J Primatol 19:83–109

De Waal FBM, van Hooff JA, Netto WJ (1976) An ethological analysis of types of agonistic interaction in a captive group of Java-monkeys (*Macaca fascicularis*). Primates 17:257–290

Deag JM, Crook JH (1971) Social behaviour and "agonistic buffering" in the wild Barbary macaques, *Macaca sylvana* L. Folia Primatol 15:183–200

Delson E (1980) Fossil macaques, phyletic relationships and a scenario of deployment. In: Lindburg D (ed) The macaques: studies in ecology, behavior and evolution. Van Nostrand Reinhold, New York, pp 10–30

Estrada A (1984) Male-infant interactions among free-raging stumptail macaques. In: Taub DM (ed) Primate paternalism. Van Nostrand Reinhold, New York, pp 56–87

Fooden J (1982) Taxonomy and evolution of the *sinica* group of macaques: 3. Species and subspecies accounts of *Macaca assamensis*. Fieldiana. Zoology (New Series) 10:1–52

Groves CP (2001) Primate taxonomy: Smithsonian series in comparative evolutionary biology. Smithsonian Institution Press, Washington, DC

Kalbitz J, Schülke O, Ostner J (2017) Triadic male-infant-male interaction serves in bond maintenance in male Assamese macaques. PLoS One 12(10):e0183981

Khanal L, Chalise MK, Hei K, Acharya BK, Kawamoto Y, Jiang X (2018) Mitochondrial DNA analyses and ecological niche modeling reveal post-LGM expansion of the Assam macaque (*Macaca assamensis*) in the foothills of Nepal Himalaya. Am J Primatol 80(3):e22748

Koirala S, Chalise MK, Katuwal H, Gaire R, Pandey B, Ogawa H (2017) Diet and activity of *Macaca assamensis* in wild and semi-provisioned groups in Shivapuri Nagarjun National Park, Nepal. Folia Primatol 88:57–74

Li C, Chao Z, Fan P (2015) Whit-cheeked macaques (*Macaca leucogenys*): a new macaque species from Modog, Southeastern Tibet. Am J Primatol 77(2):753–766

Maruhashi T, Toyoda A, Malaivijitnond S (2018) Timing and characters of Touch Baby Genital (TBG) behaviours of *Macaca arctoides* inhabiting the Khao Krapuk Khao Taomo Non-hunting Area, Petchaburi, Thailand. Abstract. Satellite international symposium on Asian primates. Kathmandu, Nepal, pp 26–27

Ogawa H (1995) Bridging behavior and other affiliative interactions among male Tibetan macaques (*Macaca thibetana*). Int J Primatol 16:707–729

Ogawa H (2006) Wily monkeys: social intelligence of Tibetan macaques. Kyoto University Press and Trans Pacific Press, Kyoto

Ogawa H, Malaivijitnond S, Hamada Y (2009) Social interactions among Assamese macaques at Wat Tham Pla, Thailand. Proceedings. The 3rd international congress on the future of animal research, Bangkok, p 29

Packer C (1980) Male care and exploitation of infants in *Papio anubis*. Anim Behav 28:512–520

Paul A (1999) The sociobiology of infant handling in primates: is the current model convincing? Primates 40:33–46

Silk JB, Samuels A (1984) Triadic interactions among *Macaca radiata*: passports and buffers. Am J Primatol 6:373–376

Sinha A, Datta A, Madhusudan MD, Mishra C (2005) *Macaca munzala*: a new species from western Arunachal Pradesh, northeastern India. Int J Primatol 26(4):977–989

Strum SC (1984) Why males use infants. In: Taub DM (ed) *Primate paternalism*. Van Nostrand Reinhold, New York, pp 20–55

Sukmak M, Malaivijitnond S, Schülke O, Ostner J, Hamada Y, Wajjwalku W (2014) Preliminary study of the genetic diversity of eastern Assamese macaques (*Macaca assamensis assamensis*) in Thailand based on mitochondrial DNA and microsatellite markers. Primates 55:189–197

Taub DM (1980) Testing the "agonistic buffering" hypothesis. Behav Ecol Sociobiol 6:187–197

Thierry B, Singh M, Kaumanns W (eds) (2004) Macaque societies: a model for the study of social organization. Cambridge University Press, Cambridge

Wada K (2005) The distribution pattern of rhesus and Assamese monkeys in Nepal. Primates 46:115–119

Wada K, Xiong C, Wang Q (1987) On the distribution of Tibetan and rhesus monkeys in southern Anhui, China. Acta Theriol Sinica 7:148–176

Zhao Q, Deng Z (1988) *Macaca thibetana* at Mt. Emei, China: I. A cross-sectional study of growth and development. Am J Primatol 16:251–260

Part IV
Living with Microbes, Parasites, and Diseases

Chapter 11
The Gut Microbiome of Tibetan Macaques: Composition, Influencing Factors and Function in Feeding Ecology

Binghua Sun, Michael A. Huffman, and Jin-Hua Li

11.1 Introduction

Microbes dominated the earth for at least 2.5 billion years before multicellular life appeared in the biosphere (Hooper and Gordon 2001; Ley et al. 2008). In fact, animals have been living and evolving in a microbial world (Margaret et al. 2013). Microbes inevitably colonized animals and coevolved through the process of interactions with their hosts (Clayton et al. 2018). Microorganisms can inhabit multiple parts of a host's body, such as the skin, oral cavity, sex organs, and the gastrointestinal (GI) tract. They are usually recognized as the microbiome of a particular host. As an essential part of the host's body, the microbiome plays an important role in host physiology by influencing nutritional intake, metabolic activity, and immune homeostasis (Turnbaugh et al. 2006; Greenblum et al. 2012; Hooper et al. 2012). Thus, the complex relationship between hosts and their microbiomes provides a unique opportunity for understanding mammalian adaptation and evolution (Hird 2017).

As the most complex and diverse ecological system of the mammalian body, the GI tract is colonized by bacteria, archaea, fungi, and viruses (Underhill and Iliev 2014). Previous studies have revealed that the gut microbiota play an important role in immune regulation, vitamin synthesis, energy acquisition, and disease risk

B. Sun
School of Resources and Environmental Engineering, Anhui University, Hefei, Anhui, China

M. A. Huffman
Primate Research Institute, Kyoto University, Kyoto, Japan
e-mail: huffman.michael.8n@kyoto-u.ac.jp

J.-H. Li (✉)
School of Resources and Environmental Engineering, Anhui University, Hefei, Anhui, China

International Collaborative Research Center for Huangshan Biodiversity and Tibetan Macaque Behavioral Ecology, Anhui, China

School of Life Sciences, Hefei Normal University, Hefei, Anhui, China
e-mail: jhli@ahu.edu.cn

J.-H. Li et al. (eds.), *The Behavioral Ecology of the Tibetan Macaque*, Fascinating Life Sciences, https://doi.org/10.1007/978-3-030-27920-2_11

reduction of the host (Turnbaugh et al. 2006; Hooper et al. 2012; Sharon et al. 2012). For example, all vertebrates lack cellulase, the enzymes, which help them to digest the fiber in their food, and are reliant upon intestinal microbes for its production (Yokoe and Yasumasu 1964; Mackie 2002). As a result, mammals maintain rich gut microbial communities, which are beneficial for the digestion of cellulose and hemicellulose abundant in the foods they ingest (Amato et al. 2015).

Nonhuman primates (NHPs) evolved in tropical forest habitats and radially spread to woodland, savanna and montane environments (Hanya et al. 2011). In order to adapt to the resultant temporal food resource shifts and to maintain ordinary functions of the body, NHPs have evolved many anatomical and behavioral strategies (Rodman and Cant 1984). For example, some primates have evolved specialized stomachs and teeth, while many others have evolved distinctive habitat use patterns, ranging behavior and social organization (Milton and May 1976; Chivers et al. 1984; Cachel 1989; Yamada and Muroyama 2010; Hanya and Chapman 2013). Generally, NHPs prefer to eat high-quality food rich in lipids, proteins, and carbohydrate, such as fruits and young leaves. However, these foods are not always available year-round (Rodman and Cant 1984; Ungar 1995; Hanya et al. 2011; Hanya and Chapman 2013). NHPs living in temperate environments can only eat mature leaves, gum and roots during winter (Fan et al. 2009). These low-quality foods usually contain high proportions of cellulose and hemicellulose, which are difficult to digest (Coley 1983; Burrows 2010). In response to ecological pressures, many primates also use behavioral strategies such as adjusting their activity budgets and foraging patterns to help overcome periods of food scarcity (Zhou et al. 2007; Starr et al. 2012; Hanya and Chapman 2013; Campera et al. 2014). However, how wild primates digest low-quality foods high in cellulose during the winter has not been well understood (Amato et al. 2014).

Although low-quality foods, including woody plants, mature leaves, fungi, mature leaves, and plant exudates are difficult to digest, more recent research related to the above has revealed that these foods can be broken down and utilized by the intestinal microbes of NHPs. For this reason, the gut microbiota may provide a way to understand how primates adapt to ecological pressures during periods of food scarcity. Since the gut microbiome is dominated by bacteria, many studies on the feeding ecology of mammals have focused solely on the bacterial microbiome (Qin et al. 2010). Other common elements, such as gut fungi, also appear to affect the health and nutrition of animals, but this too is not yet well understood (Huffnagle and Noverr 2013; Underhill and Iliev 2014; Sokol et al. 2017). Although intestinal fungi have a much smaller number of cells compared to bacteria, the evidence suggests that gut mycobiota can also play an important role in host health by affecting gut bacterial composition (Hoffmann et al. 2013), interacting with immune cells, and assisting in the metabolic activity of the host (Hajishengallis et al. 2011; Romani 2011; Iliev and Underhill 2012; Rizzetto et al. 2014). Therefore, it is important to integrate information about the role of gut bacteria and fungi when discussing the role of the gut microbiome in the evolution of animal dietary strategies.

NHPs and humans have extensive similarities in their genetic characteristics, physiology and morphology. Very important animal model systems, NHPs are extremely important for understanding human physiology, behavior, cognition,

Fig. 11.1 A female Tibetan macaque carrying a baby is foraging for food in winter (Photo credit: Qixin Zhang)

health and evolution (McCord et al. 2014; Ren et al. 2015). Therefore, compared with other laboratory animals, NHPs have an advantage in helping us to understanding the role of the gut microbiome in human health maintenance, as well as the evolutionary relationships involving the gut microbiome, host diet and evolution.

The free-ranging Tibetan macaque (*Macaca thibetana*) group at Mt. Huangshan (described in Chap. 2) presents a good opportunity for studying this, since all members of the group can be identified individually, allowing for the collection of feces from individuals of known age and sex (Zhang et al. 2010). Our study group lives in a deciduous and evergreen-broadleaf-mixed montane forest. As the temperature fluctuates across seasons (highest in summer: 34.2 °C, lowest in winter: −13.9 °C), the quality of their food also varies accordingly (see Fig. 11.1). For example, from winter to spring, Tibetan macaques experience a switch from low quality food resources (mature leaves, roots and plant stem) to high quality food resources (young leaves and flowers) (Xiong and Wang 1988; Zhao 1999). As a result, our study population is an ideal model for studying the role of the gut microbiome in primate feeding ecology and evolution.

In this chapter, we will summarize our research on the composition of the bacterial/fungal community and microbial diversity of Tibetan macaques and analysis of the factors influencing variation in the microbiome across age, sex and season in this species. Furthermore, we use this information to discuss the role of the Tibetan macaques' gut microbiome in relation to the evolution of their feeding ecology.

11.2 Gut Microbiome of Tibetan Macaque

11.2.1 Composition of Gut Bacteria

Understanding the composition of an animal's gut bacteria helps us to reveal its role in the host's feeding ecology, however, we know little about the Tibetan macaques' gut microbiota. Recently, using the methods of high-throughput sequencing and bioinformatics, Sun et al. (2016) detected 14 known bacterial phyla in this macaque's gut, the most dominant phyla (relative abundance > 1%) being Firmicutes, Bacteroidetes and Proteobacteria Spirochaetes (Sun et al. 2016). Comparing our results to those of the gut microbiome in other NHPs (Table 11.1), we found that the two most dominant phyla (Firmicutes and Bacteriodetes) were commonly reported in many other species as well (Ochman et al. 2010; Mckenney et al. 2015; Xu et al. 2015). At the genus level, the two most dominant genera present in the gut of Tibetan macaques were *Prevotella* and *Succinivibrio*. Although the genus *Prevotella* was also commonly reported elsewhere (Ochman et al. 2010; Mckenney et al. 2015; Xu et al. 2015), *Succinivibrio* has been seldom been detected as a dominant genus. The relative abundance of these main bacterial genera (relative abundance > 1%) are presented in Fig. 11.2a.

Table 11.1 The dominant phyla of gut bacteria detected in several non-human primate species

Monkey's name	Dominant phyla			Captive/ wild	Study
	Firmicutes	Bacteroidetes	Proteobacteria		
Lemur Catta	+	+		Captive	McKenney et al. (2015)
Varecia variegata	+	+		Captive	McKenney et al. (2015)
Propithecus coquereli	+	+		Captive	McKenney et al. (2015)
Rhinopithecus bieti	+	+	+	Wild	Xu et al. (2015)
Nycticebus pygmaeus		+		Wild	Xu et al. (2013)
Gorilla beringei		+	+	Wild	Ochman et al. (2010)
Gorilla gorilla		+	+	Wild	Ochman et al. (2010)
Pan troglodytes	+	+		Wild	Ochman et al. (2010)
Macaca thibetana	+	+		Wild	Sun et al. (2016)

Note: + means the relative abundance was one of the top two

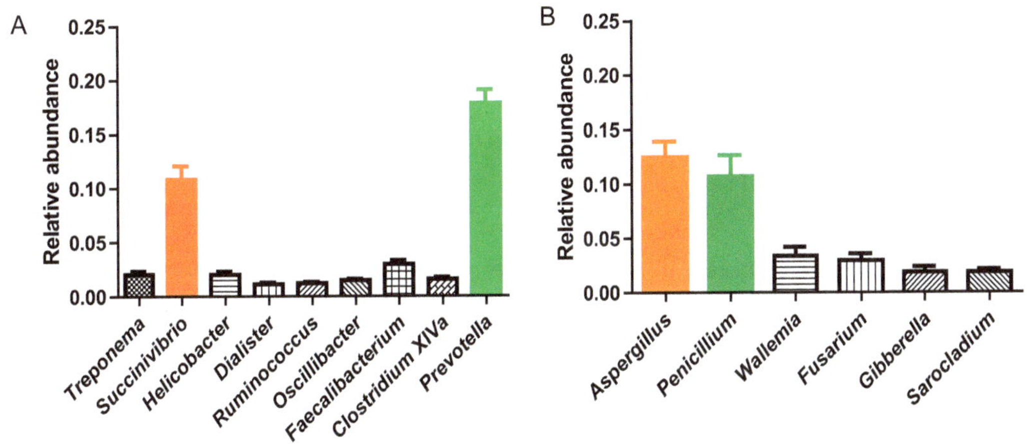

Fig. 11.2 The relative abundance of the main genera of bacteria and fungi. (**a**) The relative abundance of the major bacterial genera (relative abundance > 1%); data from Sun et al. (2016). (**b**) The relative abundance of the major fungal genera (relative abundance > 1%), data from Sun et al. (2018)

11.2.2 Composition of Gut Fungi

Previous studies on humans and mice have revealed that the mammalian gut is rich in fungi, dominated by three phyla: Ascomycota, Basidiomycota, and Zygomycota (Iliev and Underhill 2012; Qiu et al. 2015; Mar et al. 2016; Strati et al. 2016; Wheeler et al. 2016; Sokol et al. 2017). However, little is known about the gut fungi in NHPs. Using the Illlumina Miseq platform and primers of the ITS region (Bokulich and Mills 2013; Sun et al. 2018) first reported that the gut mycobiota of Tibetan macaques was dominated by two phyla Ascomycota and Basidiomycota (Sun et al. 2018), this is consistent with previous studies on mammalian gut mycobiomes. In addition, other phyla such as Zygomycota, Chytridiomycota, Glomeromycota and Rozellomycota were also detected in this macaque species. At the family level, Trichocomaceae shows the highest relative abundance (24.70%). At the genus level, the gut mycobiota of Tibetan macaques are rich in *Aspergillus* and *Penicillium* (relative abundance more than 10% respectively). Furthermore, by defining the core families and genera (both present in >90% of the samples with an average relative abundance > 0.01), Sun et al. (2018) revealed that six core taxa, namely Trichocomaceae, Nectriaceae, Davidiellaceae, *Aspergillus*, *Penicillium* and *Fusarium* can be detected in the gut of Tibetan macaques. The relative abundances of fungal genera (>1%) are presented in Fig. 11.2b. The presence of core taxa in the gut of this macaque suggests that the fungi in their guts are not random and exist as a relatively stable fungal community.

11.3 Factors Affecting the Gut Microbiome in Tibetan Macaques

11.3.1 Affects of Age, Sex, and Season on Gut Bacterial Microbiome

It is well known that many internal and external factors can affect the diversity and composition of an animal's gut microbiota, such as age, sex and seasonal change. Based on data from the Illlumina Miseq platform and linear mixed models, Sun et al. (2016) evaluated factors affecting gut microbial diversity in Tibetan macaques. The results showed that sex and age had no significant effect on the alpha diversity (including Shannon index, Chao 1, OTU richness and ACE) of the gut bacterial microbiome, as well as the beta diversity (evaluating by unweighted and weighted UniFrac distances) (Sun et al. 2016). These results differ from previous studies in humans and other vertebrates (Yatsunenko et al. 2012; Bolnick et al. 2014), but are consistent with those in chimpanzees and baboons (Degnan et al. 2012; Tung et al. 2015).

Seasonal factors on the other hand have a strong effect on the composition and diversity of the gut bacterial microbiome in Tibetan macaques (Sun et al. 2016). For example, the Shannon diversity index of the gut bacterial microbiome was significantly different between winter and spring. In addition, a significant seasonal separation of beta diversity evaluating by weighted UniFrac distances and unweighted UniFrac distances (Permanova tests, $p < 0.05$ and 0.01 respectably) were detected. Furthermore, three phyla and 20 known genera showed significant differences between the two seasons. The representative taxa rich in winter samples were Proteobacteria, Spirochaetia, *Succinivibrio*, *Clostridium* sensu stricto and *Treponema*, but Firmicutes and *Prevotella* were richest in spring samples. Differences in the composition of gut bacteria also resulted in the variations of predicted metagenomes between winter and spring. Using PICRUSt and LEfSe tests, Sun et al. (2016) found that seven KEGG pathways of Tibetan macaques' gut bacterial microbiome, including Glycan Biosynthesis and Metabolism, Amino Acid Metabolism, Signal Transduction, Cell motility, Transport and Catabolism, Neurodegenerative Diseases and Endocrine System, were significantly enriched in the winter. However six other metabolic pathways Energy Metabolism, Carbohydrate Metabolism, Cellular Processes Transcription, Signaling and Metabolism of Cofactors and Vitamins and Enzyme Families were significantly enriched in the spring (Sun et al. 2016). The potential functions of the gut microbiome in different seasons of our study group will be discussed in Sect. 11.4.

11.3.2 Gut Fungal Microbiome Affected by Age, Sex, and Season

Although previous studies in humans have revealed that age and sex can influence the composition and diversity of the gut fungal microbiome (Strati et al. 2016), little information is available for wild living primates. Using LEfSe analysis, Sun et al. (2018) first reported that only one fungal genus *Sarocladium* was enriched in the old age group. By comparing males and females, the family Mycosphaerellaceae and genus *Devriesia* were particularly enriched in females, while the phylum Ascomycota and family Tetraplosphaeriaceae were enriched in males (Sun et al. 2018). These results indicated that the effect of sex and age on the composition of gut fungi was not as strong as that found in humans. In addition, we found evidence of marked seasonal variation in the composition of this macaque species' gut mycobiota. Fifteen taxa had significant variation across the four seasons. The abundant taxa in each season were as follows, Autumn: Wallemiaceae, Hypocreaceae, *Wallemia*, *Trichoderma*; Spring: Sclerotiniaceae, Nectriaceae, *Ciboria, Fusarium, Gibberella, Sarocladium* and *Talaromyces*; Summer: Trichocomaceae and *Penicillium;* Winter: *Devriesia* and Teratosphaeriaceae. This result is the first report to find strong seasonal influences on primate gut mycobiota, indicating the potential relationship between gut mycobiota and seasonal changes in host diet.

Different from previous studies in humans, no evidence indicates that age or sex strongly affect the alpha diversity of Tibetan macaque gut mycobiota (Sun et al. 2018). This result is likely because even individuals living in the same environment have differences in their dietary preferences. In addition, using PCoA and PERMANOVA tests, we found that there existed significant variation in beta diversity among different age and sex classes. Interestingly, seasonal change can significantly affect alpha and beta diversity, as well as the composition of the gut fungal microbiome.

11.4 Functions of the Gut Microbiome in Tibetan Macaque Feeding Ecology

Feeding ecology plays an important role in understanding the evolution of NHPs (Lambert 2011). With advances in sequencing technology, many studies have revealed that the gut microbiome aids in the digestion of some food resources, especially those rich in cellulose and hemicellulose, substances which are difficult to digest. Therefore, this important adaptive mechanism has attracted increasing attention in recent years. Tibetan macaques living in a highly seasonal ecosystem with strongly seasonal changes in rainfall, temperature, food resources, and home range shifts (Xiong and Wang 1988; Zhao 1999). Previous studies have suggested that these environmental factors are closely reflected in the composition of the gut microbiome (Amato et al. 2015; Chevalier et al. 2015; Maurice et al. 2015; Wu et al. 2017). Therefore, our study subjects may provide an important model for understanding the functions of the gut microbiome in primate feeding ecology, especially for primates living in highly seasonal ecosystems.

11.4.1 Gut Bacterial Microbiome and the Feeding Ecology of Tibetan Macaques

It is well known that the gut bacterial microbiome can help hosts to digest food resources more efficiently, and they also can change according to variation in host diet across time and space, helping the host to meet its nutrient and energy requirements (Hooper et al. 2002; Donohoe et al. 2011; Koren et al. 2012; Amato et al. 2014; Chevalier et al. 2015). As a species living in a highly seasonal ecosystem, Tibetan macaques depend mostly on a plant diet for their subsistence, including leaves, grass, roots, fruits, and flowers, and they usually shift home ranges to adapt to seasonal changes in food availability (Xiong and Wang 1988; Zhao 1999). The diverse and responsive gut bacterial community of wild-living Tibetan macaques can be considered an important adaptation for their foraging lifestyle. The presence of the relatively abundant fiber-degrading bacteria of Firmicutes and Bacteroidetes in their gut can help them to utilize the heavily plant-based diet. Some well-known fiber-degrading bacteria genera detected in the gut of Tibetan macaques, such as *Succinivibrio* and *Clostridium*, suggest that the gut bacterial microbiome plays an important role in the Tibetan macaques' feeding ecology.

In seasonal ecosystems, one of the most important questions is how animals adapt to seasonal changes in food composition and availability. From winter to spring, Tibetan macaques experience a switch from low quality food resources (mature leaves, roots and plant stem) to high quality food resources (young leaves and flowers). Generally, low-quality food resources are rich in cellulose and hemicellulose, and high-quality food resources are rich in pectin, carbohydrates and simple sugars. Based on a comparative study of the Tibetan macaque gut bacterial microbiome between winter and spring (Sun et al. 2016), the gut bacterial adaptive mechanism for seasonal food type's changes in the Tibetan macaque diet is summarized in Fig. 11.3.

During winter, the representative gut bacterial taxa Proteobacteria, *Succinivibrio* and *Clostridium* significantly increased, and the predicted metagenomes of glycan biosynthesis and metabolic pathways significantly increased in winter samples (Sun et al. 2016). The genus *Succinivibrio*, which usually is detected in rumen microbial ecosystems, ferment glucose efficiently by producing acetic acid and succinic acid (de Menezes et al. 2011), as well as being beneficial to the metabolism of different types of fatty acids (Van Dyke and McCarthy 2002). Another genus *Clostridium* contains many organisms which produce cellulose and hemicellulose-digestive enzymes (Van Dyke and McCarthy 2002; Zhu et al. 2011). The significant increase

Fig. 11.3 Food type change and adaptive shifts of the gut bacterial microbiome in Tibetan macaques. The predicted metagenomes of KEGG pathway and bacterial taxa of the Tibetan macaque gut microbiome, which significantly increased in winter, are beneficial for the digestion of cellulose and hemicellulose rich foods in winter, and the other bacteria significantly increased in spring, which are beneficial for the digestion of pectin, carbohydrate and simple sugar rich foods which are abundant in spring

of the genus *Clostridium* in winter samples is very beneficial for the digestion of cellulose and dietary fiber. In addition, evidence from black howler monkeys, humans and mice reveals that Proteobacteria are highly correlated with energy acquisition (Bryant and Small 1956; Koren et al. 2012; Amato et al. 2014; Chevalier et al. 2015). In response to cold weather, primates experience an increase in energy loss (Tsuji et al. 2013), and Tibetan macaques are no exception. The increase of Proteobacteria is beneficial for coping with cold weather. In conclusion, the gut bacterial microbiome in winter increases the efficiency of dietary fiber digestion in Tibetan macaques, which helps them to meet their energy requirements.

In addition, Sun et al. (2016) found that Firmicutes and *Prevotella* were significantly enriched in the Tibetan macaque gut during spring. During this time, the diet consists mainly of young leaves, flowers and bamboo shoots, foods that are richer in digestible pectin, carbohydrates and simple sugars, compared to mature leaves (Xiong and Wang 1988; Zhao 1999; You et al. 2013). The genus *Prevotella* is associated with the digestion of carbohydrates simple sugars, pectin and hemicellulose (Amato et al. 2014). The significant increase of genes related to carbohydrate metabolism and energy metabolism pathways in spring samples also indicates that the gut bacterial community helps break down and make accessible these high-quality food resources, beneficial for macaques to recover from energy loss experienced during the cold conditions of winter.

11.4.2 Gut Fungal Microbiome and Feeding Ecology of Tibetan Macaques

Compared with the gut bacterial microbiome, there is very little data on the primate gut fungal microbiome, and the role of gut fungi in primate foraging ecology is unknown. Sun et al. (2018) first reported the diversity and composition of Tibetan macaque gut mycobiota. Three genera *Aspergillus*, *Penicillium* and *Fusarium* were detected as normal inhabitants in the gut. The two most dominant genera, were *Aspergillus* (12.46%) and *Penicillium* (10.72%). It has been reported that the anaerobic species of these two genera can produce cellulolytic and hemicellulolytic enzymes, which are beneficial to cellulosic biomass degradation (Boots et al. 2013; Liao et al. 2014; Shuji et al. 2014; Solomon et al. 2016; Trinci et al. 1994). Similarly, as robust cellulose and hemicellulose degrader, the genus *Fusarium* was reported in a recent study (Huang et al. 2015). At certain times of the year, Tibetan macaques depend on a plant diet, high in cellulose and hemicellulose (Xiong and Wang 1988; You et al. 2013; Campbell et al. 2001). Our previous studies indicated that the dominant genera of Tibetan macaque gut mycrobiota may play an important role in the digestion of these and other plant items.

In addition, the gut fungal microbiome of Tibetan macaques changed with the seasonal change in dietary food content. Fifteen taxa were detected to be significantly enriched in one of the four seasons (Sun et al. 2018). This result indicates that the gut mycrobiota can response to dietary shifts. It has been proposed that the gut mycobiota response to seasonal dietary shift is beneficial to the hosts' nutritional and reproductive needs (Noma et al. 1998; Tsuji et al. 2013). However, knowledge about the functions of the particular fungi in the mammalian gut is limited (Milton and May 1976), as are genomic databases and metabolic maps of gut fungi (Huffnagle and Noverr 2013), making it difficult to explain the relationship between mycobiota composition and primate feeding ecology. Based on the available information about the genus *Penicillium* (Shuji et al. 2014), we hypothesize that this summer season enriched genus may aid in the digestion of mature leaves, which contain a higher proportion of cellulose and hemicellulose. However, a possible explanation as to why this genus was not enriched during winter could be that the ingestion of fallen nuts in winter caused the amount of *Penicillium* to decrease due to some chemical interaction (Maria et al. 2014). Secondly, it has been reported that high diversity of gut mycobiota can improve the utilization efficiency of plant fiber consumption (Bauchop 1981; Akin et al. 1983; Denman and Mcsweeney 2010). Thus, the gut mycobiota with high alpha diversity during the winter, reported in our previous study, may be beneficial for Tibetan macaques to digest dietary fiber. In conclusion, the gut fungal mycobiota detected in Tibetan macaque provides evidence for the role of gut fungi in primate foraging ecology. More attention needs to be paid to this in future studies.

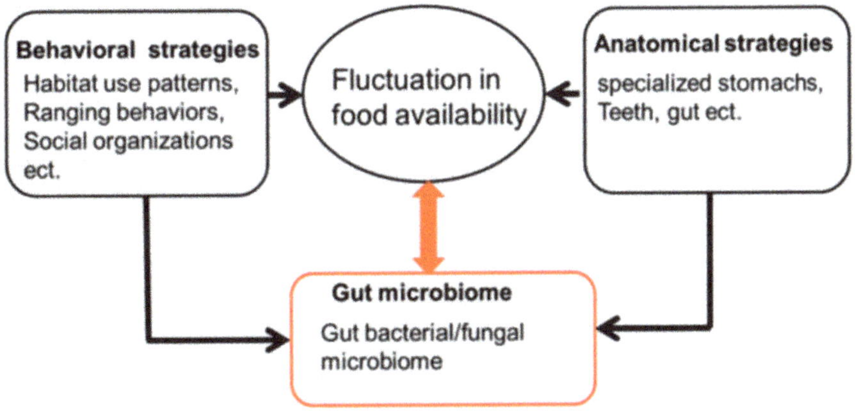

Fig. 11.4 Adaptive mechanisms of wild primates in response to fluctuations in food availability and food quality

11.5 Conclusions and Future Directions

Our studies of the Tibetan macaque gut microbiome reviewed in this chapter add important baseline data for the Tibetan macaque to our knowledge of the mammalian gut microbiome regarding difference in host age, sex and its role in feeding ecology, as well as the role of the gut microbiome in the adaptive radiation of primates. Together with anatomical and behavioral strategies, the gut microbiome is believed to be important factor in the successful evolution of NHPs, particularly as an adaption to the temporal fluctuation in food resource types and their availability. The present data on the wild Tibetan macaque gut microbiome leads us to hypothesize that gut microbiota played an important role in the adaptive radiation of primates, and other animals, which can help them to solve foraging dilemmas in seasonally fluctuating environments (Fig. 11.4). Further research is necessary to test this hypothesis on more mammalian species. In addition, mammals living in a microbial environment, and the microorganisms in the soil and plant foods can also change with seasonal fluctuations. The link between the gut microbiome and environmental microbiomes remain to be elucidated.

Acknowledgments Many thanks to the Huangshan Garden Forest Bureau for their permission and support of this work. We also gratefully acknowledge Paul A. Garber, Lori K. Sheeran and all members of our research group for their excellent work on the gut microbiome of Tibetan macaques. This chapter was supported by grants from the National Natural Science Foundation of China (No. 31870371, 31400330, 31172106), the Special Foundation for Excellent Young Talents in University of Anhui Province, China (No. 2012SQRL018ZD) and the Initial Funding for Doctoral Research in Anhui University (No. J01003229).

References

Akin DE, Gordon GL, Hogan JP (1983) Rumen bacterial and fungal degradation of *Digitaria pentzii* grown with or without sulfur. Appl Environ Microbiol 46:738–748

Amato KR, Leigh SR, Angela K, Mackie RI, Yeoman CJ, Stumpf RM, Wilson BA, Nelson KE, White BA, Garber PA (2014) The role of gut microbes in satisfying the nutritional demands of adult and juvenile wild, black howler monkeys (*Alouatta pigra*). Am J Phys Anthropol 155:652–664

Amato KR, Leigh SR, Kent A, Mackie RI, Yeoman CJ, Stumpf RM, Wilson BA, Nelson KE, White BA, Garber PA (2015) The gut microbiota appears to compensate for seasonal diet variation in the wild black howler monkey (*Alouatta pigra*). Microb Ecol 69:434–443

Bauchop T (1981) The anaerobic fungi in rumen fibre digestion. Agric Environ 6:339–348

Bokulich NA, Mills DA (2013) Improved selection of internal transcribed spacer-specific primers enables quantitative, ultra-high-throughput profiling of fungal communities. Appl Environ Microbiol 79(8):2519–2526

Bolnick DI, Snowberg LK, Hirsch PE, Lauber CL, Parks B, Lusis AJ, Knight R, Caporaso JG, Svanbäck R (2014) Individual diet has sex-dependent effects on vertebrate gut microbiota. Nat Commun 5:4500

Boots B, Lillis L, Clipson N, Petrie K, Kenny DA, Boland TM, Doyle E (2013) Responses of anaerobic rumen fungal diversity (phylum Neocallimastigomycota) to changes in bovine diet. J Appl Microbiol 114:626–635

Bryant MP, Small NJ (1956) Characteristics of two new genera of anaerobic curved rods isolated from the rumen of cattle. J Bacteriol 72:22–26

Burrows AM (2010) The evolution of exudativory in primates. Springer, New York

Cachel S (1989) Primate adaptation and evolution. Int J Primatol 10:487–490

Campbell JL, Glenn KM, Grossi B, Eisemann JH (2001) Use of local North Carolina browse species to supplement the diet of a captive colony of folivorous primates (*Propithecus* sp.). Zoo Biol 20:447–461

Campera M, Serra V, Balestri M, Barresi M, Ravaolahy M, Randriatafika F, Donati G (2014) Effects of habitat quality and seasonality on ranging patterns of collared brown lemur (*Eulemur collaris*) in littoral forest fragments. Int J Primatol 35:957–975

Chevalier C, Stojanović O, Colin DJ, Suarez-Zamorano N, Tarallo V, Veyrat-Durebex C, Rigo D, Fabbiano S, Stevanović A, Hagemann S (2015) Gut microbiota orchestrates energy homeostasis during Cold. Cell 163:1360–1374

Chivers DJ, Wood BA, Bilsborough A (1984) Food acquisition and processing in primates: concluding discussion. In: Chivers DJ, Wood BA, Alan B (eds) Food acquisition and processing in primates. Springer, New York, pp 545–556

Clayton JB, Gomez A, Amato K, Knights D, Travis DA, Blekhman R, Knight R, Leigh S, Stumpf R, Wolf T (2018) The gut microbiome of nonhuman primates: lessons in ecology and evolution. Am J Primatol 80:e22867

Coley PD (1983) Herbivory and defensive characteristics of tree species in a lowland tropical forest. Ecol Monogr 53:209–229

de Menezes AB, Lewis E, O'Donovan M, O'Neill BF, Clipson N, Doyle EM (2011) Microbiome analysis of dairy cows fed pasture or total mixed ration diets. FEMS Microbiol Ecol 78:256–265

Degnan PH, Pusey AE, Lonsdorf EV, Goodall J, Wroblewski EE, Wilson ML, Rudicell RS, Hahn BH, Ochman H (2012) Factors associated with the diversification of the gut microbial communities within chimpanzees from Gombe National Park. Proc Natl Acad Sci U S A 109:13034–13039

Denman S, Mcsweeney C (2010) Development of a real-time PCR assay for monitoring anaerobic fungal and cellulolytic bacterial populations within the rumen. FEMS Microbiol Ecol 58:572–582

Donohoe DR, Nikhil G, Xinxin Z, Wei S, O'Connell TM, Bunger MK, Bultman SJ (2011) The microbiome and butyrate regulate energy metabolism and autophagy in the mammalian colon. Cell Metab 13:517–526

Fan P, Ni Q, Sun G, Bei H, Jiang X (2009) Gibbons under seasonal stress: the diet of the black crested gibbon (*Nomascus concolor*) on Mt. Wuliang, Central Yunnan, China. Primates 50:37–44

Greenblum S, Turnbaugh PJ, Borenstein E (2012) Metagenomic systems biology of the human gut microbiome reveals topological shifts associated with obesity and inflammatory bowel disease. PNAS 109:594–599

Hajishengallis G, Liang S, Payne MA, Hashim A, Jotwani R, Eskan MA, Mcintosh ML, Alsam A, Kirkwood KL, Lambris JD (2011) Low-abundance biofilm species orchestrates inflammatory periodontal disease through the commensal microbiota and complement. Cell Host Microbe 10 (5):497–506

Hanya G, Chapman CA (2013) Linking feeding ecology and population abundance: a review of food resource limitation on primates. Ecol Res 28:183–190

Hanya G, Qarro M, Tattou MI, Fuse M, Vallet D, Yamada A, Go M, Takafumi H, Tsujino R, Agetsuma N (2011) Dietary adaptations of temperate primates: comparisons of Japanese and Barbary macaques. Primates 52:187–198

Hird SM (2017) Evolutionary biology needs wild microbiomes. Front Microbiol 8:725

Hoffmann C, Dollive S, Grunberg S, Chen J, Li H, Wu GD, Lewis JD, Bushman FD (2013) Archaea and fungi of the human gut microbiome: correlations with diet and bacterial residents. PLoS One 8:e66019

Hooper LV, Gordon JI (2001) Commensal host-bacterial relationships in the gut. Science 292 (5519):1115–1118

Hooper LV, Midtvedt T, Gordon JI (2002) How host-microbial interactions shape the nutrient environment of the mammalian intestine. Annu Rev Nutr 22:283–307

Hooper LV, Littman DR, Macpherson AJ (2012) Interactions between the microbiota and the immune system. Science 336:1268–1273

Huang Y, Busk PK, Lange L (2015) Cellulose and hemicellulose-degrading enzymes in *Fusarium* commune transcriptome and functional characterization of three identified xylanases. Enzyme Microb Technol 73–74:9–19

Huffnagle GB, Noverr MC (2013) The emerging world of the fungal microbiome. Trends Microbiol 21:334–341

Iliev ID, Underhill DM (2012) Interactions between commensal fungi and the C-type lectin receptor Dectin-1 influence colitis. Science 336:1314–1317

Koren O, Goodrich J, Cullender T, Spor A, Laitinen K, Bäckhed HK, Gonzalez A, Werner J, Angenent L, Knight R (2012) Host remodeling of the gut microbiome and metabolic changes during pregnancy. Cell 150:470–480

Lambert JE (2011) Primate nutritional ecology: feeding biology and diet at ecological and evolutionary scales. In: Campbell C, Fuentes A, MacKinnon KC, Panger M, Bearder S, Stumpf R (eds) Primates in perspective, 2nd edn. Oxford University Press, Oxford, pp 512–522

Ley RE, Micah H, Catherine L, Turnbaugh PJ, Rob Roy R, J Stephen B, Schlegel ML, Tucker TA, Schrenzel MD, Rob K (2008) Evolution of mammals and their gut microbes. Science 322 (5905):1188

Liao H, Li S, Wei Z, Shen Q, Xu Y (2014) Insights into high-efficiency lignocellulolytic enzyme production by Penicillium oxalicum GZ-2 induced by a complex substrate. Biotechnol Biofuels 7(1):162

Mackie RI (2002) Mutualistic fermentative digestion in the gastrointestinal tract: diversity and evolution. Integr Comp Biol 42:319–326

Mar RM, Pérez D, Javier CF, Esteve E, Maringarcia P, Xifra G, Vendrell J, Jové M, Pamplona R, Ricart W (2016) Obesity changes the human gut mycobiome. Sci Rep 6:21679

Margaret MFN, Hadfield MG, Bosch TCG, Carey HV, Tomislav DLO, Douglas AE, Nicole D, Gerard E, Tadashi F, Gilbert SF (2013) Animals in a bacterial world, a new imperative for the life sciences. PNAS 110:3229–3236

Maria U, Xiaoyu W, Baer DJ, Novotny JA, Marlene F, Volker M (2014) Effects of almond and pistachio consumption on gut microbiota composition in a randomised cross-over human feeding study. Br J Nutr 111:2146–2152

Maurice CF, Cl KS, Ladau J, Pollard KS, Fenton A, Pedersen AB, Turnbaugh PJ (2015) Marked seasonal variation in the wild mouse gut microbiota. ISME J 9:2423–2434

McCord AI, Chapman CA, Weny G, Tumukunde A, Hyeroba D, Klotz K, Koblings AS, Mbora DN, Cregger M, White BA (2014) Fecal microbiomes of non-human primates in Western Uganda reveal species-specific communities largely resistant to habitat perturbation. Am J Primatol 76:347–354

Mckenney EA, Melissa A, Lambert JE, Vivek F (2015) Fecal microbial diversity and putative function in captive western lowland gorillas (*Gorilla gorilla gorilla*), common chimpanzees (*Pan troglodytes*), Hamadryas baboons (*Papio hamadryas*) and binturongs (*Arctictis binturong*). Integr Zool 9:557–569

Milton K, May ML (1976) Body weight, diet and home range area in primates. Nature 259:459–462

Noma N, Suzuki S, Izawa K (1998) Inter-annual variation of reproductive parameters and fruit availability in two populations of Japanese macaques. Primates 39:313–324

Ochman H, Worobey M, Kuo CH, Ndjango JBN, Peeters M, Hahn BH, Hugenholtz P (2010) Evolutionary relationships of wild hominids recapitulated by gut microbial communities. PLoS Biol 8:8

Qin J, Li R, Raes J, Arumugam M, Burgdorf KS, Manichanh C, Nielsen T, Pons N, Levenez F, Yamada T (2010) A human gut microbial gene catalogue established by metagenomic sequencing. Nature 464(7285):59–65

Qiu X, Zhang F, Yang X, Wu N, Jiang W, Li X, Li X, Liu Y (2015) Changes in the composition of intestinal fungi and their role in mice with dextran sulfate sodium-induced colitis. Sci Rep 5:10416

Ren T, Grieneisen LE, Alberts SC, Archie EA, Wu M (2015) Development, diet and dynamism: longitudinal and cross-sectional predictors of gut microbial communities in wild baboons. Environ Microbiol 18(5):1312–1325

Rizzetto L, De FC, Cavalieri D (2014) Richness and diversity of mammalian fungal communities shape innate and adaptive immunity in health and disease. Eur J Immunol 44:3166–3181

Rodman PS, Cant JGH (1984) Anatomy and behavior. (Book reviews: adaptations for foraging in nonhuman primates). Science 226:1187–1188

Romani L (2011) Immunity to fungal infections. Nat Rev Immunol 11:275–288

Sharon G, Turnbaugh PJ, Elhanan B (2012) Metagenomic systems biology of the human gut microbiome reveals topological shifts associated with obesity and inflammatory bowel disease. PNAS 109:594–599

Shuji T, Takashi K, Tetsuo K (2014) Complex regulation of hydrolytic enzyme genes for cellulosic biomass degradation in filamentous fungi. Appl Microbiol Biotechnol 98:4829–4837

Sokol H, Leducq V, Aschard H, Pham HP, Jegou S, Landman C, Cohen D, Liguori G, Bourrier A, Nionlarmurier I (2017) Fungal microbiota dysbiosis in IBD. Gut 66:1039–1048

Solomon KV, Haitjema CH, Henske JK, Gilmore SP, Borges-Rivera D, Lipzen A, Brewer HM, Purvine SO, Wright AT, Theodorou MK (2016) Early-branching gut fungi possess a large, comprehensive array of biomass-degrading enzymes. Science 351:1192–1195

Starr C, Nekaris KAI, Leung L (2012) Hiding from the moonlight: luminosity and temperature affect activity of Asian nocturnal primates in a highly seasonal forest. PLoS One 7:e36396.1–e36396.8

Strati F, Paola MD, Stefanini I, Albanese D, Rizzetto L, Lionetti P, Calabrò A, Jousson O, Donati C, Cavalieri D (2016) Age and gender affect the composition of fungal population of the human gastrointestinal tract. Front Microbiol 7:1227

Sun B, Xi W, Bernstein S, Huffman MA, Xia DP, Gu Z, Rui C, Sheeran LK, Wagner RS, Li J (2016) Marked variation between winter and spring gut microbiota in free-ranging Tibetan Macaques (*Macaca thibetana*). Sci Rep 6:26035

Sun B, Gu Z, Wang X, Huffman MA, Garber PA, Sheeran LK, Zhang D, Zhu Y, Xia DP, Li JH (2018) Season, age, and sex affect the fecal mycobiota of free-ranging Tibetan macaques (*Macaca thibetana*). Am J Primatol 80(7):e22880

Trinci APJ, Davies DR, Gull K, Lawrence MI, Nielsen BB, Rickers A, Theodorou MK (1994) Anaerobic fungi in herbivorous animals. Mycol Res 98:129–152

Tsuji Y, Hanya G, Grueter CC (2013) Feeding strategies of primates in temperate and alpine forests: comparison of Asian macaques and colobines. Primates 54:201–215

Tung J, Barreiro LB, Burns MB, Grenier JC, Lynch J, Grieneisen LE, Altmann J, Alberts SC, Blekhman R, Archie EA (2015) Social networks predict gut microbiome composition in wild baboons. elife 4:e05224

Turnbaugh PJ, Ley RE, Mahowald MA, Magrini V, Mardis ER, Gordon JI (2006) An obesity-associated gut microbiome with increased capacity for energy harvest. Nature 444:1027–1031

Underhill DM, Iliev ID (2014) The mycobiota: interactions between commensal fungi and the host immune system. Nat Rev Immunol 14:405–416

Ungar PS (1995) Fruit preferences of four sympatric primate species at Ketambe, Northern Sumatra, Indonesia. Int J Primatol 16:221–245

Van Dyke M, McCarthy A (2002) Molecular biological detection and characterization of Clostridium populations in municipal landfill sites. Appl Environ Microbiol 68:2049–2053

Wheeler ML, Limon JJ, Bar AS, Leal CA, Gargus M, Tang J, Brown J, Funari VA, Wang HL, Crother TR (2016) Immunological consequences of intestinal fungal dysbiosis. Cell Host Microbe 19(6):865–873

Wu Q, Wang X, Ding Y, Hu Y, Nie Y, Wei W, Ma S, Yan L, Zhu L, Wei F (2017) Seasonal variation in nutrient utilization shapes gut microbiome structure and function in wild giant pandas. Proc Biol Sci 284:20170955

Xiong C, Wang Q (1988) Seasonal habitat used by Thibetan Monkeys. Acta Theriol Sinica 8:176–183

Xu B, Xu W, Yang F, Li J, Yang Y, Tang X, Mu Y, Zhou J, Huang Z (2013) Metagenomic analysis of the pygmy loris fecal microbiome reveals unique functional capacity related to metabolism of aromatic compounds. PLoS One 8(2):e56565

Xu B, Xu W, Li J, Dai L, Xiong C, Tang X, Yang Y, Mu Y, Zhou J, Ding J (2015) Metagenomic analysis of the *Rhinopithecus bieti* fecal microbiome reveals a broad diversity of bacterial and glycoside hydrolase profiles related to lignocellulose degradation. BMC Genomics 16:174

Yamada A, Muroyama Y (2010) Effects of vegetation type on habitat use by crop-raiding Japanese macaques during a food-scarce season. Primates 51:159

Yatsunenko T, Rey FE, Manary MJ, Trehan I, Dominguez-Bello MG, Contreras M, Magris M, Hidalgo G, Baldassano RN, Anokhin AP (2012) Human gut microbiome viewed across age and geography. Nature 486(7402):222–227

Yokoe Y, Yasumasu I (1964) The distribution of cellulase in invertebrates. Comp Biochem Physiol 13:323–338

You SY, Yin HB, Zhang SZ, Tian-Ying JI, Feng XM (2013) Food habits of *Macaca thibetana* at Mt. Huangshan,China. J Biol 30(5):64–67

Zhang M, Li J, Zhu Y, Wang X, Wang S (2010) Male mate choice in Tibetan macaques Macaca thibetana at Mt. Huangshan, China. Curr Zool 56:213–221

Zhao Q-K (1999) Responses to seasonal changes in nutrient quality and patchiness of food in a multigroup community of Tibetan macaques at Mt. Emei. Int J Primatol 20:511–524

Zhou Q, Wei F, Huang C, Li M, Ren B, Luo B (2007) Seasonal variation in the activity patterns and time budgets of *Trachypithecus francoisi* in the Nonggang Nature Reserve, China. Int J Primatol 28:657–671

Zhu L, Wu Q, Dai J, Zhang S, Wei F (2011) Evidence of cellulose metabolism by the giant panda gut microbiome. PNAS 108:17714–17719

Chapter 12
Medicinal Properties in the Diet of Tibetan Macaques at Mt. Huangshan: A Case for Self-Medication

Michael A. Huffman, Bing-Hua Sun, and Jin-Hua Li

12.1 Introduction

Life history strategies include growth, maintenance, and reproduction (Gadgil and Bossert 1970), all of which are dependent upon a proper diet for metabolic functions. Animal feeding strategies are based on finding and consuming a balance of the most essential nutritional elements, carbohydrates, fats, proteins, trace elements, and vitamins, while at the same time avoiding the negative impacts of secondary metabolites in plants (Lambert 2011; Simpson et al. 2004). These secondary metabolites protect plants from predation by an array of insect and vertebrate herbivores that prey upon them by reducing palatability and digestibility (Freeland and Janzen 1974; Glander 1982; Rosenthal and Berenbaum 1992). Nonetheless, this does not always inhibit animals from ingesting such plants in tolerable amounts for purposes other than nutrition.

The idea that animals may ingest plants for their medicinal value was first suggested by Janzen (1978), based on a variety of anecdotal reports from the wild. The study of primate self-medication then began in earnest as a scientific discipline in the mid-to-late 1980s with observations of chimpanzees in the wild (see Huffman 2015). Nonetheless, it is widely documented that humans have traditionally seen animals as a source of knowledge about the use of plants for their medicinal value

M. A. Huffman (✉)
Primate Research Institute, Kyoto University, Kyoto, Japan
e-mail: huffman.michael.8n@kyoto-u.ac.jp

B.-H. Sun
School of Resources and Environmental Engineering, Anhui University, Hefei, Anhui, China

J.-H. Li
School of Resources and Environmental Engineering, Anhui University, Hefei, Anhui, China

International Collaborative Research Center for Huangshan Biodiversity and Tibetan Macaque Behavioral Ecology, Anhui, China

School of Life Sciences, Hefei Normal University, Hefei, Anhui, China
e-mail: jhli@ahu.edu.cn

J.-H. Li et al. (eds.), *The Behavioral Ecology of the Tibetan Macaque*, Fascinating Life Sciences, https://doi.org/10.1007/978-3-030-27920-2_12

(Engel 2002; Huffman 1997, 2002, 2007). Humans can learn from watching sick wild primates, because we share the same evolutionary history, possess a common physiology, and have lived together under similar environmental conditions for much of our species' history. It has been argued that we have inherited many of the same ways to combat common diseases in the environment (Huffman 2016). Indeed, recent archeological and biochemical evidence supports the idea by showing that one of our closest extinct ancestors, *Homo neanderthalensis*, also used medicinal plants still widely in use today by modern humans (Hardy et al. 2012, 2013; Huffman 2016).

Self-medication research focuses on understanding how animals respond to illness and how these behaviors can be transmitted across generations (Huffman 1997). It has also been argued to be a bio-rational for the exploration and exploitation of novel secondary plant compounds and new insights into how they can be used for the management of health in humans and livestock (Huffman et al. 1998; Krief et al. 2005; Petroni et al. 2016). At the proximate level, self-medication may be driven by the necessity to maintain physiological homeostasis to stay in relatively good condition (Foitova et al. 2009; Forbey et al. 2009).

Currently, the majority of evidence for self-medication in animals comes from the study of how they deal with parasite or pathogen-induced illness (see Huffman 1997, 2011). While some parasitic infections likely go unnoticed, when homeostasis is disrupted or threatened, it is expected to be in the best interest of the host to actively respond in ways to alleviate discomfort. However, there is no reason why self-medication should be limited to parasitosis, since animals are faced with a wide variety of health homeostatic challenges brought upon by such factors as reproductive events, climatic extremes, or other seasonal events (Carrai et al. 2003; Huffman 1997, 2011; Ndagurwa 2012). The ability of a species to defend itself against life-threatening conditions provides a significant adaptive advantage and thus should be present throughout the entire animal kingdom.

In 1997, the concept of "medicinal foods" was formally introduced to primatology, adding the extra element of passive prevention of disease based on the presence of plants in the diet that contain noticeable bioactive, physiology-modifying properties, from which the animals ingesting them could potentially benefit (Huffman 1997). This term was borrowed from the human ethnopharmacological literature (Etkin 1996). For example, among the Hausa of Nigeria, 30% of the wild plant food species they ingest are also used as medicine. Interestingly, of the species used by these people to treat symptoms of malaria, 89% are also eaten as food (Etkin and Ross 1983). Many food items eaten by primates and other mammals have also been shown to contain a variety of secondary metabolites with medicinal properties (roughly 15–25% of any population's food plant species list), suggesting that animals may benefit from the periodic ingestion, in small amounts of these plants (sifaka *Propithecus verreauxi verreauxi*, Carrai et al. 2003; gorillas *Gorilla gorilla* and *G. beringei*, Cousins and Huffman 2002; chimpanzees *Pan troglodytes*, Huffman 1997, 2003; Japanese macaques *Macaca fuscata*, Huffman and MacIntosh 2012; MacIntosh and Huffman 2010; various ungulate species, Mukherjee et al. 2011; lemurs *Eulemur fulvus*, Negre et al. 2006; wooly spider monkeys *Brachyteles arachnoides*, Petroni et al. 2016). The secondary compound rich content of some foods in the diet may play a significant role in the maintenance of health. Two examples associated with risks to parasite infection illustrate this point.

The medicinal diet of chimpanzees in the Mahale M group of Tanzania was examined (Huffman 1998) by conducting a database search using the African ethnomedicine literature. From 172 chimpanzee food species, 43 (22%) items were found to be used to treat parasitic- or gastrointestinal-related illnesses by humans. It was also common for some species to have multiple ethnomedicinal uses. While not all 43 species may have been ingested by chimpanzees in such a way as to benefit from these potential medicinal properties, 33% (20/63) of the plant parts ingested (leaf and stem = 75%, bark 15%, seed = 5%, fruit = 5%) from 16 of these species corresponded to the parts utilized by humans specifically for the treatment of intestinal parasites and gastrointestinal illness.

The medicinal diet and parasite richness of ten Japanese macaque troops were examined (MacIntosh and Huffman 2010). The study troops were selected to represent the species entire distribution, ranging from the extreme cold temperate zone of Shimokita peninsula down to the subtropical island of Yakushima. A total of 1664 plant part items (range, 56–408) from 694 species were the target of an extensive literature search for potential antiparasitic activity in the diet. Of all these ingested items, 198 (from 135 species) were found to have reported antiparasitic properties. The proportion of these antiparasitic items ranged from 12 to 18% across these ten troops. A further 167 plant items (133 species) exhibited medicinal properties not related to parasitic infection or gastrointestinal symptoms. Because nematode species richness is negatively associated with latitude among Japanese macaques (Gotoh 2000), it was predicted that the proportion of antiparasitic items in their diets would also follow the same pattern (MacIntosh and Huffman 2010). A tendency was noted for the proportion of antiparasitic food items to decrease with increasing latitude, and a strong positive statistically significant relationship between the proportions of antiparasitic items ingested and nematode species richness was found. The proportion of medicinal items unrelated to parasite activity showed no such relationship with either latitude or with nematode species richness, supporting the hypothesis that the medicinal diet was somehow influenced by the degree of parasite pressure, relative species richness, and potentially parasite load.

From these and other studies noted above, a pattern is emerging, by which primates and other mammals incorporate certain food items with medicinal properties into their diet. In primates the information thus far shows a strong connection between medicinal food consumption and parasite infection and or reproductive events. Both factors present significant challenges to the survival and fitness of an individual. Knowing what immediate homeostasis challenges that impact individuals of a group can help to better understand how the medicinal components of their diet may work in an animal's favor.

Currently, limited details are available about health and diseases affecting Tibetan macaques (*Macaca thibetana*). In a case study of the troop fission event of YA and YB, the sudden mass death of 17 individuals over a 1-week period was mentioned (Li et al. 1996). Disease was suggested to be responsible, but no diagnosis was given. The speed and widespread effect could be suggestive of an aggressive viral infection, whose virulence was perhaps exacerbated by high stress levels.

Zhu et al. (2012) report the presence of nine intestinal nematode species in Tibetan macaques, with special note of the zoonotic *Gongylonema pulchrum*, having the highest prevalence rate of 31.58%, followed by *Trichuris trichiura* (25.00%), *Oesophagostomum apiostomum* (23.68%), *Ancylostoma duodenale* (hookworm) (14.47%), *Trichostrongylus* sp. (13.16%), and other species of lower prevalence. These parasites are known to be responsible for mild to severe pathogenesis (Brack 1987). Among them, *O. apiostomum* is noted to be responsible for perhaps the severest pathogenesis, in particular among previously infected (pre-immunized) individuals. In these cases, encysted larvae are trapped in the intestinal mucosa by elevated immune response, causing the larvae to die inside the cyst, leading to inflammation, necrosis, and hemorrhaging, and in severe cases leading to secondary bacterial infections, necrosis of the intestinal mucosa, weight loss, weakness, and mortality (Brack 2008). *O. stephanostomum*, a sister species infecting chimpanzees, and other great apes, is equally pathogenic and is associated with self-medicative behaviors used in the therapeutic treatment with *Vernonia amygdalina* by chimpanzees with high-level infections during the rainy season (Huffman 1997; Huffman and Caton 2001). Furthermore, evidence for the possible role of medicinal foods for controlling *O. stephanostomum* and other infections has also been suggested (Huffman 1997).

While more work is clearly needed to understand the different factors affecting the homeostasis of Tibetan macaques (*Macaca thibetana*) possibly leading to self-medication, we take this opportunity to evaluate their diet for its potential medicinal value. We predict there is a proportion of the diet containing plants with medicinal value, and it is our goal to highlight that potential for future research, in order to better understand the role of diet in health maintenance and self-medication in Tibetan macaques.

12.2 Materials and Methods

The subjects of this investigation are Tibetan macaques living in the Valley of the Monkeys (118°11′ W, 30°29′ N), Mt. Huangshan, Anhui Province, China. The site is situated at an elevation of approximately 1840 m above sea level, and the year is divided into four seasons: winter (December to February), spring (March to May), summer (June to August), and autumn (September to November). Average temperatures range from around 0 °C in mid-winter to around 25 °C in summer (Fig. 12.1). Peaks in rainfall occur during the months of May through July, tapering off in autumn and winter. (Further details of the study site are presented in other chapters of this book.)

Analysis of the diet of Yulinkeng 1 (YA1) troop was based on a plant food list previously published by You et al. (2013). At the time of that study, the troop consisted of 28 individuals: 12 adults (4 males, 8 females), 10 juveniles (6 males,

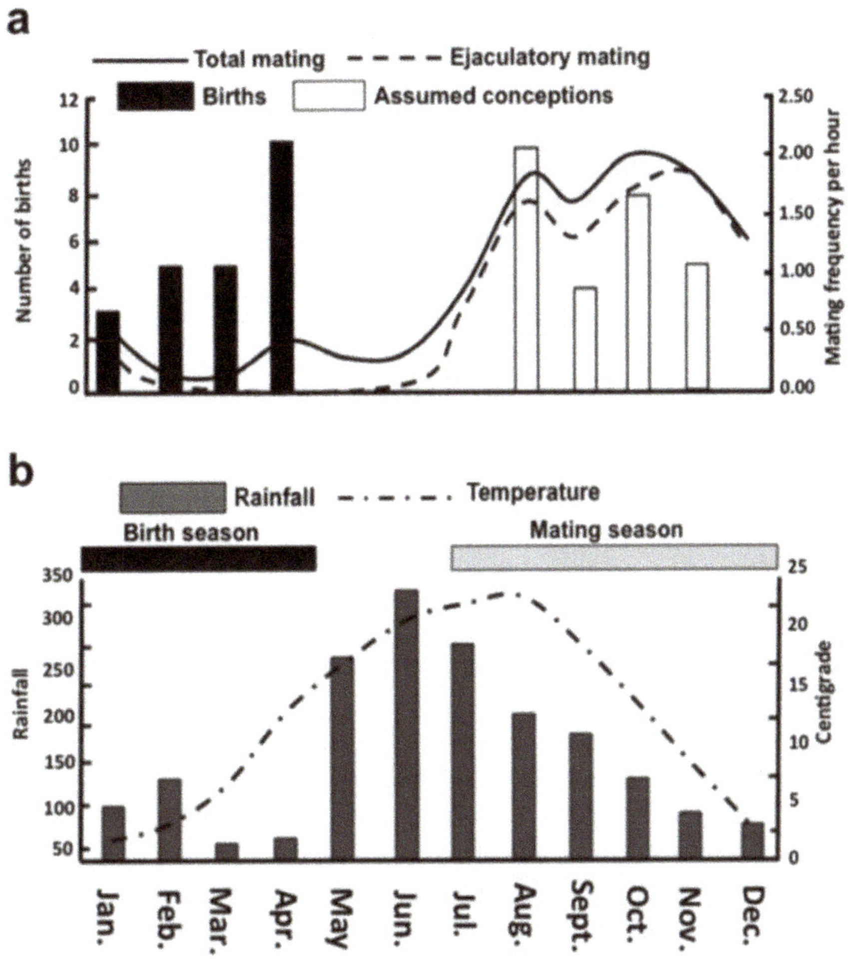

Fig. 12.1 Annual distribution of mating, births, and conceptions of Tibetan macaques (**a**) and monthly rainfall and temperature (**b**) at Mt. Huangshan. These figures were redrawn, modified, and combined from previously published material for illustrative purposes (Li et al. 2005). The meteorological data presented here is the average of 10 years (1983–1992)

4 females), and 6 infants (5 males, 1 female). Direct observations of the troop's feeding behavior in the rugged terrain of Mt. Huangshan are challenging, and to overcome these difficulties, the investigators produced a food list based on the analysis of 81 fecal samples collected over 10-day blocks, one block each during winter, spring, summer, and autumn between November 2011 and October 2012 (You et al. 2013). While the study covered just 1 year, it provides an important basis for an evaluation of the medicinal potential of food items ingested by the troop during that period.

The method used to quantify the relative contribution of each plant species (part) took a microhistopathological approach to identify ingested food remains in the feces. The method is a validated, well-established, procedure developed by Sparks and Malechek (1968) and has been used to quantify the botanical composition of diets in a variety of free-ranging, sometimes difficult to directly monitor, domestic and wild animals in their natural habitats (e.g., mule deer *Odocoileus hemionus*,

Anthony and Smith 1974; musk deer *Moschus leucogaster*, Green 1987; domestic goats *Capra aegagrus hircus*, Mellado et al. 1991; cattle *Bos taurus*, sheep *Ovis aries*, Angora goats, Alipayo et al. 1992; wild turkeys *Meleagris gallopavo*, Rumble and Anderson 1993; Yunnan snub-nosed monkeys *Rhinopithecus bieti*).

The method requires the assembly of histopathological plant tissue reference key slides prepared from identified plant species in the habitat. These keys are then used to microscopically identify plant parts in the feces based on each species' unique cell tissue structure characteristics (see Ahmed et al. 2015). At Huangshan, a total of 224 plant species (73 families) in the habitat were identified and histopathological reference keys were made (You et al. 2013). They calculated the relative density (RD) of each species (and family) present in the feces using the following formula, where:

$$\text{RD} = \text{the density of a plant particle/the total density of all plant particles} \times 100.$$

We evaluated the medicinal properties of each plant species in the resulting list for medicinal value using online database sources [Traditional Chinese Medicinal Plants (Duke and Ayensu 1985); Find Me A Cure (2018); Herbpathy Data Base (2018); Plants For A Future Data Base (1996–2012); World Agroforestry Data Base (2018)], followed up with Google Scholar article searches by plant species and/or active compound name, focusing on the plant items (leave, fruits, stems, bark, roots, flowers) ingested by YA1 troop members.

Data sorting and descriptive statistics were carried out using Microsoft® Excel® for Mac 2011 (Ver. 14.7.2.). Statistical analyses for chi-square were carried out using an online calculator: https://www.socscistatistics.com/tests/chisquare2/Default2.aspx. Statistical significance was set at 0.05.

12.3 Results and Discussion

12.3.1 Plant Food Species and Their Relative Density (RD) Values

To put the medicinal foods ingested by members of the troop into perspective, we first describe the overall trends of their feeding habits as revealed in the dietary analysis. The plant food species consumed by YA1 troop members are listed in descending order of RD by family and species in Table 12.1. The species RD values are summarized seasonally, with each season totaling 100%. Fifty species (61 different items, 26 families) from across the entire study period were analyzed here. Of the 61 items ingested, leaves accounted for 78%, by far the largest proportion of plant parts found, followed by fruits 11%, stems 3%, buds 2%, seeds 2%, young shoots 2%, and flowers 2%.

Table 12.1 Plant food species, part(s) eaten, and seasonal variation of use in the YA1 troop of Tibetan macaques at Mt. Huangshan

Family and species	Parts Consumed	Relative density (RD) Winter	Spring	Summer	Autumn
Fagaceae					
Castanpopsis eyrei (Champion ex Bentham) Tutcher	Leaf, fruit	14.78%	9.17%	10.49%	14.48%
Lithocarpus glaber (Thunb.) Nakai	Leaf, fruit	6.60%	6.62%	4.27%	6.46%
Quercus myrsinaefolia Blume	Leaf, fruit	6.06%	7.38%	7.41%	9.37%
Quercus glauca Thunb.	Leaf, fruit	2.50%	3.12%	5.05%	7.69%
Quercus glandulifera Blume	Leaf, fruit	0	1.22%	5.41%	2.08%
Quercus aliena Blume	Leaf, fruit	0	0	0	0.60%
Lauraceae					
Litsea coreana H. Léveillé	Leaf	12.09%	8.08%	9.55%	4.66%
Machilus leptophylla Handel-Mazzetti	Leaf	4.53%	2.53%	6.84%	7.11%
Phoebe sheareri (Hemsley) Gamble in Sargent	Leaf	4.04%	2.88%	5.12%	5.55%
Machilus thunbergii Siebold & Zuccarini	Leaf, bud	0.66%	13.50%	2.20%	1.68%
Lindera aggregata (Sims) Kostermans	Leaf	0	0.17%	0.30%	0.07%
Poaceae					
Zea mays L. (corn)	Seed	8.94%	7.13%	6.76%	4.06%
Carex tristachya Thunberg in Murray	Leaf	2.08%	2.82%	4.78%	1.37%
bamboo	Young shoots	3.47%	2.88%	2.65%	2.16%
Ericaceae					
Rhododendron ovatum (Lindley) Planchon ex Maximowicz	Leaf	3.37%	3.97%	1.54%	2.65%
Vaccinium bracteatum Thunberg in Murray	Leaf	2.05%	1.34%	2%	1.06%
Rhododendron sp.	Leaf, flower	1.27%	0.87%	2.07%	1.22%
Vaccinium mandarinorum Diels	Leaf		1.40%	1.30%	0.68%
Hamamelidaceae					
Loropetalum chinense (R. Brown) Oliver	Leaf, stem	5.18%	4.09%	3.51%	2.81%
Distylium myricoides Hemsley	Leaf, stem	1.31%	0.35%	1.26%	1.61%
Liquidambar formosana Hance	Leaf	0	0.23%	0	0
Theaceae					
Camellia cuspidata (Kochs) H. J. Veitch	Leaf	5.18%	4.76%	2.84%	3.81%
Eurya alata Kobuski	Leaf	1.09%	0.64%	1.44%	0.37%
Eurya muricata Dunn	Leaf	0	0.58%	0	1.45%
Eurya nitida Korthals	Leaf	0	0.23%	3.58%	2.08%

(continued)

Table 12.1 (continued)

Family and species	Parts Consumed	Relative density (RD)			
		Winter	Spring	Summer	Autumn
Leguminosae					
Millettia dielsiana Harms. ex Diels.	Leaf	2.36%	2.53%	1.88%	1.84%
Lespedeza bicolor Turczaninow	Leaf	0	0.35%	1.44%	0
Saxifragaceae					
Itea omeiensis C. K. Schneider in Sargent	Leaf	2.09%	1.34%	1.09%	0.22%
Taxaceae					
Torreya grandis Fortune ex Lindley	Leaf	1.37%	0.64%	0	0
Aquifoliaceae					
Ilex purpurea Hassk.	Leaf	0.52%	0.29%		0.68%
Myrtaceae					
Syzygium buxifolium Hooker & Arnott	Leaf	0.49%	0.06%	0.75%	0.15%
Cephalotaxaceae					
Cephalotaxus fortunei Hooker	Leaf	0.40%	0	0	0
Taxodiaceae					
Cunninghamia lanceolata (Lambert) Hooker	Leaf, fruit	7.60%	0.11%	0.91%	2.08%
Ranunculaceae					
Thalictrum aquilegifolium L.	Leaf	0	0.11%	0.53%	0
Anacardiaceae					
Toxicodenddron sylvestre (Siebold & Zucc.) Kuntze	Leaf	0	0.17%	0.83%	0.30%
Araliaceae					
Hedera nepalensis var. *sinensis* (Tobler) Rehder	Leaf	0	0.29%	0	1.76%
Styracaceae					
Pterostyrax corymbosus Siebold & Zuccarini	Leaf	0	0.23%	0	0
Tiliaceae					
Tilia oliveri Szyszyłowicz	Leaf	0	0.52%	1.13%	0
Berberidaceae					
Epimedium davidii Franchet	Leaf	0	0.23%	0	0
Gesneriaceae					
Conandron ramondioides Siebold & Zuccarini	Leaf	0	0	0	0.30%

(continued)

Table 12.1 (continued)

Family and species	Parts Consumed	Relative density (RD) Winter	Spring	Summer	Autumn
Araceae		0	1.33%	0	0
Pinellia cordata Dunn.	Leaf	0	0.11%	0	0
Grassleaf Sweetfalg	Leaf	0	1.22%	0	0
Lardizabalaceae		0	0.46%	0	2.05%
Akebia trifoliata (Thunberg) Koidzumi	Leaf	0	0.06%	0	0.83%
Akebia quinata (Houttuyn) Decaisne	Leaf	0	0.23%	0	1.22%
Sargentodoxa cuneata (Oliver) Rehder & E. H. Wilson in Sargent	Leaf	0	0.17%	0	0
Rosaceae					
Rhaphiolepis indica (Linnaeus) Lindley	Leaf	0	2.88%	1.07%	1.92%
Photinia davidsoniae Hook.	Leaf	0	0.58%	0	0.15%
Cucurbitaceae					
Thladiantha nudiflora Hemsley	Leaf	0	0.46%	0	0
Actinidiaceae					
Actinidia lanceolata Dunn.	Leaf	0	0	0	3.47%
Menispermaceae					
Menispermum dauricum Candolle	Leaf	0	1.40%	0	0
Food species consumed		18	35	25	30
Medicinal food species consumed		7	11	6	7

List is modified from You et al. (2013). RD percentages are summarized per season. $n = 50$ species
Medicinal food species consumed are highlighted in grey

Seasonal variation was noted in the number of plant species ingested from their total repertoire (Fig. 12.2), but these differences were not statistically significant ($\chi^2 = 6.93$, $df = 3$, $p > 0.05$). The largest number of species consumed was in spring ($n = 46$), followed by autumn ($n = 37$), summer ($n = 31$), and winter ($n = 25$). With the exception of one species, *Cunninghamia lanceolata* (Taxodiaceae) whose leaves and fruit were both consumed, all ingested fruit items (nuts, acorns) come from the Fagaceae family. Combined, use of leaf and fruit items in this family had the highest RD values of all plant food species identified, and none of these were classified as medicinal foods. The highest of these RD values were recorded in autumn and summer and involved the consumption of ripening and fallen ripe fruits, the main food items sought after in these two seasons. The second largest RD value family was Lauraceae, with five species, whose leaves and buds were consumed. Only one of the five species in this family, *Litsea coreana,* was classified as a medicinal food.

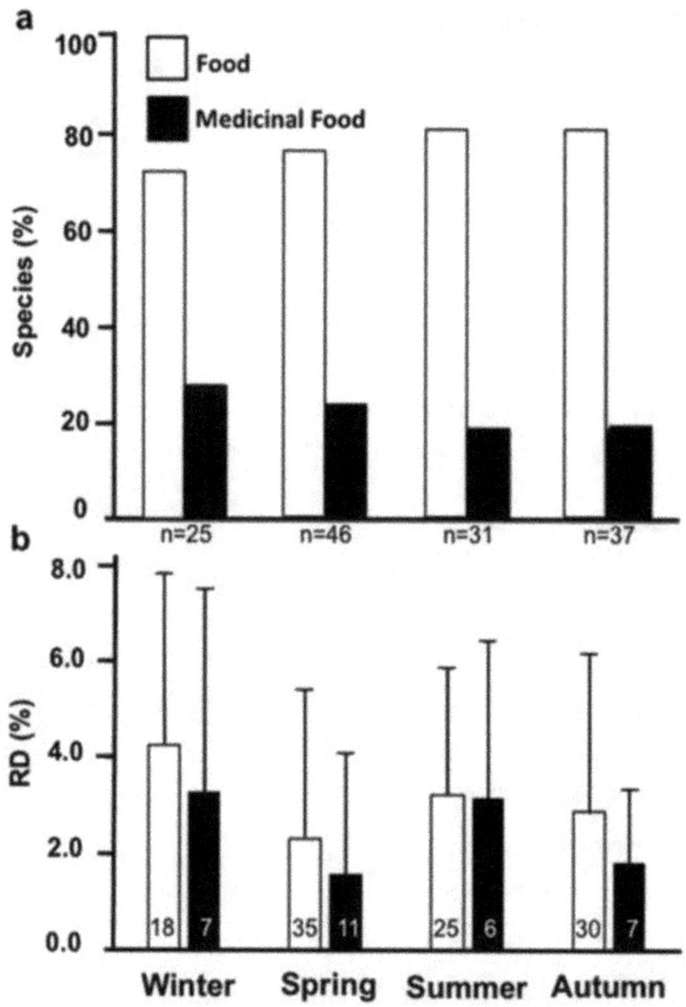

Fig. 12.2 Seasonal variation in the ingestion of food and medicinal food species (**a**) and average RD values of food and medicinal food species by YA1 troop. (**b**) Mean RD values calculated from individual species RD values per season presented in Table 12.1. Data derived from You et al. (2013)

12.3.2 Medicinal Foods in the Diet

Across the entire study period, 12 species (13 items: 12 leaves, 1 stem) in the diet had notable medicinal activity in the plant part ingested by members of YA1 troop (Table 12.1). Our literature search revealed an array of pharmacological properties of high medicinal value including antiparasitic, antiprotozoal, antibacterial, antifungal, antiviral, anti-dysentery, anti-enteritis, anticancer (antitumor), anti-inflammation, antirheumatic, antidiabetic, cardiovascular protective, neuroprotective, osteoprotective, reproductive stimulant, antidepressant, immunosuppressant, and diaphoretic; treatment for ulcers, wounds, skin disease, weakness, and dizziness; and other health-protecting and health-promoting activities. Associated with these activities are a variety of physiologically active secondary plant metabolites (Appendix).

The following are detailed descriptions of the 12 plant species' seasonality of use, life form, pharmacological properties, and prescribed uses by humans.

***Cephalotaxus fortunei* (Leaf, Used in Winter) Evergreen Tree** Abietane diterpenoids synthesized by suspension-cultured cells displayed a wide spectrum of biological activities including antiparasitic, antibacterial, antifungal, and antiviral properties. These compounds were found to be effective against pathogens such as *Mycobacterium tuberculosis* and *Staphylococcus aureus*, including methicillin-resistant (MRSA) strains and biofilm infection of *S. aureus* (Neto et al. 2015). Abietane diterpenes isolated from aerial parts of *Plectranthus barbatus* showed remarkable activity with acceptable selectivity against the blood parasites *Plasmodium falciparum* (malaria) and *Trypanosoma brucei* (sleeping sickness in animals). Nonspecific antiprotozoal activity was also detected and is likely due to the compound's high cytotoxicity (Mothana et al. 2014). This plant also demonstrates significant anticancer activity (Duke and Ayensu 1985).

***Epimedium davidii* (Leaf Used in Spring) Perennial Herb** Compounds isolated from the leaves include Icariside II (Baohuoside I) and Baohuoside II, III, V, and VI (Ma et al. 2011). These compounds possess therapeutic activities such as osteoprotective effect, neuroprotective effect, cardiovascular protective effect, anticancer effect, anti-inflammation effect, immunoprotective effect, and enhancement of reproductive function (sexual health for males and females) (Li et al. 2015). Treatment with the compound icariin is reported to significantly increase epididymal sperm counts and testosterone levels of male rats (Chen et al. 2014) and significantly increases erectile function in castrated Wistar rats by increasing the percentage of smooth muscle and inducible nitric oxide synthase in the corpus cavernosum (Liu et al. 2005). Active ingredients administered at a dose of 5 g/kg^{-1} with an ig volume of 5 mL kg^{-1} alleviated the impact of high-intensity exercise on serum testosterone, maintaining it at normal physiological levels. It can also promote protein synthesis, inhibit degradation of amino acid and protein, and increase hemoglobin and glycogen reserves in rats receiving exercise training (Zhou et al. 2013).

A series of tests on rats and mice using icariin have also reported significant stress reduction and antidepressant properties via downregulation of glucocorticoid receptor activity and regulation of hippocampal neuroinflammation are associated with this plant species (Pan et al. 2010; Wu et al. 2011; Gong et al. 2016). The leaves of this species are also noted for their anticancer activity (Frohne and Pfänder 1984), immunosuppressive action, and inhibitory properties of lymphocyte activation (Ma et al. 2004).

***Hedera nepalensis* (Leaf Used in Spring and Autumn), Evergreen Climber or Creeper Vine** The leaves and berries are reported to have cathartic, diaphoretic, and stimulant properties. A decoction is used to treat skin diseases. Significant cancer chemo-preventative and cytotoxic properties have been demonstrated (Jafri et al. 2016). It is also reported to be an important folk medicine for the treatment of diabetes. No specific mention of *H. nepalensis* is given, but a closely related species *Hedera helix* is relevant to this discussion, as it has a similar chemical makeup and has been noted to be poisonous in large doses, but the leaves are eaten without observed side effects in wild mammals. Leaves contain hederagenin, a saponic glycoside, which can cause gastrointestinal nervous system disturbances, breathing difficulties, and coma if eaten in large amounts (Plants For A Future.com).

***Ilex purpurea* (Leaf Used in Winter, Spring, and Autumn) Evergreen Tree** The plant is considered to be one of the 50 fundamental herbs in Chinese medicine. It is reported to have antitumor properties, and an extract of the leaves is made into a solution and used for treating burns and ulcers in the lower extremities. The ashes of burnt leaves are used as a dressing for skin ailments and infected wounds (Duke and Ayensu 1985).

***Lespedeza bicolor* (Leaf Used in Spring and Summer) Deciduous Shrub** Leaves contain flavonoids, alkaloids, terpenes, organic acids, and stigmasterols. Potential for antioxidant, anticancer, and antibacterial activity is also noted (Ullah 2017) and is traditionally used for coughs and colds, kidney and urethra problems, fever, headache, weakness, and dizziness (Chang et al. 2017).

***Liquidambar formosana* (Leaf Used in Spring) Deciduous Tree** The leaves are used in the treatment of cancerous growths (Duke and Ayensu 1985).

***Litsea coreana* (Leaf Used Year Round) Evergreen Tree** Exhibits notable bioactivities, such as hepatoprotection, hyperglycemia, anti-inflammation, antioxidation, and antibacterial properties through multiple molecular mechanisms. These compounds augment immunoglobulin M and G values and show significant inhibitory effects on the pathogens *Bacillus anthracis, Proteusbacillus vulgaris, Staphylococcus aureus*, and *Bacillus subtilis*. Anti-HSV-1 activity and anti-gastric carcinoma and anti-colon carcinoma HT-29 activity have been reported. Leaves contain polysaccharides, polyphenols, essential oils, and numerous flavonoids (Jia et al. 2017).

***Loropetalum chinense* (Leaf, Stem Used Year Round) Evergreen Woody Shrub** The leaves are crushed and pulverized for external application on wounds (Find Me A Cure.com), while a decoction of the whole plant is used to treat coughing in tuberculosis patients and is a treatment against dysentery and enteritis. This species also promotes wound healing and possesses antibacterial, anti-inflammatory, antioxidant, and antitumoral activity, as well as adjusts fat metabolism, and protects from cardiovascular disease (Zhou et al. 2014).

***Millettia dielsiana* (Leaf Used Year Round) Deciduous Shrub** A decoction or tincture of leaves is used to treat hookworm, roundworm (nematodes), and filarial infections and is a treatment for amenorrhea, metrorrhagia, anemia, traumatic injuries, and rheumatoid arthritis (Anonymous 1977; Yeung 1985). Stem extracts exhibit anti-enterovirus activity and is effective against Coxsackie virus B3, Coxsackie virus B5, Poliovirus I, Echovirus 9, and Echovirus 29 (Guo et al. 2006). It also aids in the prevention of cardiovascular and cerebrovascular disease. It possesses high antioxidant activity (Gan et al. 2010). Seven isoflavones have been isolated from the stem and identified as 6-ethoxyca lpogonium isoflavone A, durmillone, ichthynone, jamaicin, toxicaro l isoflavone, barbigerone, and genistein (Gong et al. 2007).

***Sargentodoxa cuneata* (Leaf Used in Spring) Deciduous Climbing Shrub** The stem is anthelmintic, antibacterial, antirheumatic, carminative, diuretic, and tonic (Usher 1974; Anonymous 1977; Yeung 1985; Duke and Ayensu 1985). The mashed leaves are plastered onto sores (Duke and Ayensu 1985). A decoction or tincture is used in the treatment of anemia, traumatic injuries, rheumatoid arthritis, hookworm disease, roundworm, and filariasis (Anonymous 1977; Yeung 1985). Stem extracts exhibited anti-enterovirus activity against anti-Coxsackie virus B3, Coxsackie virus B5, Polio virus I, Echovirus 9, and Echovirus 29 (Guo et al. 2006).

***Syzygium buxifolium* (Leaf Used Year Round) Evergreen Shrub** The juice of macerated leaves are taken to reduce fever (Duke and Ayensu 1985). Leaf powder is rubbed on the skin of smallpox patients for its cooling effect (World Agroforestry.org).

***Vaccinium bracteatum* (Leaf Used Year Round) Evergreen Shrub** Leaves contain isoorientin, orientin, vitexin, isovitexin, isoquercitrin, quercetin-3-O-α-L-rhamnoside, and chrysoeriol-7-O-β-D-glucopyranoside. Radical scavenging activity and protection against KBrO3-mediated kidney damage have been demonstrated (Zhang et al. 2014a, b). Anticancer and anti-inflammatory activity was demonstrated (Landa et al. 2014).

12.3.3 Seasonality of Medicinal Food Ingestion

In total, 76% of the 50 species (61 food items from 26 families) reported in YA1 troop's diet showed no evidence of significant toxicity or pharmacological activity, supporting the assumption that most plants in the diet are indeed selected for their nutritional value. However, the remaining 24% ($n = 12$) of the species are considered medicinal foods, with potential health-promoting properties.

There was no statistically significant seasonal difference in the number of medicinal foods in the diet ($\chi^2 = 0.9371$, $p > 0.05$, $df = 3$; Fig. 12.2). However, more medicinal food species were consumed in winter (28%) and spring (24%) and then summer (19%) or autumn (19%). All the consumed medicinal plant items were leaves. Seven of these species are evergreen trees, shrubs, or creepers, so leaves of these species are available year-round. The remaining five species were deciduous trees or shrubs, with leaves only in spring and summer months. These results suggest a broad based potential for access to medicinal food across the year. While the number of potential medicinal species ingested is not significantly different across seasons, the particular species ingested varied between seasons in some cases and present some interesting patterns related to their possible seasonal benefits. The three following categories of health concern extracted from our analysis provide some important areas for future investigation.

12.3.4 Antiparasitic Properties

Previous reports of medicinal food ingestion and therapeutic self-medication in response to parasite infections point out the relationship between seasonality of reinfection and plant ingestion as evidence for the context of plant use (Huffman et al. 1997, 1998). While no clear-cut seasonal trend for medicinal food ingestion was apparent in the YA1 troop diet, the relationship between medicinal properties of five specific medicinal food species and their seasonality of use provides information for future detailed investigation. *C. fortunei, L. bicolor, L. coreana, L. chinese,* and *S. cuneata* are reported to possess significant broad-spectrum antiparasitic, antibacterial, and antiviral activity. One of the highest RD values of all medicinal foods was assigned to *L. coreana* (RD = 12.09%) consumption in winter. However, two of the most intriguing medicinal food candidates with wide-spectrum antiparasitic properties are *C. fortunei* and *S. cuneat*a, even though both had low RD values (0.40 and 0.17, respectively). Ingestion was restricted to winter in the former and spring in the later.

Winter and spring, the mating and birth seasons, respectively, could be potentially key seasons for investigating the parasite infection status of individuals before and after the ingestion of these plant species. One of the three zoonotic parasites identified in YA1 troop is *O. apiostomum* (Zhu et al. 2012). Infection in Japanese macaques by *O. apiostomum* occurs more frequently during winter months (MacIntosh et al. 2010), and self-medication in response to a sister species, *O. stephanostomum*, occurs in chimpanzees (Huffman 1997). The reduction of worm burden and temporary relief from related gastrointestinal upset has been linked to the ingestion of the bitter pith of *V. amygdalina* (Huffman et al. 1993, 1996a). In vitro pharmacological assays of *V. amygdalina* demonstrate a broad range of antiparasitic activities (e.g., Ohigashi et al. 1994; Oyeyemi et al. 2018).

For respiratory viruses, seasonal changes in humidity improve viral survival and increase opportunities for infection (Altizer et al. 2006). A general survey of disease prevalence in YA1 troop is necessary to provide further insights about the ecology of infection dynamics that could lead to important links with their medicinal diet. *L. bicolor* is associated with possible respiratory health, and a number of other species could inhibit respiratory viruses from establishment (Chang et al. 2017).

12.3.5 Reproductive Modulation

Throughout history, humans have utilized a number of plant hormones to suppress or enhance their reproductive and sexual activity (Lewis and Elvin-Lewis 1977). The reproductive behavior of male and female Tibetan macaques has been studied from a variety of perspectives including endocrinology, behavior, and seasonal variation (e.g., Li et al. 2005; Xia et al. 2018; Zhao 1993). Tibetan macaques at Mt. Huangshan exhibit high levels of sexuality inside and sometimes outside of the mating season (Li et al. 2007, see Fig. 12.2 above; Xia et al. 2010). Could there be something in the diet that stimulates or enhances Tibetan macaque sexual behavior?

One medicinal food consumed by YA1 troop in particular deserves attention in this respect. Used as an aphrodisiac in Chinese traditional medicine, *E. davidii* is known as horny goat weed or rowdy lamb herb (Ma et al. 2011), suggesting an origin for the use of this plant from watching the behavior of animals. Experimental studies have demonstrated significant enhancement of reproductive function, including increased sperm count, testosterone levels, and enhanced erectile function in rats (Chen et al. 2014; Li et al. 2015; Liu et al. 2005), but ingestion of *E. davidii* by macaques at Mt. Huangshan is limited to spring, the birth season of this troop. The seasonal timing of the consumption of this plant is preceded by a decline in ejaculatory mating frequency at the end of the mating season in late winter. For a few months in spring, non-ejaculatory mating continues at low levels with a slight peak around April (Fig. 12.1). Could the ingestion of this plant be having some effect on their reproductive activity? There is precedence in the literature to believe there might be.

Ingestion of plant hormones by animals has been found to have a number of other influences on reproductive behavior (e.g., Berger et al. 1977; Starker 1976; Sadlier 1969). Wasserman et al. (2012) report the seasonal influence of estrogenic plant consumption on hormonal and behavioral fluctuations in red colobus monkeys (*Procolobus rufomitratus*) in Uganda. Peaks in the consumption of young leaves (*Millettia dura*) with high estrogen levels coinciding with both fecal estradiol and cortisol levels in the feces. This was associated with increased levels of both copulation and aggressive interactions. In a study by Whitten (1983), the timing of onset, duration, and ending of seasonal mating behavior in female vervet monkeys (*Cercopithecus aethiops*) were closely correlated with the availability and ingestion of the flowers of *Acacia elatior* (Mimosaceae). Later Garey et al. (1992) analyzed the flowers of this species and found them to be estrogenic. Garey and colleagues determined that the amount of flowers consumed by vervet monkeys could provide adequate exogenous estrogen to stimulate the onset of mating activity. *Sargentodoxa cuneata* is only ingested during the spring birth season and has been found to assist in menstrual regulation among people, as a prophylactic against amenorrhea or metrorrhagia (Anonymous 1977; Yeung 1985). Therefore, including this species in the diet could have some sex steroid-like properties that have a role in modifying female reproductive status after birth.

Phytoestrogens present in the diet of many primates have been proposed to affect birth spacing, influence the sex of offspring, and regulate fertility. Glander (1980) proposed that inter-annual variation in birth spacing of howler monkeys (*Alouatta palliata*) was due to inter-annual variation in food quality. That is births were concentrated seasonally in years when secondary compound concentration in plant foods were low (high food quality) and spread across the year when concentrations were high (low food quality).

An in-depth examination of sex and reproduction in the Gombe chimpanzees by Wallis (1995, 1997) noted significant seasonal patterns, and multiple reproductive parameters including conception, anogenital swelling, infant mortality, and fertility. Wallis proposed that intensive foraging on seasonally available plant foods containing phytoestrogens mediate these fertility factor (Wallis 1992, 1994, 1997).

Hence, both male and female Tibetan macaque reproductive biology and behavior might be influenced by the inclusion in the diet of plants that have been shown in other species to influence reproductive hormones.

12.3.6 Stress Reduction

Stress disrupts health homeostasis, affecting reproductive function, overall health, and well-being of animals. Stress can be induced by both environmental and social factors, leading to physiological and behavioral imbalances (e.g., Takeshita et al. 2013, 2014; Wooddell et al. 2016). For example, primates living in seasonally cold habitats have a number of behavioral means for ameliorating cold stress, such as staying warm by sleeping and resting site selection, huddling, and, in one unique case, taking therapeutic hot spring baths (e.g., Hori et al. 1977; Zhang and Watanabe 2007; Kelley et al. 2016; Takeshita et al. 2018).

For stress induced by social interactions relating to social instability, dominance interactions, intergroup encounters, competition for food or mates, etc., affiliative behaviors such as grooming, reconciliation, consolation, and nonreproductive sexual behavior have been reported to be mechanisms of physiological reduction of stress in socially living species. Primates in particular have received wide attention (e.g., Aureli et al. 2002; Berry and Kaufer 2015; Carter et al. 2008; Fraser et al. 2008).

Tibetan macaques are classified as having a despotic, strongly linear, dominance style (Berman et al. 2004). They are well known for their kin-biased affiliation, tolerance, post-conflict reconciliation, nonreproductive mating, and the use of infants as a buffer to reduce tension between adults: "bridging" behavior (e.g., Bauer et al. 2014; Berman et al. 2004, 2007; Li et al. 2007; Ogawa 1995a, b). This suite of behaviors is linked to stress reduction through conflict buffering, suggestive of an undercurrent of social stress in their daily lives. Schenepel (2015) recorded high levels of agonistic and submissive behaviors around the provisioning area of YA1 troop compared to non-provision areas in the forest. Within the context of our study, we pose the question: "Do Tibetan macaques also have a dietary choice that aids in stress reduction?"

E. davidii is a prime candidate. Experimental evidence from several studies on hormonal and behavioral stress amelioration has been reported in relation to the administration of icariin, a major flavonoid isolated from *E. davidii*, in stress-induced rats and mice (Pan et al. 2010; Wu et al. 2011; Liu et al. 2015; Gong et al. 2016). Wu et al. (2011) demonstrated that icariin markedly decreased stress-induced downregulation of glucocorticoid receptors in mice subjected to "social defeat" by conspecifics. The compound also displays antidepressant activity and is used in Chinese traditional medicine for this purpose.

To the best of our knowledge, a dietary strategy for stress reduction has not yet received attention in the animal self-medication literature. Further investigation of the context of stress and the ingestion of plants like *E. davidii* may allow us to expand our knowledge of the role of diet in this area as well (Table 12.2).

Table 12.2 Medicinal properties of the 12 candidate "medicinal foods" in the diet of YA1 troop of Tibetan macaques at Mt. Huangshan

Species (part ingested) [season of use] form	Medicinal properties (see text for references and further details)
Cephalotaxus fortunei (leaf) [winter] evergreen, tree	Antiparasitic antibacterial, antifungal, and antiviral properties. Effective against pathogenic *Mycobacterium tuberculosis* and *Staphylococcus aureus*, including methicillin-resistant (MRSA) strains and biofilm infection of *S. aureus*. Contains abietane diterpenes showing remarkable activity against *Plasmodium falciparum* (malaria), *Trypanosoma brucei* (sleeping sickness in animals). Nonspecific antiprotozoal activity likely due to high cytotoxicity. Cancer prevention properties
Epimedium davidii (leaf) [spring] herbaceous, perennial	Enhancement of reproductive function (erectile, sperm count, testosterone levels). Stress reduction and antidepressant properties via downregulation of glucocorticoid receptor activity and regulation of hippocampal neuroinflammation. Anticancer activity, immunosuppressive action, and inhibition of lymphocyte activation. Osteoprotective effect, neuroprotective effect, cardiovascular protective effect, anti-inflammation effect, and immunoprotective effect
Hedera nepalensis (leaf) [spring, autumn] evergreen climber or creeper vine	Purgative action, sweat-inducing and stimulant properties. Used to treat skin diseases. Cancer preventative and cytotoxic properties. Important folk medicine for the treatment of diabetes. Contains saponins and is toxic. Ingestion induces gastrointestinal nervous system disturbances
Ilex purpurea (leaf) [winter, spring, autumn] evergreen tree	One of the 50 fundamental herbs in Chinese medicine. Antitumor properties. Used for treating burns, ulcers in the lower extremities
Lespedeza bicolor (leaf) [spring, summer] deciduous shrub	Antioxidant, anticancer, and bactericidal activity. Traditionally used for coughs and colds, kidney and urethra problems, fever, headache, weakness, and dizziness. Contains flavonoids, alkaloids, terpenes, organic acids, and stigmasterols
Liquidambar formosana (leaf) [spring] deciduous tree	Treatment of cancerous growths
Litsea coreana (leaf) [year-round] evergreen tree	Augments immunoglobulin M and G values and shows significant inhibitory effects on pathogenic *Bacillus anthracis, Proteusbacillus vulgaris, Staphylococcus aureus*, and *Bacillus subtilis*. Anti-HSV-1 activity, anti-gastric carcinoma and colon carcinoma HT-29 activity. Exhibits antibacterial, hepatoprotective, hyperglycemic, anti-inflammatory, and antioxidation activity. Contains polysaccharides, polyphenols, essential oils, and numerous flavonoids
Loropetalum chinense (leaf, stem) [year-round] evergreen woody shrub	An external application to wounds. A treatment for coughing in patients with tuberculosis, dysentery, and enteritis. Promotes wound healing, possesses antibacterial activity. Anti-inflammatory and antioxidant activity. Adjusts fat metabolism. Contains antitumoral activity and protects from cardiovascular disease

(continued)

Table 12.2 (continued)

Species (part ingested) [season of use] form	Medicinal properties (see text for references and further details)
Millettia dielsiana (leaf) [year-round] deciduous shrub	Used for prevention of cardiovascular and cerebrovascular disease. Significant antioxidant activity
Sargentodoxa cuneata (leaf) [spring] deciduous climbing shrub	Anthelmintic (hookworm, roundworm, filariasis) and antibacterial activity. Stem extracts exhibit anti-enterovirus activity against anti-Coxsackie virus B3, Coxsackie virus B5, Poliovirus I, Echovirus 9, and Echovirus 29. Also possess antirheumatic, carminative, diuretic, and tonic properties. Treatment for sores, amenorrhea, metrorrhagia (irregular uterine bleeding), traumatic injuries, rheumatoid arthritis, and anemia
Syzygium buxifolium (leaf) [year-round] evergreen shrub	Taken as a febrifuge (reduce fever). Cooling effect when rubbed on the bodies of smallpox patients
Vaccinium bracteatum (leaf) [year-round] evergreen shrub	Radical scavenging activity and protection against KBrO3-mediated kidney damage. Anticancer and anti-inflammatory activity. Contains isoorientin, orientin, vitexin, isovitexin, isoquercitrin, quercetin-3-O-α-L-rhamnoside, and chrysoeriol-7-O-β-D-glucopyranoside

12.4 Future Research

This study was designed to evaluate the potential for self-medication in Tibetan macaques. At present we cannot completely rule out the possibility that macaques consumed some or all of these plants described above only to meet some micronutrient deficiency, or that the amounts consumed were insufficient to bring about physiological change. However, the evidence presented in this chapter is compelling enough to warrant further research.

Where do we go from here? In order to demonstrate therapeutic self-medication, there are four basic requirements: (1) identify the disease or symptom(s) being treated, (2) distinguish the use of a therapeutic agent from that of everyday food items, (3) demonstrate a positive change in health condition following self-medicative behavior, and (4) provide evidence for plant activity and or direct pharmacological analysis of compounds extracted from these therapeutic agents (Huffman 2010). The pharmacological activity reported here for 12 species is the first stage of fulfilling requirement (4). This suggests that ingesting these plants may elicit significant physiological benefit if ingested in sufficient amounts.

The broad spectrum of confirmed pharmacological activity reported here suggests several avenues of research to pursue in the future and reason to believe that Tibetan macaques self-medicate. Depending on the pharmacological activities of the plant in question, future work needs to attempt to directly link the context of use with the health status of the individual. This requires longitudinal investigation with attention to (1) seasonal influences (e.g., birth, mating, infection seasonality), (2) presence or absence of large-scale disruptive social influences likely to induce psycho-physiological stress (e.g., troop fission or the death of a leader or principle care-giver), (3) reproductive state (e.g., pregnancy, estrus, reproductive history), and (4) age and health status. Monitoring of identified individuals representative of all age-sex classes, recording reinfection seasonality, infection intensity and behavioral indicators of poor health, weakened body condition, poor appetite, plant food selection patterns, etc. are required. These data will help to provide the necessary context of medicinal plant use to strengthen the case for self-medication in Tibetan macaques (e.g., see Huffman et al. 1996a, b; Huffman and Caton 2001; Alados and Huffman 2000; MacIntosh et al. 2011; Burgunder et al. 2017). Only recently has the effect of seasonal dietary change on micro- and mycobiota composition of Tibetan macaques been investigated (Sun et al. 2016, 2018).

In closing, it should be noted that there are still other medicinal properties in the medicinal diet that have not been discussed in detail. They need to be looked at more closely in the future. These include the possible roles of anti-inflammation, immunoprotection, antibiotic, antibacterial, and antiviral properties, in the passive protection or treatment of seasonal afflictions brought on by cold-damp or hot-humid weather conditions. Do troop members ingest these items more in some seasons than others? Noteworthy too about these understudied properties of the diet are the widespread anticancer (antitumoral), osteoprotective, cardiovascular protective, and neuroprotective effects. Do older members of the troop ingest items with these properties more often than younger ones?

In the aging Western society today, such diseases form the core of many of our health problems, and it has been argued that the cause of this is due to the change in our diets, shifting towards more processed foods and away from more natural food sources (Johns 1990). The properties of the Tibetan macaque diet may provide us with important insights into the long-term dietary strategy of primates occurring at the interface of food and medicine.

Acknowledgments MAH is grateful to the organizers of the International Symposium "Nonhuman Primates: Insights into Human Behavior and Society" held in Mt. Huangshan from July 21 to the 25, 2017, for being able to attend and exchange valuable scientific ideas and cultivate friendships and collaborations. This chapter is one result of that opportunity. MAH also thanks Yamato Tsuji and Massimo Bardi for their statistical advice and comments on the manuscript. BH SUN is very grateful to the Huangshan Garden Forest Bureau for their permission and support to his study. BH SUN also wishes to thank Xiaojuan Xu and Jayue Sun for their support during the writing of this manuscript.

Appendix: Plant Secondary Metabolites in Plant Items Ingested by Tibetan Macaques at Mt. Huangshan

Abietane (diterpenoid)
Alkaloids
Essential oils
Flavonoids
Icariin (prenylated flavonol glycoside)
Icariside II (Baohuoside I) and Baohuoside II, III, V, VI
Isoflavones
i6-ethoxyca lpogonium isoflavone A, durmillone, ichthynone, jamaicin, toxicaro l
Isoflavone, barbigerone, genistein
Organic acids
Polysaccharides
Polyphenols
Saponic glycosides (hederagenin)
Stigmasterols
Terpenes

References

Ahmed T, Khan A, Chandan P (2015) Photographic key for the identification of some plant of Indian Trans-Himalaya. Not Sci Biol 7:171–176

Alados CL, Huffman MA (2000) Fractal long-range correlations in behavioural sequences of wild chimpanzees: a non-invasive analytical tool for the evaluation of health. Ethology 106 (2):105–116

Alipayo D, Valdez R, Holechek JL et al (1992) Evaluation of microhistological analysis for determining ruminant diet botanical composition. J Range Manag 45:148–152

Altizer S, Dobson A, Hosseini P, Pascual M, Rohani P (2006) Seasonality and the dynamics of infectious diseases. Ecol Lett 9:467–484

Anonymous (1977) A barefoot doctors manual. Running Press, Philadelphia

Anthony RG, Smith NS (1974) Comparison of rumen and fecal analysis to describe deer diets. J Wildl Manag 38:535–540

Aureli F, Cords M, van Schaik CP (2002) Conflict resolution following aggression in gregarious animals: a predictive framework. Anim Behav 64:325–343

Bauer B, Sheeran LK, Matheson MD, Li J-H, Wagner RS (2014) Male Tibetan macaques' (*Macaca thibetana*) choice of infant bridging partners. Zool Res 35(3):222–230

Berger PJ, Sanders EH, Gardner PD, Negus NC (1977) Phenolic plant compounds functioning as reproductive inhibitors in *Microtus montanus*. Science 195:575–577

Berman CM, Ionica CS, Li J-H (2004) Dominance style among *Macaca thibetana* on Mt. Huangshan, China. Int J Primatol 25(6):1283–1312

Berman CM, Ionica CS, Li J-H (2007) Supportive and tolerant relationships among male Tibetan macaques at Huangshan, China. Behaviour 144:631–661

Berry AK, Kaufer D (2015) Stress, social behavior, and resilience: insights from rodents. Neurobiol Stress 1:116–127

Brack M (1987) Agents transmissible from simians to man. Springer, Berlin

Brack M (2008) Oesophagostomiasis. EAZWV transmissible disease fact sheet no 116. https://c.ymcdn.com/sites/www.eazwv.org/resource/resmgr/Files/Transmissible_Diseases_Handbook/Fact_Sheets/116_Oesophagostomiasis.pdf. Accessed 19 Jun 2018

Burgunder J, Pafco B, Petrelkova KJ, Modry D, Hashimoto C, MacIntosh AJJ (2017) Complexity in behavioral organization and strongylid infection among wild chimpanzees. Anim Behav 129:257–268

Carrai V, Borgognini-Tarli SM, Huffman MA et al (2003) Increase in tannin consumption by sifaka (*Propithecus verreauxi verreauxi*) females during the birth season: a case for self-medication in prosimians? Primates 44(1):61–66

Carter CS, Grippo AJ, Pournajafi-Nazarloo H, Ruscio MG, Porges SW (2008) Oxytocin, vasopressin and sociality. Prog Brain Res 170:331–336

Chang N, Luo Z, Li D et al (2017) Indigenous uses and pharmacological activity of traditional medicinal plants in Mount Taibai, China. Evid Based Complement Alternat Med 2017:8329817. https://doi.org/10.1155/2017/8329817

Chen M, Hao J, Yang Q et al (2014) Effects of icariin on reproductive functions in male rats. Molecules 19(7):09502–09514

Cousins D, Huffman MA (2002) Medicinal properties in the diet of gorillas – an ethnopharmacological evaluation. Afr Study Monogr 23:65–89

Duke JA, Ayensu ES (1985) Medicinal plants of China. Reference Publications, Algonac. isbn:0-917256-20-4

Engel C (2002) Wild health. Houghton Mifflin, Boston

Etkin NL (1996) Medicinal cuisines: diet and ethnopharmacology. Int J Pharmacog 34(5):313–326

Etkin NL, Ross PJ (1983) Malaria, medicine, and meals: plant use among the Hausa and its impact on disease. In: Romanucci-Ross L, Moerman DE, Tancredi LR (eds) The anthropology of medicine: from culture to method. Praeger, New York, pp 231–259

Find Me A Cure (2018). https://findmeacure.com. Accessed 30 Apr 2018

Foitova I, Huffman MA, Wisnu N et al (2009) Parasites and their effect on orangutan health. In: Wish SA, Utami SS, Setia TM et al (eds) Orangutans—ecology, evolution, behavior and conservation. Oxford University Press, Oxford, pp 157–169

Forbey J, Harvey A, Huffman MA et al (2009) Exploitation of secondary metabolites by animals: a behavioral response to homeostatic challenges. Integr Comp Biol 49(3):314–328

Fraser ON, Stahl D, Aureli F (2008) Stress reduction through consolation in chimpanzees. PNAS 105(25):8557–8562

Freeland WJ, Janzen DH (1974) Strategies in herbivory by mammals: the role of plant secondary compounds. Am Nat 108(961):269–289

Frohne D, Pfänder JA (1984) Colour atlas of poisonous plants. Timber Press, Portland

Gadgil M, Bossert WH (1970) Life historical consequences of natural selection. Am Nat 104:1–24

Gan R-Y, Xu X-R, Song F-L et al (2010) Antioxidant activity and total phenolic content of medicinal plants associated with prevention and treatment of cardiovascular and cerebrovascular diseases. J Med Plant Res 4(22):2438–2444

Garey J, Markiewicz L, Gurpide E (1992) Estrogenic flowers, a stimulus for mating activity in female vervet monkeys. In: XIVth Congress of the International Primatological Society Abstracts, Strassbourg, p 210

Glander KE (1980) Reproduction and population growth in free-ranging mantled howling monkeys. Am J Phys Anthropol 53:25–36

Glander KE (1982) The impact of plant secondary compounds on primate feeding behavior. Yrbk Phys Anthropol 25:1–18

Gong T, Wang H-Q, Chen R-Y (2007) Isoflavones from vine stem of *Millettia dielsiana*. Zhongguo Zhong Yao Za Zhi 32(20):2138–2140

Gong M-J, Han B, Wang S-M, Liang S-W (2016) Icariin reverses corticosterone- induced depression-like behavior, decrease in hippocampal brain-derived neurotrophic factor (BDNF) and metabolic network disturbances revealed by NMR-based metabonomics in rats. J Pharm Biomed Anal 123:63–73
Gotoh S (2000) Regional differences in the infection of wild Japanese macaques by gastrointestinal helminth parasites. Primates 41(3):291–298
Green MJ (1987) Diet composition and quality in Himalayan musk deer based on faecal analysis. J Wildl Manag 51:880–892
Guo J-P, Pang J, Wang X-W et al (2006) *In vitro* screening of traditionally used medicinal plants in China against enteroviruses. World J Gastroenterol 12(25):4078–4081
Hardy K, Buckley S, Collins MJ et al (2012) Neanderthal medics? Evidence for food, cooking and medicinal plants entrapped in dental calculus. Naturwissenschaften 99:617–626
Hardy K, Buckley S, Huffman MA (2013) Neanderthal self-medication in context. Antiquity 87:873–878
Herbpathy Data Base (2018). https://herbpathy.com. Accessed 27 Apr 2018
Hori T, Nakayama T, Tokura H, Hara F, Suzuki M (1977) Thermoregulation of Japanese macaque living in a snowy mountain area. Jpn J Physiol 27:305–319
Huffman MA (1997) Current evidence for self-medication in primates: a multidisciplinary perspective. Yrbk Phys Anthropol 40:171–200
Huffman MA (2002) Animal origins of herbal medicine. In: Fleurentin J, Pelt J-M, Mazars G (eds) From the sources of knowledge to the medicines of the future. IRD Editions, Paris, pp 31–42
Huffman MA (2003) Animal self-medication and ethno-medicine: exploration and exploitation of the medicinal properties of plants. Proc Nutr Soc 62:371–381
Huffman MA (2007) Animals as a source of medicinal wisdom in indigenous societies. In: Bekoff M (ed) Encyclopedia of human-animal relation, vol 2. Greenwood Publishing Group, Westport, CT, pp 434–441
Huffman MA (2010) Self-medication: passive prevention and active treatment. In: Breed MD, Moore J (eds) Encyclopedia of animal behavior, vol 3. Academic, Oxford, pp 125–131
Huffman MA (2011) Primate self-medication. In: Campbell C, Fuentes A, MacKinnon K et al (eds) Primates in perspective. University of Oxford Press, Oxford, pp 563–573
Huffman MA (2015) Chimpanzee self-medication: a historical perspective of the key findings. In: Nakamura M, Hosaka K, Itoh N, Zamma K (eds) Mahale chimpanzees—50 years of research. Cambridge University Press, Cambridge, pp 340–353
Huffman MA (2016) An ape's perspective on the origins of medicinal plant use in humans. In: Hardy K, Kubiak-Martens L (eds) Wild harvest: plants in the hominin and pre-agrarian human worlds. Oxbow Books, Oxford, pp 55–70
Huffman MA, Caton JM (2001) Self-induced increase of gut motility and the control of parasitic infections in wild chimpanzees. Int J Primatol 22:329–346
Huffman MA, MacIntosh AJJ (2012) Plant-food diet of the Arashiyama Japanese macaques and its potential medicinal value. In: Leca J-B, Huffman MA, Vasey PL (eds) The monkeys of Stormy Mountain: 60 years of primatological research on the Japanese macaques of Arashiyama. Cambridge University Press, Cambridge, pp 356–431
Huffman MA, Gotoh S, Izutsu D, Koshimizu K, Kalunde MS (1993) Further observations on the use of the medicinal plant, *Vernonia amygdalina* (Del) by a wild chimpanzee, its possible effect on parasite load, and its phytochemistry. Afr Stud Monogr 14(4):227–240
Huffman MA, Koshimizu K, Ohigashi H (1996a) Ethnobotany and zoopharmacognosy of *Vernonia amygdalina*, a medicinal plant used by humans and chimpanzees. In: Caligari PDS, Hind DJN (eds) Compositae: biology and utilization, vol 2. Royal Botanical Gardens, Kew, pp 351–360
Huffman MA, Page JE, Sukhdeo MVK et al (1996b) Leaf-swallowing by chimpanzees, a behavioral adaptation for the control of strongyle nematode infections. Int J Primatol 17(4):475–503
Huffman MA, Gotoh S, Turner LA et al (1997) Seasonal trends in intestinal nematode infection and medicinal plant use among chimpanzees in the Mahale Mountains National Park, Tanzania. Primates 38(2):111–125

Huffman MA, Ohigashi H, Kawanaka M et al (1998) African great ape self-medication: a new paradigm for treating parasite disease with natural medicines? In: Ebizuka Y (ed) Towards natural medicine research in the 21st century. Elsevier Science BV, Amsterdam, pp 113–123

Jafri L, Saleem S, Kondrytuk TP et al (2016) *Hedera nepalensis* K. Koch: a novel source of natural cancer chemopreventive and anticancerous compounds. Phytother Res 30(3):447–453. https://doi.org/10.1002/ptr.5546

Janzen DH (1978) Complications in interpreting the chemical defenses of trees against tropical arboreal plant-eating vertebrates. In: Montgomery GG (ed) The ecology of arboreal folivores. Smithsonian Institution Press, Washington, DC, pp 73–84

Jia X, Li P, Wan J, He C (2017) A review on phytochemical and pharmacological properties of *Litsea coreana*. Pharm Biol 55(1):1368–1374. https://doi.org/10.1080/13880209.2017.1302482

Johns T (1990) With bitter herbs they shall eat it. University of Arizona Press, Tucson

Kelley EA, Jablonski NG, Chaplin G, Sussman RW, Kamilar JM (2016) Behavioral thermoregulation in *Lemur catta*: the significance of sunning and huddling behaviors. Am J Primatol 78:745–754

Krief S, Hladik CM, Haxaire C (2005) Ethnomedicinal and bioactive properties of the plants ingested by wild chimpanzees in Uganda. J Ethnopharmacol 101:1–15

Lambert JE (2011) Primate nutritional ecology, feeding biology and diet at ecological and evolutionary scales. In: Campbell CJ, Fuentes A, MacKinnon KC et al (eds) Primates in perspective. Oxford University Press, Oxford, pp 512–522

Landa P, Skalova L, Bousova I et al (2014) *In vitro* anti-proliferative and anti-inflammatory activity of leaf and fruit extracts from *Vaccinium bracteatum* Thunb. Pak J Pharm Sci 27(1):103–106

Lewis WH, Elvin-Lewis MPF (1977) Medical botany. Wiley, New York

Li J-H, Wang Q, Han D (1996) Fission in a free-ranging Tibetan macaque troop at Huangshan Mountain, China. Chin Sci Bull 41(16):1377–1381

Li J-H, Yin HB, Wang QS (2005) Seasonality of reproduction and sexual activity in female Tibetan macaques (*Macaca thibetana*) at Huangshan, China. Acta Zool Sin 51(3):365–375

Li J-H, Yin HB, Zhou L-Z (2007) Non-reproductive copulation behavior among Tibetan macaques at Huangshan, China. Primates 48:64–72

Li C, Li Q, Mei Q et al (2015) Pharmacological effects and pharmacokinetic properties of icariin, the major bioactive component in Herba Epimedii. Life Sci 126:57–68

Liu WJ, Xin ZC, Xin H et al (2005) Effects of icariin on erectile function and expression of nitric oxide synthase isoforms in castrated rats. Asian J Androl 7:381–388

Ma A, Qi S, Xu D et al (2004) Baohuoside-1, a novel immunosuppressive molecule, inhibits lymphocyte activation in vitro and in vivo. Transplantation 78(6):831–838

Ma H, He X, Yang Y et al (2011) The genus *Epimedium*, an ethnopharmacolgical and phytochemical review. J Ethnopharmacol 143:519–541

MacIntosh AJJ, Huffman MA (2010) Towards understanding the role of diet in host-parasite interactions in the case of Japanese macaques. In: Nakagawa F, Nakamichi M, Sugiura H (eds) The Japanese macaques. Springer, Tokyo, pp 323–344

MacIntosh AJJ, Hernandez A, Huffman MA (2010) Host age, sex, and reproductive seasonality affect nematode parasitism in wild Japanese macaques. Primates 51:353–364

MacIntosh JJJ, Alados CI, Huffman MA (2011) Fractal analysis of behavior in a wild primate: behavioural complexity in health and disease. J R Soc Interface 8(63):1497e1509

Mellado M, Foote RH, Rodriquez A et al (1991) Botanical composition and nutrient content of diets selected by goats grazing on desert grassland in northern Mexico. Small Rumin Res 6:141–150

Mothana RA, Al-Said MS, Al-Musayeib NM et al (2014) *In vitro* antiprotozoal activity of abietane diterpenoids isolated from *Plenctranthus barbatus* Andr. Int Mol Sci 15:8360–8371. http://www.mdpi.com/1422-0067/15/5/8360. Accessed 27 Apr 2018

Mukherjee JR, Chelladurai V, Ronald J et al (2011) Do animals eat what we do? Observations on medicinal plants used by humans and animals of Mudanthurai Range, Tamil Nadu. In: Kala CP (ed) Medicinal plants and sustainable development. Nova Science, New York, pp 179–195

Ndagurwa HGT (2012) Bark stripping by chacma baboons (*Papio hamadryas ursinus*) as a possible prophylactic measure in a pine plantation in eastern Zimbabwe. Afr J Ecol 51:164–167

Negre A, Tarnaud L, Roblot JF et al (2006) Plants consumed by *Eulemur fulvus* in Comoros Islands (Mayotte) and potential effects on intestinal parasites. Int J Primatol 27(6):1495–1517

Neto I, Faustino C, Rijo P (2015) Antimicrobial abietane diterpenoids against resistant bacteria and biofilms. In: Méndez-Vilas A (ed) The battle against microbial pathogens: basic science, technological advances and educational programs, Microbiology book series. Formatex Research Center, Badajoz

Ogawa H (1995a) Triadic male-female-infant relationships and bridging behavior among Tibetan macaques. Folia Primatol 64:153–157

Ogawa H (1995b) Recognition of social relationships in bridging behavior among Tibetan macaques (*Macaca thibetana*). Am J Primatol 35:305–310

Ohigashi H, Huffman MA, Izutsu D, Koshimizu K, Kawanaka M, Sugiyama H, Kirby GC, Warhurst DC, Allen D, Wright CW, Phillipson JD, Timmon-David P, Delmas F, Elias R, Balansard G (1994) Toward the chemical ecology of medicinal plant use in chimpanzees: the case of *Vernonia amygdalina*, a plant used by wild chimpanzees possibly for parasite-related diseases. J Chem Ecol 20(3):541–553

Oyeyemi IT, Akinlabi AA, Adewumi A, Aleshinloye AO, Oyeyemi OT (2018) *Vernonia amygdlina*: a folkloric herb with anthelmintic properties. Beni-Suef Univ J Basic Appl Sci 7:43–49

Pan Y, Wang F-M, Quang L-Q, Zhang D-M, Kong L-D (2010) Icariin attenuates chronic mild stress-induced dysregulation of the LHPA stress circuit in rats. Psychoneuroendocrinology 35:272–283

Petroni LM, Huffman MA, Rodriguez E (2016) Medicinal plants in the diet of woolly spider monkeys (*Brachyteles arachnoides*, E. Geoffroy, 1806)—a bio-rational for the search of new medicines for human use? Rev Bras Farm 27(2):135–142

Plants For A Future Data Base (1996–2012). https://pfaf.org/user/Default.aspx. Accessed 27 Apr 2018

Rosenthal GA, Berenbaum MR (1992) Herbivores: their interactions with secondary plant metabolites. Academic, San Diego

Rumble MA, Anderson SH (1993) Evaluating the microscopic fecal technique for estimating hard mast in turkey diets. USDA Forest Service Research Paper RM-310. https://www.fs.fed.us/rm/pubs_rm/rm_rp310.pdf

Sadlier RM (1969) The ecology of reproduction in wild and domestic mammals. Methuer, London

Schenepel BL (2015) Provisioning and its effects on the social interactions of Tibetan Macaques (Macaca Thibetana) at Mt. Huangshan, China. All Master's Theses. 399. https://digitalcommons.cwu.edu/etd/399

Simpson SJ, Sibly RM, Lee KP, Behmer ST, Raubenheimer D (2004) Optimal foraging when regulating intake of multiple nutrients. Anim Behav 68:1299–1311

Sparks DR, Malechek JC (1968) Estimating percentage dry weight in diets using a microscope technique. J Range Manage 21:264–265

Starker LA (1976) Phytoestrogens: adverse effects on reproduction in California quail. Science 191:98–100

Sun B, Wang X, Bernstein S, Huffman MA, Dong-Po Xia D-P, Gu Z, Chen R, Sheeran LK, Wagner RS, Li J (2016) Marked variation between winter and spring gut microbiota in free-ranging Tibetan Macaques (*Macaca thibetana*). Sci Rep 6:26035. https://doi.org/10.1038/srep26035

Sun B-H, Gu Z, Wang X, Huffman MA, Garber PA, Sheeran LK, Zhang D, Zhu Y, Xia D-P, Li J-H (2018) Season, age, and sex affect the fecal mycobiota of free-ranging Tibetan macaques (*Macaca thibetana*). Am J Primatol 80:e22880. https://doi.org/10.1002/ajp.22880

Takeshita RSC, Huffman MA, Bercovitch FB, Mouri K, Shimizu K (2013) The influence of age and season on fecal dehyroepiandrosterone-sulfate (DHEAS) concentrations in Japanese macaques (*Macaca fuscata*). Gen Comp Endocrinol 19:39–43

Takeshita RSC, Bercovitch FB, Huffman MA, Mouri K, Garcia C et al (2014) Environmental, biological, and social factors influencing fecal adrenal steroid concentrations in female Japanese macaques (*Macaca fuscata*). Am J Primatol 76:1084–1093

Takeshita RSC, Bercovitch FB, Kinoshita K, Huffman MA (2018) Beneficial effects of hot springs bathing on stress levels in Japanese macaques. Primates 59(3):215–225

Ullah S (2017) Methanolic extract from Lespedeza bicolor: potential candidates for natural antioxidant and anticancer agent. J Tradit Chin Med 37(4):444–451

Usher G (1974) A dictionary of plants used by man. Macmillan, London

Wallis J (1992) Socioenvironmental effects on timing of first postpartum cycles in chimpanzees. In: Nishida T, McGrew WC, Marler P, Pickford M, de Waal F (eds) Topics in primatology, Human origins, vol 1. University of Tokyo Press, Tokyo, pp 119–130

Wallis J (1994) Socioenvironmental effects on full anogenital swellings in adolescent female chimpanzees. In: Roeder JJ, Thierry B, Anderson JR, Herrenschmidt N (eds) Current primatology, Social development, learning and behavior, vol 2. University of Louis Pasteur, Strasbourg, pp 25–32

Wallis J (1995) Seasonal influence on reproduction in chimpanzees of Gombe National Park. Int J Primatol 16:435–451

Wallis J (1997) A survey of reproductive parameters in the free-ranging chimpanzees of Gombe National Park. J Reprod Fertil 109:297–307

Wasserman MD, Chapman CA, Milton K, Gogarten JF, Wittwer DJ, Zeigler TE (2012) Estrogenic plant consumption predicts red colobus monkey hormonal state and behavior. Horm Behav 62:553–562

Whitten PL (1983) Flowers, fertility and females. Abstract. Am J Phys Anthropol 60:269–270

Wooddell LJ, Kaburu SSK, Rosenberg KL et al (2016) Matrilineal behavioral and physiological changes following the removal of a non-alpha matriarch in rhesus macaques. PLoS One 11(6): e0157108

World Agroforestry Data Base (2018). http://www.worldagroforestry.org/treedb/. Accessed 27 Apr 2018

Wu J, Du J, Xu C, Le J, Xu Y, Liu B, Dong J (2011) Icariin attenuates social defeat-induced downregulation of glucocorticoid receptor in mice. Pharmacol Biochem Behav 98:273–278

Xia D-P, Li J-H, Zhu Y, Sun B-H, Sheeran LK, Matheson MD (2010) Seasonal variation and synchronization sexual behaviors in free-ranging male macaques (*Macaca thibetana*) at Huangshan, China. Zool Res 31(5):509–515

Xia D-P, Wang X, Zhang Q-X, Sun B-H, Sun L, Sheeran LK, Li J-H (2018) Progesterone levels in seasonally breeding, free-ranging male *Macaca thibetana*. Mamm Res 63:99–106

Yeung H-C (1985) Handbook of Chinese herbs and formulas. Institute of Chinese Medicine, Los Angeles

You S-Y, Yin H-B, Zhang S-Z et al (2013) Food habits of *Macaca thibetana*. China J Biol 30 (5):64–67

Zhang P, Watanabe K (2007) Extra-large cluster formation by Japanese macaques (*Macaca fuscata*) on Shodoshima Island, Central Japan, and related factors. Am J Primatol 69:1119–1130

Zhang J, Chu C-J, Li X-L et al (2014a) Isolation and identification of antioxidant compounds in *Vaccinium bracteatum* Thunb. by UHPLC-Q-TOF LC/MS and their kidney damage protection. J Funct Foods 11:62–10

Zhang Y, Ren F-X, Yang Y et al (2014b) Chemical constituents and biological activities of *Loropetalum chinense* and *Loropetalum chinense* var. *rubrum*: research advances. J Int Pharm Res 41(3):307–312

Zhao Q-K (1993) Sexual behavior of Tibetan macaques at Mt. Emei, China. Primates 34 (4):431–444

Zhou H-T, Cao J-M, Lin Q et al (2013) Effect of *Epimedium davidii* on testosterone content, substance metabolism an exercise capacity in rats receiving exercise training. Chin Pharm J 48 (1):25–29

Zhou J, Wang Y-S, Wu Z-G (2014) Study of medicinal value of *Loropetalum chinense*. CJTCMP 29(7):2283–2286

Zhu Y, Ji H, Li J-H, Xia D-P, Sun B-H, Xu Y-R, Kyes RC (2012) First report of the wild Tibetan macaque (*Macaca thibetana*) as a new primate host of *Gongylonema pulchrum* with high incidence in China. J Anim Vet Adv 11(24):4514–4518

Chapter 13
Primate Infectious Disease Ecology: Insights and Future Directions at the Human-Macaque Interface

Krishna N. Balasubramaniam, Cédric Sueur, Michael A. Huffman, and Andrew J. J. MacIntosh

13.1 Introduction

The expansion of human populations has increased interactions and conflict between humans and nonhuman primates (hereafter primates) throughout their range. Assessing the causal factors and thereby mitigating such conflict pose a major challenge for anthropologists, primatologists, and conservation biologists. This is because human-primate interactions are spatiotemporally variable in form and frequency (reviewed in Dickman 2012; Paterson and Wallis 2005). For instance, some of these interactions include (1) human-induced changes to primate habitat that lead to the fragmentation and decline of primate populations [e.g., Zanzibar red colobus monkeys (*Procolobus kirkii*): Siex 2005], (2) increased crop-raiding by primates leading to transactional costs on humans [e.g., Buton macaques (*Macaca ochreata*): Priston et al. 2012], (3) human-primate competition for space and resources [e.g., chimpanzees (*Pan troglodytes*): Hockings et al. 2012], (4) injuries to both humans and primates on account of direct aggression [e.g., rhesus macaques (*Macaca mulatta*): Southwick and Siddiqi 1994, 1998, 2011], and (5) primate-induced damage to human property and landscapes that generate transactional or opportunity costs to humans (Barua et al. 2013).

In comparison to such readily discernible negative effects, one outcome of conflict that is subtler and hence often goes undetected or unchecked is the

K. N. Balasubramaniam (✉)
Department of Population Health and Reproduction, School of Veterinary Medicine, University of California at Davis, Davis, CA, USA

C. Sueur
IPHC, UMR 7178, Université de Strasbourg, CNRS, Strasbourg, France
e-mail: cedric.sueur@iphc.cnrs.fr

M. A. Huffman · A. J. J. MacIntosh
Primate Research Institute, Kyoto University, Kyoto, Japan
e-mail: huffman.michael.8n@kyoto-u.ac.jp

J.-H. Li et al. (eds.), *The Behavioral Ecology of the Tibetan Macaque*, Fascinating Life Sciences, https://doi.org/10.1007/978-3-030-27920-2_13

acquisition and transmission of infectious diseases (Barua et al. 2013; Wolfe et al. 2007). Our shared evolutionary histories, along with physiological and behavioral similarities, make many primate species natural reservoirs of human parasites (Fiennes 1967; Nunn and Altizer 2006; Tutin 2000). Likewise, the acquisition of parasites from humans has led to disease outbreaks among free-living primates (Kaur and Singh 2009; Kaur et al. 2008, 2011; Nunn and Altizer 2006). From an ecological standpoint, free-living primates may acquire parasites from humans in many ways. For example, increasing epidemiological assessments continue to establish the sharing of parasites between humans and populations of socioecologically flexible primates like baboons and macaques which have become increasingly reliant on human-provisioned food or garbage in areas of overlap (Engel and Jones-Engel 2011; Engel et al. 2008; Jones-Engel et al. 2005). Humans may also indirectly influence primate exposure to parasites by altering the environment, which may potentially subdivide primate populations and change their behavioral and foraging strategies (Chapman et al. 2005, 2006a; Huffman and Chapman 2009). Third, wild primates may also sometimes be exposed to "spillovers" of parasites from international travelers during ecotourism and biological field research (Carne et al. 2017; Engel et al. 2008; Jones-Engel et al. 2005; Marechal et al. 2011; Muehlenbein and Ancrenaz 2010). Such a wide range of potential disease acquisition and transmission routes make human-primate interfaces hot spots for emerging infectious diseases (EIDs) (Nunn et al. 2008; Wolfe et al. 2007). This is especially significant in the light of the growing call for a global, transdisciplinary strategy to deal with zoonoses in both humans and animals (the One Health, hereafter OH, concept: Destoumieux-Garzon et al. 2018; Zinsstag et al. 2011, 2015). Finally, human activities like agricultural and urban land development, tourism, and provisioning, aside from directly influencing exposure as stated above, may also influence variation in the susceptibility of primates to parasites once exposed, for example, by altering levels of stress and immune function (Chapman et al. 2006b; Marechal et al. 2011, 2016; Muehlenbein and Ancrenaz 2010).

In this chapter, we focus on how human-macaque interfaces, being hot spots for the transmission of a diverse array of parasites, present opportunities for human-primate infectious disease ecology research. We first briefly outline the significance and primary objectives behind research on primate infectious disease ecology, highlighting the greater focus to date on research implementing such approaches to study wild primates in comparison to research at human-primate interfaces. We next reveal how macaques, and more broadly the variable nature of human-macaque interfaces, present opportunities to study human-primate disease transmission from a socioecological perspective (Engel and Jones-Engel 2011; Jones-Engel et al. 2005; Nahallage and Huffman 2013). We then provide a detailed account of previous studies we extracted from the online Global Mammal Parasite Database (Nunn and Altizer 2005; Stephens et al. 2017) that have detected parasites at human-macaque interfaces. Finally, we demonstrate how the implementation of novel conceptual frameworks like the Coupled Natural and Human Systems (An and Lopez-Carr 2012; Destoumieux-Garzon et al. 2018; Liu et al. 2007) and One Health concepts (Destoumieux-Garzon et al. 2018; Zinsstag et al. 2011, 2015), as well as the implementation of cutting-edge methodological approaches like Social Network Analyses (e.g., Drewe and Perkins 2015; Pasquaretta et al. 2014; Rushmore et al. 2017; VanderWaal and Ezenwa 2016) and community-level bipartite or multimodal Networks (e.g., Dormann et al. 2017; Gomez et al. 2013; Latapy et al. 2008), can address some of the critical gaps in these studies to offer key future directions for epidemiological research at these interfaces.

13.2 Primate Infectious Disease Ecology

An infectious disease is a disorder that is caused by an infectious agent, or in ecological terms a "parasite," that causes pathology in its host (MacIntosh 2016). In the ecological realm, a "parasite" is considered any organism that lives within (or on) another "host" organism, at some cost to the latter (MacIntosh 2016). For the remainder of the chapter, we deal with enteric parasites or "endoparasites" (hereafter just "parasites"), which live within the body of the host organism. These typically fall under seven major types of organisms. Five of these, specifically bacteria, viruses, rickettsia, prions, and fungi, are conventionally pathogenic microorganisms. The last two, protozoa and helminths, include both pathogenic and non-pathogenic species. All parasites typically disrupt the normal, homeostatic functioning of the body, both directly as a result of their own activity and indirectly by stimulating the host's immune system to produce a defensive response. They may do so by their sheer presence, by competing with host cells and symbiotic microbes, and, in extreme cases, by releasing toxins that increase the severity of diseases. Depending on their ecologies or life histories, parasites may enter hosts via their exposure to contaminated environmental sources such as food, water, and soil (e.g., enteric bacterial pathogens: Kilonzo et al. 2011; Sinton et al. 2007). They may also spread rapidly through host populations via mechanisms such as (1) direct host-to-host contact (e.g., respiratory viruses), (2) the sharing of common, contaminated environmental space or resources (e.g., enteric bacteria such as *Salmonella* sp., *Shigella* sp.), (3) exchange of body fluids (e.g. blood-borne pathogens like HIV and HPV), or via (4) vector-borne transmission [e.g., mosquitoes spreading malarial parasites (*Plasmodium* sp.)] (summarized in Engel and Jones-Engel 2011; Nunn and Altizer 2006).

Infectious disease ecology is a subfield that deals with the evolutionary and environmental factors that influence the exposure, acquisition, and transmission dynamics of parasites within and (more recently) between human and animal populations (Grenfell and Dobson 1995; Hudson et al. 2002). As we have now entered the Anthropocene epoch, human influence on the environment has generated an increased awareness of the importance of both public health and the conservation of natural ecosystems. So it is not surprising that over the last two decades in particular we have seen an incredible surge in research related to infectious disease ecology and evolution (reviewed in Huffman and Chapman 2009; Kappeler et al. 2015; MacIntosh and Frias 2016; Nunn 2012; Nunn and Altizer 2006), with interdisciplinary approaches drawing on theory and methods from several biological sciences including anthropology, evolutionary genetics, behavioral ecology, epidemiology, network theory, and statistics.

Nonhuman primates have served as especially useful model host systems in these endeavors (summarized in Huffman and Chapman 2009; Nunn 2012; Nunn and Altizer 2006). In addition to sharing evolutionary histories and, increasingly, ecological space with humans, primates also exhibit diverse forms of social systems, characterized by heterogeneity in group composition and size, dispersal patterns, foraging strategies, mating systems, and social structures (Hinde 1976; Kappeler and Van Schaik 2002; Sterck 1998; Thierry 2007a). For these reasons, they are physiological, ecological, and behavioral model host systems for infectious disease research (MacIntosh 2016). There is now consensus among scientists that the evolutionary, ecological, and social diversity of free-living primates is impacted by (or indeed impact) the risk of acquisition and transmission of parasites (Sueur et al. 2018).

Broadly, empirical research on primate infectious disease ecology to date has had five major foci. First, in studies related to (1) *parasite-host co-evolution*, evolutionary anthropologists have attempted to establish links between the phylogenetic relationships of parasites and their primate hosts (MacIntosh and Frias 2016; Nunn 2011; Nunn and Altizer 2006; Petrášová et al. 2011; Vallo et al. 2012). Second, studies on (2) *primate parasite socioecology*, in addition to the relative role(s) of resource abundance, predation pressure, and infanticidal risk, have also begun to examine the role of parasites in shaping the evolution of primate group sizes and social network structure (Chapman et al. 2009; Nunn et al. 2011; meta-analyses by Griffin and Nunn 2012; Nunn et al. 2015; Patterson and Ruckstuhl 2013; Rifkin et al. 2012). Conversely, the idea that group-living and social structure may also impact the diversity and prevalence of parasites in hosts (Drewe and Perkins 2015; VanderWaal and Ezenwa 2016) has led to such socioecological approaches to also focus on the identification of potential "super spreaders" or "social bottlenecks" of infection (Balasubramaniam et al. 2016, 2018; Duboscq et al. 2016; Griffin and Nunn 2012; MacIntosh et al. 2012; Romano et al. 2016). Other studies have used agent-based models to predict the prevalence and transmission of parasites through artificial primate groups and networks (Griffin and Nunn 2012; Nunn et al. 2015). Yet social life does not always equate to disease transmission or threats to homeostasis. Indeed, studies on both captive and wild primates that assess the links between (3) *infection risk and sociality, stress, and immune function* have tested the opposite paradigm, i.e., that possessing strong, diverse social connections, rather than increasing the risk of pathogenic acquisition, may function to socially buffer some primates against infection (Balasubramaniam et al. 2016; Duboscq et al. 2016; Sapolsky et al. 2000; Young et al. 2014). More research has focused on the impact of (4) *parasites in primate conservation and management*—while some deal with the implications of introduced species on the prevalence and diversity of parasites in indigenous primates (Petrášová et al. 2010, 2011), other research has attempted to quantify differences in parasite richness or diversity in primates living in varying degrees of human influence or in relation to their threatened status(es) (Bublitz et al.

2015; Chapman et al. 2006a; Gillespie et al. 2005; Goldberg et al. 2007; Kowalewski et al. 2011). Finally, emerging lines of research have focused on (5) *primate counter-strategies*, including avoidance behaviors to minimize exposure to parasites (Amoroso et al. 2017; Poirotte et al. 2017, 2019; Sarabian and MacIntosh 2015; Sarabian et al. 2017), and self-medication that removes or minimizes the impact of an infection on the host (Huffman 2016).

To date, much of the empirical work related to primate infectious disease ecology has focused on wild or red-listed primate populations (reviewed in Frias and MacIntosh 2018). Aside from habitat loss and fragmentation (Hussain et al. 2013), red-listed populations also face the risk of extinction on account of infectious diseases transmitted from humans or livestock (reviewed in Frias and MacIntosh 2018). In comparison, less research has focused on the relationship between host socioecology and transmission of parasites between humans and free-living primates at overlapping interfaces (Kaur and Singh 2009). This is despite the wide recognition that humans and primates strongly influence each other's biology, behavior, and health (Fuentes 2012; Fuentes and Hockings 2010) and that such human-primate interfaces are also potential sources of EIDs (Jones-Engel et al. 2005; Nunn et al. 2008; Wolfe et al. 2007).

13.3 Human-Macaque Interfaces

The genus *Macaca* is the most diverse, geographically widespread, and ecologically successful group of primates (Cords 2013; Thierry 2007a, b). They constitute 23 extant species, which range from North Africa in the West (Barbary macaques: *M. sylvanus*), across the Indian subcontinent [e.g., rhesus macaques (*M. mulatta*), bonnet macaques (*M. radiata*)], China [e.g., rhesus macaques (*M. mulatta*), Tibetan macaques (*M. thibetana*)], and Southeast Asia [e.g., long-tailed macaques (*M. fascicularis*), Sulawesi macaque species (e.g., *M. nigra*, *M. tonkeana*)], and up to Japan in the Far East [Japanese macaques (*M. fuscata*)] (Thierry 2007a, b). Across this range, their ecological flexibility is evidenced by the fact that macaque species, and indeed populations of the same species, inhabit a wide variety of habitats, from tropical rainforests to snowcapped mountains and from dry scrub forests to urbanized human settlements (Cords 2013; Gumert et al. 2011; Thierry 2007a, b).

In nature, all macaque species show broadly similar social organization [but see Sinha et al. (2005) for an exception]—they live in multi-male multi-female social groups in which females are philopatric and males disperse from their natal groups (Cords 2013; Thierry 2013). At the same time, they show a remarkable degree of inter- and intraspecific variation in the structure of social relationships, ranging from despotic, nepotistic societies with steep dominance hierarchies and modular, centralized, and kin-biased social networks (e.g., rhesus macaques, Japanese macaques) to tolerant or egalitarian societies characterized by shallower dominance relationships and dense, decentralized, and well-connected social networks (e.g., Sulawesi

Fig. 13.1 Macaques at human-macaque interfaces, specifically (**a**) rhesus macaques in Himachal Pradesh, Northern India; (**b**) long-tailed macaques in Kuala Lumpur, Malaysia; (**c**) toque macaques in Colombo, Sri Lanka; (**d**) bonnet macaques in Kerala, Southern India. Photo credits: (**a**), (**b**) and (**d**): K. N. Balasubramaniam; (**c**): M. A. Huffman

crested macaques: Balasubramaniam et al. 2012; Thierry et al. 2008; Sueur et al. 2011a).

More pertinently, macaques also vary in the extent to which they show adaptive or maladaptive responses to human disturbance and anthropogenic landscapes, i.e., along a spectrum of overlap at human-macaque interfaces (Priston and McLennan 2013; Radhakrishna and Sinha 2011; Radhakrishna et al. 2013). At the upper end of this spectrum lie rhesus and long-tailed macaques (Fig. 13.1a, b). Large populations of these "weed" species, categorized as "Least Concern" by IUCN since 2010 (IUCN 2019), gravitate toward and even preferentially exploit human settlements (Jaman and Huffman 2013; Southwick and Siddiqi 1994, 2011; Southwick et al. 1983). Long-tailed macaques are even listed among the IUCN Invasive Species Specialist Group's (ISSG) top 100 invasive species in the world (Lowe et al. 2000). Thus, they inhabit a variety of human-macaque interfaces: from buffer zones of ecotourism in national parks, to agricultural fields bordering rural villages, to urbanized cities like Delhi, Dhaka, and Kuala Lumpur (Fig. 13.1a, b). Other species like bonnet macaques and toque macaques (*M. sinica*) are not far behind, with both wild and semi-urban populations that inhabit the smaller town-, temple-, and university campus-interfaces of Southern India and Sri Lanka, respectively (Huffman et al. 2013a; Nahallage and Huffman 2013; Nahallage et al. 2008; Radhakrishna et al. 2013; Ram et al. 2003; Sinha et al. 2005) (Fig. 13.1c, d). Yet some of these species, like toque macaques, remain listed as "vulnerable" or "endangered" on account of the negative effects of ecotourism and habitat loss throughout their range (IUCN 2019). Finally, some less ecologically flexible species like lion-tailed macaques (*M. silenus*), Tibetan macaques (*M. thibetana*), and Sulawesi crested macaques (*M. nigra*) are still exposed to the negative impact of human activity in the form of habitat loss affecting their socioecology (Kumara et al. 2014; Singh et al. 2001), ecotourism-related stressors and mortality rates (Berman et al. 2007; Marechal et al. 2011), and hunting for bush-meat impacting mortality rates (Kyes et al. 2012; Palacios et al. 2012; Riley 2007; Riley and Fuentes 2011). Indeed, many of these species are classified as being "endangered" or "critically endangered" as a result (IUCN 2019).

The rise of *ethnoprimatology* as a subfield of biological anthropology has occurred simultaneously with the rise of primate infectious disease ecology. Specifically, ethnoprimatology is related to understanding how humans and primates impact each other's niche construction, behavioral biology, and health-related outcomes (Dore et al. 2017; Fuentes 2012; Fuentes and Hockings 2010). Unsurprisingly, human-macaque interfaces in North Africa, India, Sri Lanka, and Southeast Asia have been the primary foci of most ethnoprimatological research, with some more recent studies in Africa also having been conducted on baboons (Fehlmann et al. 2016; Kaplan et al. 2011; Hoffman and O'Riain 2012), chimpanzees (Hockings et al. 2012), and lemurs (Loudon et al. 2017). To date, this work has revealed that the nature, frequency, and severity of interactions and conflict at human-macaque interfaces vary broadly by context (Radhakrishna and Sinha 2011; Radhakrishna et al. 2013). For instance, across the Indian subcontinent, China, and Southeast Asia, conflict is heavily influenced by whether macaques also play more positive roles with resident and/or visiting human communities, e.g., monkeys as religious symbols, pets, trade commodities, or tourist attractions (Jones-Engel et al. 2004; Nahallage and Huffman 2013; Radhakrishna et al. 2013). At the same time, some intrinsic characteristics of macaques, including the age-sex class, personalities of individuals, and/or species-typical adaptive responses, have also been shown to influence interface interactions (Beisner et al. 2014; Fuentes 2006; Marechal et al. 2011; Sha et al. 2009). Such variation in human- and macaque-specific features across interfaces generates a broad variety of direct and indirect interactions, such as (1) human provisioning of macaques, (2) macaques using anthropogenic landscape features (e.g., buildings, fences, water tanks), (3) mutual contact- and non-contact aggression, (4) the exchange of body fluids like blood and saliva, (5) the fragmentation of macaque populations on account of the loss of natural habitat, (6) the hunting and consumption of macaques by humans as bush-meat, and (7) the use of macaques as pets, trade commodities, or tourist attractions (Fuentes et al. 2011; Gumert et al. 2011; Hussain et al. 2013; Jones-Engel et al. 2004; Radhakrishna et al. 2013; Riley and Fuentes 2011; Riley 2003). Naturally, the dynamic nature of such environments provides myriad mechanisms for the acquisition and transmission of parasites (Engel and Jones-Engel 2011; Nunn 2012).

13.4 Parasites at Human-Macaque Interfaces

To extract and review previous studies that report parasites among free-living macaque populations at human-macaque interfaces, we relied on the Global Mammal Parasite Database (or GMPD, Version 2.0: Stephens et al. 2017). The GMPD is a compilation of studies that report disease-causing organisms—bacteria, viruses, protozoa, helminths, and fungi—isolated from wild or free-living populations of some of the major mammalian taxa, specifically ungulates, carnivores, and primates (Nunn and Altizer 2005; Stephens et al. 2017). The database now contains 24,000 records, from over 2700 literature sources including journal articles, books and book chapters, and reports at conference proceedings. Records may be filtered on the basis of different parasite or host-specific characteristics, such as taxonomic categories, geographic location, and mode of transmission.

A search of the GMPD database filtered by host genus (macaques) and type of parasite (bacteria, viruses, helminths, and protozoa) revealed 570 records from across 80 different studies. Figure 13.2 indicates the distribution of these records by study period. Aside from the general geographic location, the GMPD does not offer more specific filtering options that aid in the classification of studies in accordance with socioecological conditions under which they were conducted. So, we manually screened for "human-macaque interface" studies as those among the above studies that reported one or more of the following: (1) the direct transmission of these agents between humans and macaques in either direction (e.g., contact

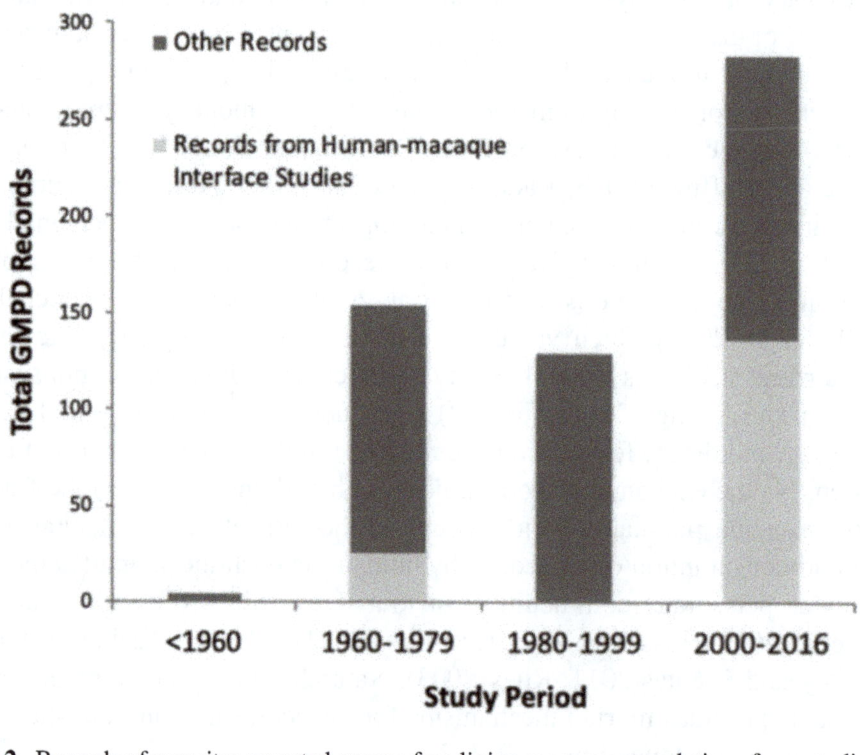

Fig. 13.2 Records of parasites reported among free-living macaque populations from studies in the Global Mammal Parasite Database (GMPD)

aggression or provisioning, pet macaques released into free-living populations), (2) the possible or potential transmission of such agents, or (3) the indirect impact of humans or anthropogenic factors (e.g., habitat fragmentation, livestock, macaque foraging on provisioned food) being identified to have influenced the acquisition of these parasites among macaques. Since this chapter primarily deals with the socioecological impact of human-macaque interfaces on disease risk, we also did not include studies on the phylogenetic co-evolutionary roots of parasites and their primate hosts.

These criteria led to the extraction of 161 (out of 570) macaque records, the vast majority of which were dated post 2000 (130 out of 161: Fig. 13.2). These records were spread across nine cited studies conducted on six different species of macaques. In Table 13.1, we summarize information from these studies, providing details on the parasites isolated, host macaque species, geographic location, prevalence and the number of individuals sampled (where the information is available), and the type of acquisition reported or speculated. Unsurprisingly, viral agents dominate this list with 88 of the 161 entries (or 55%) from 4 citations. During the last two decades, several zoonotic viruses have been described and studied in nonhuman primates in Africa and Asia from evolutionary and virulence perspectives (e.g., Ebola in great apes, reviewed in Leendertz et al. 2017; respiratory viruses in macaques, reviewed below). From a socioecological standpoint, studies on macaques have revealed strong associations between the frequency of intense human-macaque contact behaviors that involve the exchange of body fluids (e.g., aggressive bites and scratches) and the prevalence of respiratory viruses. Early work showed that wild-caught rhesus macaques in Northern India that had the highest degrees of exposure to human contact were also the most likely to show blood serum antibodies against human respiratory viruses (Shah and Southwick 1965). Later work in Nepal revealed correlations between the frequency of intense contact interactions with humans like aggressive bites and scratches and the seroprevalence of respiratory viruses such as the simian foamy virus (SFV), simian type-D virus, Cercopithecine herpesvirus-1 (CHV-1), and simian virus-40 (SV-40) (Jones-Engel et al. 2006) among rhesus macaques living in human settlements. In comparison, the Barbary macaques of Gibraltar, which engage in less intense aggression and have lower rates of contact bites in comparison to the Asian rhesus populations, showed a markedly lower seroprevalence (or were even seronegative) of these same respiratory viruses (Engel et al. 2008). More recently, humans traveling with performing "pet" rhesus macaques were found to indirectly influence the genetic structure and translocation of macaque SFV across rhesus populations in Bangladesh (Feeroz et al. 2013). Finally, other work not included in the GMPD has recorded the prevalence of macaque-borne viral pathogens like SFV and retroviruses among humans that regularly come into contact with these populations (Jones-Engel et al. 2005).

Gastrointestinal protozoa (36 entries out of 161, or 22%) and helminths (26 entries out of 161, or 16%) were the next most commonly reported parasites among the human-macaque interface studies examined. In nature, these are among the most commonly occurring parasites in wild primates (Huffman and Chapman 2009; Nunn and Altizer 2006). Yet, a few ecological assessments of human-perturbed landscapes

Table 13.1 Summary of studies extracted from the GMPD that have reported parasites at human-macaque interfaces

Citations	GMPD entries	Parasite	Parasite genus	Host species	Location	Lat.	Long.	Prev.	Type of acquisition (reported or speculated)
Huffman et al. (2013a)	7990	Bacteria	*Escherichia coli*	*Macaca sinica*	Sri Lanka	7	81	0.25	Direct: human-macaque interactions
Huffman et al. (2013a)	7992	Helminth	*Trichuris* sp.	*Macaca sinica*	Sri Lanka	7	81	0.22	Indirect: habitat fragmentation
Huffman et al. (2013a)	7991	Protozoa	*Entamoeba histolytica*	*Macaca sinica*	Sri Lanka	7	81	0.07	Direct: human-macaque interactions
Hussain et al. (2013)	7873	Helminth	*Ancylostoma* sp.	*Macaca silenus*	Southern India	10.4	77.02	0.26	Indirect: habitat fragmentation
Hussain et al. (2013)	7880	Helminth	*Diphyllobothrium* sp.	*Macaca silenus*	Southern India	10.4	77.02	0.02	Indirect: habitat fragmentation
Hussain et al. (2013)	7875	Helminth	*Strongyloides* sp.	*Macaca silenus*	Southern India	10.4	77.02	0.32	Indirect: habitat fragmentation
Hussain et al. (2013)	7877	Helminth	*Trichuris* sp.	*Macaca silenus*	Southern India	10.4	77.02	0.32	Indirect: habitat fragmentation
Hussain et al. (2013)	7881	Protozoa	*Balantidium* sp.	*Macaca silenus*	Southern India	10.4	77.02	0.06	Indirect: habitat fragmentation
Ekanayake et al. (2006)	5945–5948; 6090	Helminth	*Enterobius* sp.	*Macaca sinica*	Sri Lanka	7.9	81	0.51	Indirect: soil/livestock
Ekanayake et al. (2006)	5961–5964; 6094	Helminth	*Balantidium coli*	*Macaca sinica*	Sri Lanka	7.9	81	0.26	Indirect: soil/livestock
Ekanayake et al. (2006)	5969–5972; 6096	Protozoa	*Chilomastix* sp.	*Macaca sinica*	Sri Lanka	7.9	81	0.12	Indirect: soil/livestock
Ekanayake et al. (2006)	5937–5940	Protozoa	*Cryptosporidium* sp.	*Macaca sinica*	Sri Lanka	7.9	81	0.29	Indirect: soil/livestock
Ekanayake et al. (2006)	5965–5968; 6095	Protozoa	*Entamoeba coli*	*Macaca sinica*	Sri Lanka	7.9	81	0.30	Indirect: soil/livestock

Ekanayake et al. (2006)	5973–5976, 6097	Protozoa	*Entamoeba hartmanni*	*Macaca sinica*	Sri Lanka	7.9	81	0.27	Indirect: soil/livestock
Ekanayake et al. (2006)	5977–5980; 6098	Protozoa	*Entamoeba histolytica*	*Macaca sinica*	Sri Lanka	7.9	81	0.30	Indirect: soil/livestock
Ekanayake et al. (2006)	5957–5960; 6963	Protozoa	*Iodamoeba* sp.	*Macaca sinica*	Sri Lanka	7.9	81	0.27	Indirect: soil/livestock
Ekanayake et al. (2006)	5949–5952; 6091	Helminth	*Strongyloides* sp.	*Macaca sinica*	Sri Lanka	7.9	81	0.29	Indirect: soil/livestock
Ekanayake et al. (2006)	5953–5956; 6092	Helminth	*Trichuris* sp.	*Macaca sinica*	Sri Lanka	7.9	81	0.09	Indirect: soil/livestock
Engel et al. (2008)	7131; 7302; 7308; 7314; 7321	Virus	*Betaretrovirus Mason-Pfizer monkey virus*	*Macaca sylvanus*	Gibraltar	36	−5.6	0.00	Direct: human-macaque interactions
Engel et al. (2008)	7128; 7304; 7310; 7315; 7320	Virus	*Deltaretrovirus STLV 1*	*Macaca sylvanus*	Gibraltar	36	−5.6	0.00	Direct: human-macaque interactions
Engel et al. (2008)	7130; 7301; 7307; 7313; 7319	Virus	*Lentivirus SIV* sp.	*Macaca sylvanus*	Gibraltar	36	−5.6	0.00	Direct: human-macaque interactions
Engel et al. (2008)	7127; 7303; 7309; 7316; 7323	Virus	*Simplexvirus Herpes simplex virus 1*	*Macaca sylvanus*	Gibraltar	36	−5.6	0.00	Direct: human-macaque interactions
Engel et al. (2008)	7132; 7300; 7306; 7312; 7318	Virus	*Spumavirus Simian foamy virus*	*Macaca sylvanus*	Gibraltar	36	−5.6	0.59	Direct: human-macaque interactions
Feeroz et al. (2013)	7833–7845; 8102–8107	Virus	*Spumavirus Simian foamy virus*	*Macaca mulatta*	Bangladesh			0.92	Direct: human-macaque interactions, geographic isolation
Jones-Engel et al. (2006)	3729; 5422; 5458; 5521; 5581	Virus	*Cytomegalovirus Cercopithecine herpes-virus 8*	*Macaca mulatta*	Nepal	27.7	85.3	0.95	Indirect: habitat fragmentation, human provisioning

(continued)

Table 13.1 (continued)

Citations	GMPD entries	Parasite	Parasite genus	Host species	Location	Lat.	Long.	Prev.	Type of acquisition (reported or speculated)
Jones-Engel et al. (2006)	3730; 5423; 5459; 5522; 5582	Virus	*Polyomavirus SV-40*	*Macaca mulatta*	Nepal	27.7	85.3	0.90	Direct: human-macaque interactions
Jones-Engel et al. (2006)	3731;5424; 5560; 5523; 5583	Virus	*Simplexvirus Cercopithecine herpes-virus 1*	*Macaca mulatta*	Nepal	27.7	85.3	0.64	Direct: human-macaque interactions
Jones-Engel et al. (2006)	3732; 5425; 5461; 5524; 5584	Virus	*Spumavirus Simian foamy virus*	*Macaca mulatta*	Nepal	27.7	85.3	0.97	Direct: human-macaque interactions
Lee et al. (2011)	7157; 7164	Protozoa	*Plasmodium coatneyi*	*Macaca fascicularis*	Borneo			0.45	Indirect: anopheline vector
Lee et al. (2011)	7160; 7161	Protozoa	*Plasmodium cynomolgi*	*Macaca fascicularis*	Borneo			0.48	Indirect: anopheline vector
Lee et al. (2011)	7159; 7162	Protozoa	*Plasmodium fieldi*	*Macaca fascicularis*	Borneo			0.02	Indirect: anopheline vector
Lee et al. (2011)	7158; 7163	Protozoa	*Plasmodium inui*	*Macaca fascicularis*	Borneo			0.80	Indirect: anopheline vector
Lee et al. (2011)	7156; 7165	Protozoa	*Plasmodium knowlesi*	*Macaca fascicularis*	Borneo			0.89	Indirect: anopheline vector
Shah and Southwick (1965)	3135; 5519; 5579	Virus	*Alphavirus Chikungunya*	*Macaca mulatta*	Northern India	28	80	0.00	Direct: human-macaque interactions
Shah and Southwick (1965)	5508; 5568	Virus	*Dengue virus group Dengue 1*	*Macaca mulatta*	Northern India	28	80	0.00	Direct: human-macaque interactions
Shah and Southwick (1965)	5516; 5576	Virus	*Enterovirus poliovirus 1*	*Macaca mulatta*	Northern India	28	80	0.00	Direct: human-macaque interactions

Shah and Southwick (1965)	5517; 5577	Virus	*Enterovirus poliovirus 2*	*Macaca mulatta*	Northern India	28	80	0.13	Direct: human-macaque interactions
Shah and Southwick (1965)	5518; 5578	Virus	*Enterovirus poliovirus 3*	*Macaca mulatta*	Northern India	28	80	0.00	Direct: human-macaque interactions
Shah and Southwick (1965)	5507; 5567	Virus	*Flavivirus Japanese encephalitis*	*Macaca mulatta*	Northern India	28	80	0.00	Direct: human-macaque interactions
Shah and Southwick (1965)	5514; 5574	Virus	*Pneumovirus Human respiratory syncytial virus*	*Macaca mulatta*	Northern India	28	80	0.00	Direct: human-macaque interactions
Shah and Southwick (1965)	5510; 5570	Virus	*Polyomavirus SV-40*	*Macaca mulatta*	Northern India	28	80	0.60	Direct: human-macaque interactions
Shah and Southwick (1965)	5511; 5571	Virus	*Respirovirus Human parainfluenza virus 1*	*Macaca mulatta*	Northern India	28	80	0.05	Direct: human-macaque interactions
Shah and Southwick (1965)	5513; 5573	Virus	*Respirovirus Human parainfluenza virus 3*	*Macaca mulatta*	Northern India	28	80	0.51	Direct: human-macaque interactions
Shah and Southwick (1965)	5512; 5572	Virus	*Rubulavirus Human parainfluenza virus 2*	*Macaca mulatta*	Northern India	28	80	0.01	Direct: human-macaque interactions
Shah and Southwick (1965)	5509; 5569	Virus	*Simplexvirus Herpes simplex virus 1*	*Macaca mulatta*	Northern India	28	80	0.44	Direct: human-macaque interactions
Wenz-Mücke et al. (2013)	7862; 7866	Helminth	*Oesophagostomum* sp.	*Macaca fascicularis*	Thailand	16.2	103.1	0.07	Indirect: habitat fragmentation, human provisioning

(continued)

Table 13.1 (continued)

Citations	GMPD entries	Parasite	Parasite genus	Host species	Location	Lat.	Long.	Prev.	Type of acquisition (reported or speculated)
Wenz-Mücke et al. (2013)	7858; 7863	Helminth	*Strongyloides fuelleborni*	*Macaca fascicularis*	Thailand	16.2	103.1	0.41	Indirect: habitat fragmentation, human provisioning
Wenz-Mücke et al. (2013)	7859; 7864	Helminth	*Trichuris* sp.	*Macaca fascicularis*	Thailand	16.2	103.1	0.54	Indirect: habitat fragmentation, human provisioning

have revealed that anthropogenic factors may indirectly influence their acquisition among macaques [but see Lane et al. (2011) who report a decrease in such acquisition]. In Sri Lanka, for instance, the prevalence of both gastrointestinal protozoan parasites like *Cryptosporidium* sp., *Entamoeba* sp., and *Balantidium coli* and nematodes like *Enterobius* sp. and *Strongyloides* sp. was more common among toque macaques in more human-disturbed than pristine environments (Ekanayake et al. 2006). Further, Huffman et al. (2013a) speculate that increased human impact may in part be responsible for why the prevalence of helminth parasites was lower among toque macaques sampled at lower altitudes. Increased contact with anthropogenic landscapes and human-provisioned food was strongly linked to the prevalence of *Strongyloides fuelleborni* in wild long-tailed macaques in Thailand (Wenz-Mücke et al. 2013). Among populations of critically endangered lion-tailed macaques in Southern India, the anthropogenic fragmentation of their natural habitat was positively correlated to the diversity of gastrointestinal helminths and protozoan parasites (Hussain et al. 2013).

We extracted 10 records (6% of 161 records) of the protozoan parasite *Plasmodium* sp., the causative agent of malaria, all from a single study that surveyed wild populations of long-tailed and pig-tailed macaques (*Macaca nemestrina*) in Borneo (Lee et al. 2011). This study revealed especially high prevalence of three *Plasmodium* species—*P. knowlesi*, *P. cynomolgi*, and *P. inui*—among the macaque populations. They also revealed that *P. knowlesi*, previously hypothesized as having been transmitted to humans via anopheline vectors from overlapping macaque populations, was derived from an ancestral malarial parasite that existed before humans came to Southeast Asia. Since then, high prevalence of *P. knowlesi* has been detected among free-living macaques at vegetation mosaics and forest fragments in other parts of Southeast Asia, including Indonesia, Cambodia, Laos, and Vietnam (Huffman et al. 2013b; Zhang et al. 2016). Malaria is now widely recognized as being a threat at human-primate interfaces (Singh et al. 2004). In addition to thriving macaque populations acting as natural reservoirs for these parasites, the fragmented mosaic landscapes of Southeast Asia are also highly conducive to the proliferation of mosquito vector complexes like *Anopheles dirus* and *A. leucosphyrus*, which may transmit malaria into otherwise infection-naive macaque and human populations (Moyes et al. 2014).

The least reported type of parasite was bacterial, with only one record speculating that anthropogenic factors may be responsible for the prevalence of *Escherichia coli* in toque macaques (Huffman et al. 2013a). This was more broadly reflective of the general lack of studies that have focused on the detection of bacterial pathogens in free-living primates (Kaur and Singh 2009; Nunn and Altizer 2006). Bacterial pathogens like *Salmonella* sp., *Shigella* sp., and *E. coli* O157:H7 routinely cause acute diarrheal infection among humans and domestic livestock (Gorski et al. 2011; Rwego et al. 2008; Sinton et al. 2007; Suleyman et al. 2016). They have been previously isolated from wild primate populations in Africa that live in human-perturbed, fragmented habitats (chimpanzees: McLennan et al. 2017; lemurs of Madagascar: Bublitz et al. 2015). Since they strongly overlap and rely heavily on anthropogenic resources, urban and semi-urban macaques may be natural reservoirs of these agents, with the potential to disseminate them into overlapping human and critically endangered wildlife populations. A preliminary study at human-rhesus macaque interfaces in Northern India revealed that anthropogenic factors, such as rates of human-macaque aggression and provisioning, were positively correlated with the prevalence of enteric bacteria like *Salmonella* sp. and *E. coli* O157:H7 (Beisner et al. 2016). This finding should lead to future assessments of the relative prevalence of enteric bacteria among humans, livestock, and other overlapping macaque populations.

Our GMPD search yielded no studies on parasites in Tibetan macaques, which is somewhat surprising. There is scope for future work to focus on parasite transmission at human-Tibetan macaque interfaces. At both Mt. Emei and Mt. Huangshan, China, where they have been best studied (Zhao 1996; Li 1999), wild Tibetan macaque groups are indeed exposed to anthropogenic factors, particularly tourism. At Mt. Emei, a Buddhist community that is visited by tourists for its temples, there is no regulation of tourist-macaque interactions. Tourists regularly hand-provision the macaques, and there have been reports of tourists suffering fatal injuries from macaque attacks (Zhao 2005). Such intense and frequent contact presents scope for the transmission of parasites. On the other hand, a primate tourism program that is currently in place at Mt. Huangshan restricts the scope for macaque-tourist interactions. This program was laid down following a period between 1994 and 2004, when a group of Tibetan macaques at Mt. Huangshan was "managed" for tourist activity by restricting their home range (Berman et al. 2007). Studies on this group have revealed that intragroup aggression, attacks on infants, and infant mortality rates were all much higher during periods when the group's home range was restricted for tourist viewing than in periods prior to such activity (Berman et al. 2007). Later work in the mid-2000s that was conducted following the period of severe range restriction revealed that the macaques showed increased self-directed behaviors (e.g., self-scratching, yawning, body shake) as well as stress-coping social behaviors (e.g., allogrooming, body contact) when they were closer to tourists, in comparison to when there were no tourists present (reviewed in Matheson et al. 2013). Such tourist activity, now more controlled, may have presented or may continue to present a stressful environment to Tibetan macaques that may heighten the acquisition and transmission of parasites.

13.5 The Future of Human-Macaque Disease Ecology

Our review of studies on human-macaque interfaces reveals the detection and confirmation of a range of parasites. Yet many of these studies, based on either symptomatic or mortality-based evidence of pathogenic infection in either humans or macaques, have inferred that disease transmission has occurred without ever having established that transmission did occur (VanderWaal and Ezenwa 2016; VanderWaal et al. 2014). In other words, little or no work has assessed the precise mechanisms and pathways of parasite transmission at human-macaque interfaces. In this section, we illustrate how implementing (A) the conceptual frameworks of Coupled Natural and Human Systems and One Health, in combination with (B) cutting-edge network analytical techniques, may significantly enhance our current knowledge of infectious disease transmission at human-macaque interfaces.

(A) Unifying Conceptual Frameworks: Coupled Systems and One Health
Conflict at human-macaque interfaces may be spatiotemporally variable in form and frequency and may affect parasite transmission in dynamic and sometimes unpredictable ways. In this light, one of the biggest challenges facing research on infectious disease ecology at these, and indeed all human-wildlife interfaces, is the lack of a consensual theoretical or conceptual framework applicable across multiple types of systems.

One framework that may prove useful in this regard stems from the broader conceptualization that human interactions with nature and the environment may be viewed as dynamic, coupled systems (Liu et al. 2007). Since its proposition, the Coupled Natural and Human Systems (or CNHS) approach has presented a significant advancement in our understanding of human impact on abiotic and (more recently) biotic factors. Traditionally, studies examining the interactions between humans and natural phenomena have been largely reductionist in nature (summarized in Liu et al. 2007). They have adopted principles from biology, anthropology, geography, and environmental sciences, with an almost exclusive focus on how a single component of the human system may influence a given property of a natural system or vice versa. Further, they have tended to focus on short-term effects rather than conduct long-term assessments of the feedback effects of such interactions on both human and natural systems. Expanding significantly on these assessments, the CNHS approach explicitly acknowledges that aspects of human systems and natural systems are coupled or interlinked and must therefore be assessed as a collective whole. Multiple, dynamic components of human and natural systems are expected to influence the nature and types of interactions at interfaces, with such impact being expected to reciprocally impact long-term indicators of the overall stability, sustainability, and health of both human populations and natural components.

In its short history, studies implementing CNHS frameworks have primarily focused on the impact of humans on *inanimate, abiotic* factors (e.g., landscape ecology, climatic conditions: (Foley et al. 2005; Postel et al. 1996) and their long-term effects (e.g., via environmental degradation, natural disasters: Dilley et al. 2005) on human population dynamics and ecology (Liu et al. 2007)). In comparison, fewer studies have tackled the relationship between humans and *animate* natural systems like wildlife populations (Dickman 2010, 2012). The well-documented nature of interactions and conflict at human-macaque interfaces (reviewed above) offer opportunities to address this gap. Or conversely, the CNHS framework maybe useful to assess the mechanistic processes through which the variant nature of human-macaque interfaces may favor or inhibit the transmission of parasites across human and macaque systems.

As we allude to earlier, macaques may acquire parasites in many ways, including direct physical contact with humans or during social interactions with infected conspecifics, changes in foraging strategies induced by anthropogenic landscapes, or human-induced stressors increasing macaques' susceptibility to infection. Implementing the CNHS approach would entail examining the relative likelihood (s) of these mechanisms and indeed whether specific (suites of) attributes of the human system (e.g., community type, history of interactions with macaques, visitors versus tourists) or the macaque system (e.g., age-sex class, group size, species-typical social style) are linked with the degree to which one type of interface interaction may be expected to prevail over another in influencing parasite acquisition. More tellingly, we reckon that the CNHS framework would finally take research on human-macaque infectious disease research beyond mere descriptions of parasites at interfaces. Expanding on these findings, a CNHS approach would naturally lead to more long-term assessments of the impact of parasite diversity and distribution on indicators of macaque and human population health (e.g., symptomatic evidence for disease outbreaks, stress-induced illness), reproductive success (e.g., the number and fitness of offspring), and survival (e.g., population demographics and infant mortality rates).

Closely related to the CNHS framework is the One Health (OH) concept (derived from the "One Medicine" concept: Schwabe 1984), or the idea that addressing the challenges surrounding human health issues cannot be dissociated from environmental health (or EcoHealth) or from veterinary medical practices associated with treating wild and domestic animals (Destoumieux-Garzon et al. 2018; Zinsstag et al. 2011, 2015). The OH concept stems from the acknowledgment that the impact of human population expansion on the environment generates negative health outcomes, such as the occurrence of chronic, non-infectious diseases in humans, human and animal exposure to environmental toxins and emerging pollutants like plastics (Kannan et al. 2010; Waters et al. 2016), as well as the emergence of infectious diseases at human-wildlife interfaces (Gomez et al. 2013; Hudson et al. 2002; Nunn et al. 2008; Wolfe et al. 2007). So, it constitutes a global strategy highlighting the need for a holistic, transdisciplinary approach in dealing with the health of humans, animals, and ecosystems (the One Health Initiative).

Since its proposition more than a decade ago, proponents of OH approaches have (with varying degrees of success) proposed to deal with some of the barriers facing infectious disease research (summarized in Destoumieux-Garzon et al. 2018). We highlight three as being particularly relevant to human-macaque interfaces. The first is a resolution of the extent to which the factors that influence human health outcomes overlap with those that influence the health of natural ecosystems. Human-macaque interfaces are useful to conduct such assessments. This is because many (although not all) social and environmental factors that may potentially drive parasite transmission from humans to macaques, such as direct physical contact, contaminated food or water sources, and the exchange of body fluids, are also likely to transmit agents from macaques to humans (Engel and Jones-Engel 2011; Engel et al. 2008; Jones-Engel et al. 2005; Kaur and Singh 2009).

A second barrier is related to the promotion of interdisciplinary projects that combine veterinary medical assessments to detect and diagnose infectious diseases, with ecological and evolutionary approaches to understand the relationships between parasites and their hosts (MacIntosh and Frias 2016; Nesse et al. 2010; Nunn and Altizer 2006). Among all the primates, macaques (particularly rhesus macaques and long-tailed macaques) continue to be the most common genus used in captivity as models for biomedical research (Hannibal et al. 2017; Phillips et al. 2014). Further, as we review above, infectious disease research among free-living macaque populations have had variant foci, ranging from the detection and diagnosis of parasites (Engel and Jones-Engel 2011; Engel et al. 2008; Jones-Engel et al. 2004, 2005), through establishing co-evolutionary links between parasites and macaque hosts (Huffman et al. 2013b), to assessing the social and environmental underpinnings of parasite prevalence and transmission (Duboscq et al. 2016; MacIntosh et al. 2010, 2012; Romano et al. 2016). The OH concept, along with our above-stated argument that human-macaque interfaces are functionally interdependent, coupled systems, would provide a means to bring such diverse foci under a single, unifying framework (Destoumieux-Garzon et al. 2018).

Finally, a third direction involves placing a strong emphasis on complementing strong medical and theoretical knowledge, with current advancements in methodological and data analytical approaches. Below we elaborate on how one set of approaches—Network Analyses—may be especially significant for future research on infectious disease ecology at human-macaque interfaces.

(B) Network-Based Analytical Approaches

In the last two or three decades, network-based analytical techniques have revolutionized infectious disease epidemiology and ecology (Craft 2015; Craft and Caillaud 2011; Drewe and Perkins 2015; Godfrey 2013; Keeling 2005; Moore and Newman 2000; Newman 2002; VanderWaal and Ezenwa 2016). From a biological perspective, networks are reconstructions of entities (nodes) that are connected to each other based on one or more shared characteristics (edges) (Fig. 13.1a–c). For instance, *animal social and spatial networks* capture relationships between individuals in a social group linked together based on the frequency with which they interact or the degree of spatial overlap, respectively (reviewed in Brent et al. 2011; Croft et al. 2008; Farine and Whitehead 2015; Kasper and Voelkl 2009; Krause et al. 2007; Lusseau and Newman, 2004; Newman 2004; Sueur et al. 2011b; Wey et al. 2008) (Fig. 13.3a). *Bipartite or multimodal networks* add a level of complexity by distinguishing two or more components or layers of organization within a system, such that distinctions can be made between the edges that link nodes within the same layer to nodes across layers (Dormann et al. 2017; Kane and Alavi 2008; Latapy et al. 2008) (Fig. 13.3b, c). Third, links based on the degree of genotypic similarity of gastrointestinal microbes isolated from potentially interacting hosts within the same time frame may be used to reconstruct *microbial sharing or transmission networks* (VanderWaal and Ezenwa 2016; VanderWaal et al. 2014). Finally, in the absence or sparsity of real data, mathematical *agent-based models and artificial networks* have proven exceptionally useful in modeling the transmission of parasites (Bente et al. 2009; Griffin and Nunn 2012; Nunn 2009; Romano et al. 2016). Such heterogeneity in connectedness within and between the components of socioecological systems may strongly influence the likelihood of parasite acquisition and transmission.

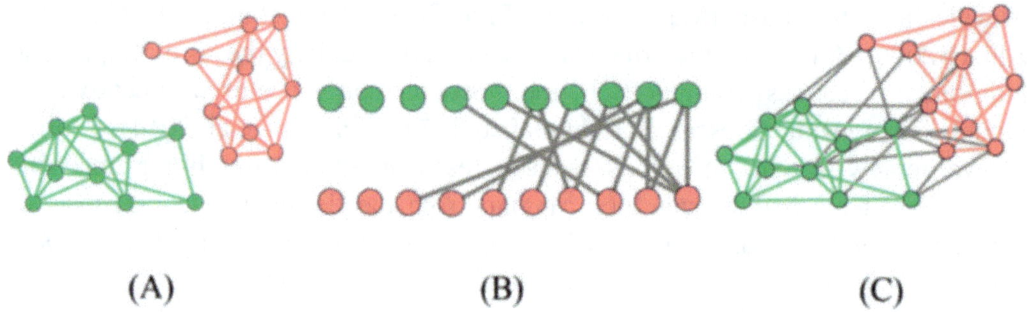

Fig. 13.3 Hypothetical social (**a**), bipartite (**b**), and multimodal (**c**) networks of the same individuals. Green nodes might represent macaques and pink ones might be humans. Green links are interactions between macaques, pink links are interactions between humans, and gray links are interactions between macaques and humans (interspecies)

Not surprisingly, network approaches, particularly social network analysis, have already found a wide range of applications in infectious disease epidemiology (Craft 2015; Craft and Caillaud 2011; Drewe and Perkins 2015; Godfrey 2013; Keeling 2005; VanderWaal and Ezenwa 2016). We briefly review these applications and related studies below. Recognizing the relative dearth in the implementation of network approaches at human-macaque interfaces, we also highlight some context (s) in which they may be implemented in human-macaque infectious disease ecological research.

Social Networks and Parasite Acquisition In humans and other animals, heterogeneity in space use overlap or contact social behavior may strongly influence the acquisition of parasites (reviewed in Drewe and Perkins 2015; Kappeler et al. 2015; Silk et al. 2017). Such heterogeneity can be modeled using social network analysis (Brent et al. 2011; Croft et al. 2008; Farine and Whitehead 2015; Krause et al. 2007; Sueur et al. 2011b). The first applications of social network approaches in the context of disease transmission were focused on humans, particularly in the spread of sexually transmitted diseases (or STDs) (Klovdahl 1985) and later following the detection of the severe acute respiratory syndrome (SARS) outbreak in 2003 (Meyers 2007). Since then, social networks across a wide range of taxa have been used to identify central or well-connected individuals, which may be potential "super spreaders" of parasites [e.g., lizards (*Egernia stokesii*): Godfrey et al. 2009; spider monkeys (*Ateles geoffroyi*): Rimbach et al. 2015; meerkats (*Suricata suricatta*): Drewe 2010; Japanese macaques: MacIntosh et al. 2012; reviewed in Drewe and Perkins 2015; VanderWaal and Ezenwa 2016]. In wild Japanese macaques, for instance, high-ranking individuals with more direct and indirect connections, or *eigenvector centrality* (Newman 2006) in their social grooming networks, were also shown to have greater species richness and infection intensities of nematode parasites (MacIntosh et al. 2012). Yet having strong and diverse social connections, rather than increasing parasite acquisition owing to contact-mediated transmission, may actually decrease the likelihood of such acquisition by mitigating stressors or enhancing immune function (e.g., Balasubramaniam et al. 2016; Cohen et al. 2015;

Hennessy et al. 2009). Consistent with this "social buffering hypothesis," work on captive rhesus macaques revealed that individuals with the strongest and most diverse social grooming and huddling connections were also the least prone to the acquisition of environmental bacterial pathogens (Balasubramaniam et al. 2016). Finally, aside from inter-individual differences in contact patterns, the higher order structure of social networks may also influence parasite transmission in contrasting ways. For instance, increased community modularity or substructuring in social networks, on account of individuals interacting more with subsets of preferred partners (Fushing et al. 2013; Newman 2006; Whitehead and Dufault 1999), may enhance parasite transmission within subgroups while presenting "social bottlenecks" to the group-wide spread of parasites (Griffin and Nunn 2012; Huang and Li 2007; Nunn et al. 2011, 2015; Romano et al. 2018; Salathe and Jones 2010). On the other hand, dense, well-connected networks, with a higher efficiency (the inverse of the number of shortest paths in the network), may facilitate rather than hinder the rapid transmission of parasites (Drewe and Perkins 2015; Griffin and Nunn 2012; Pasquaretta et al. 2014).

Such dynamic relationships between social networks and parasite acquisition suggest that broader socioecological contexts may determine the circumstances under which social life may be beneficial versus detrimental to infectious disease risk. The variant nature of the human-macaque interface may present such contexts and thereby influence parasite acquisition by altering the structure and connectedness of macaque social networks. For instance, higher frequencies of interactions with humans and/or changes to macaque movement or foraging behavior in landscapes altered by anthropogenic disturbance, by constraining the time available for macaques to engage in social interactions (Dunbar 1992; Kaburu et al. 2019; Marty et al. 2019), may lead to more modular, substructured social networks which may present bottlenecks to parasite transmission. On the other hand, such interactions or anthropogenic changes may present environmental stressors to the macaques (e.g., Barbary macaques: Carne et al. 2017; Marechal et al. 2011, 2016), in which individuals possessing strong and diverse social networks may benefit by being socially buffered against infection. In summary, future research may focus on establishing the precise mechanism(s) of super spreading, social bottlenecking, or stress-induced acquisition of parasites, through which social or contact network connectedness may influence parasite transmission dynamics at human-macaque interfaces.

Bipartite and Multimodal Networks Social networks have proven exceptionally useful to model the acquisition and transmission of parasites. Yet by themselves, they are somewhat limited in not capturing heterogeneity at higher organizational scales, for instance, across interactions between different components of a social or ecological system. This may be the reason why most epidemiological studies implementing network approaches to model heterogeneity in contact patterns have focused on either human systems or more recently wildlife systems (reviewed above), but almost never at the human-wildlife interface. Bipartite and multimodal networks, which establish connections between different interlinked components of a system, may prove especially useful in this regard (Dormann et al. 2017; Kane and Alavi 2008; Latapy et al. 2008; Finn et al. 2019). Recently, bipartite networks are beginning to feature in ecological and evolutionary research (reviewed in Bascompte and Jordano 2014; Cagnolo et al. 2011; Dormann et al. 2017), as illustrated by their being used to model marine food webs (Rezende et al. 2009), mutualistic interactions between flowers and seed-dispersing animal pollinators (Spiesman and Gratton 2016; Stang et al. 2009; Vazquez et al. 2009), and, more pertinently, host-parasitoid relationships (Laliberte and Tylianakis 2010; Poulin et al. 2013). For networks that combine links both within and across system components, some researchers have coined the term "multimodal networks," aka "multilayer" or "multislice networks" (Kane and Alavi 2008; Finn et al. 2019).

Their potential to model the connections of complex systems make bipartite and multimodal networks highly relevant tools for infectious disease research at human-wildlife interfaces. Yet to our knowledge, they have not been extensively used in this context. In the context of primates and EIDs, Gomez et al. (2013) used a combination of bipartite network construction and social network analytical tools to identify primate species that are likely to harbor EIDs. They constructed a bipartite network that linked each primate species with each parasite isolated from them. From this, they projected a unipartite "social" network in which primate species as nodes were linked to each other by edges weighted by the number of parasites they shared. They then revealed that species that were highly central in this "primate-parasite" network were also the most likely to function as "super spreaders" of EIDs to humans or at least most likely to share those EIDs with humans, suggesting potential conservation implications as well. However, only one macaque species—the toque macaque—was among the top 10 most central primates in this network. So, the extent to which macaques pose threats as transmitters of EIDs remains unclear [and more generally reflective of a knowledge gap in infectious disease research in East and Southeast Asia (Hopkins and Nunn 2007)], though this is undoubtedly likely to vary across contexts.

As reviewed in the previous section, humans and macaques engage in a variety of interactions at interfaces, some forms of which have already been linked to the acquisition and transmission of parasites (summarized in Table 13.1). To assess the mechanistic bases of such transmission, we recommend that future studies, implementing a CNHS approach, focus on constructing bipartite and multimodal networks based on intra- and interspecies interactions and spatial overlap at human-macaque interfaces. For such networks, the choice of system components may also be informed by the characteristics of the parasite studied (Craft 2015). For instance, the transmission of RNA respiratory and retroviruses between humans and macaques may require intense contact events such as bites and scratches and the exchange of body fluids like blood and saliva. Yet their fast mutation rates and short generation times may make them difficult to detect. So, multimodal networks of mild and severe contact aggressive interactions among and between humans and macaques over shorter durations of time may be useful for these purposes. On the other hand, the transmission of gastrointestinal helminths, protozoa, and enteric bacteria may require just subtle interactions such as human provisioning of macaques, acquisition from soil and anthropogenic surfaces, and macaque social grooming and contact huddling interactions, any or all of which would involve fecal-oral transmission (Balasubramaniam et al. 2016; Beisner et al. 2014, 2016; MacIntosh et al. 2012). Given that enteric bacteria may also survive longer in anthropogenic environments such as human-contaminated food, water sources and substrates (Sinton et al. 2007), moist soil (Kilonzo et al. 2011), and livestock (Craft 2015; Rwego et al. 2007; VanderWaal et al. 2014), unraveling their transmission routes may involve the construction of "multipartite" networks connecting contact patterns and spatial overlap between these biotic and abiotic factors. An even higher level of complexity may be required to detect the transmission routes of vector-borne malarial parasites like *Plasmodium knowlesi* (Abkallo et al. 2014; Huffman et al. 2013b; Lee et al. 2011). These may depend heavily on geospatial variation in the distribution and overlap of humans, reservoir macaques, and other host wildlife populations, anopheline vectors, and a host of environmental factors that may be conducive to the completion of both vector and pathogen life histories (Lee et al. 2011; Moyes et al. 2014).

Microbial Transmission Networks More recently, the phylogenetic relationships of symbiotic gut microbes isolated from animal hosts have been used to construct microbial transmission networks (VanderWaal and Ezenwa 2016). Such networks offer special advantages to detecting the potential transmission pathways of parasites that spread through the fecal-oral route (Sears et al. 1950, 1956; Tenaillon et al. 2010). Symbiotic microbes like gastrointestinal *E. coli* are present in almost every individual, have shared evolutionary histories with intestinal pathogens, and are typically acquired via fecal-oral routes (Caugant et al. 1981; Sears et al. 1950, 1956). So, if two individuals have genotypically similar or identical strains of *E. coli*, they are likely to have either shared the strain via fecal-oral transmission, which may occur either through direct social contact or through using a common environmental source (Chiyo et al. 2014; Springer et al. 2016; VanderWaal et al. 2013, 2014). Further, they rarely (if ever) alter the behavior of the host (VanderWaal et al. 2014), which allows researchers to potentially detect subtle transmission events that may signal the potential for a more devastating outbreak. Limited research to date has revealed strong links between the degree of dyadic similarity in *E. coli* and the frequency of animal space use overlap and/or social contact patterns (Balasubramaniam et al. 2018; Chiyo et al. 2014; Springer et al. 2016; VanderWaal et al. 2013, 2014). Recently, a study on captive rhesus macaques established that macaques in the same social network communities were more likely to share strains of *E. coli* among themselves than they were to macaques from other communities (Balasubramaniam et al. 2018). Previous assessments at human-primate interfaces have revealed that the population genetic structure of gut *E. coli* from great ape populations living in human-perturbed habitats was more similar to *E. coli* from humans and livestock than they were to bacteria. Such findings encourage future efforts to construct microbial transmission networks between humans and overlapping macaque populations, which may serve as models to gauge the potential transmission pathways of parasites with greater precision. However, the lack of discernible mortality or sickness behaviors associated with acquiring non-pathogenic microbes makes them models, rather than accurate forecasters, for parasite transmission.

Agent-Based Modeling When data on biological systems are either unavailable or incomplete, mathematical models offer ways of dealing with such inadequacies. In simple terms, a model may be thought of as a simplified version of a study system used to better understand it (Epstein and Axtell 1996; Minsky 1965). Thus, the complexity of a model has to be lower than that of the study system; otherwise its usefulness is lost. The main advantage of a computational model is that it can be tested infinitely by recreating or simulating a situation in the same way many (thousands of) times by adding, removing, or varying parameters and measuring emergent characteristics. These results are then usually compared with findings from empirical data to confirm or reject the tested hypotheses and accept (at least temporarily) the one for which the simulations best explain the data (Epstein and Axtell 1996; Minsky 1965). Furthermore, outcomes from these models may reveal complex or unexpected effects which may be different from, or even go undetected, based on a priori predictions. This might in turn provide a stronger basis to make other predictions, including those related to the relative role(s) of different predictive factors, in future assessments of biological systems (Epstein and Axtell 1996; Minsky 1965).

Unsurprisingly, the bottom-up approaches of simulated models have made them exceptionally useful tools to understand the acquisition and transmission of parasites. The first mathematical models of parasite transmission did not implement social network approaches: virtual individuals moved and interacted randomly in their environments (Wilensky and Stroup 1999). From individual characteristics (e.g., age, sex, hierarchical rank) and basic interaction rules (social contact, spatial proximity, conflicts, etc.), these tried to assess how the more global phenomena of parasite prevalence and outbreak potential emerge in a system (Romano et al. 2016; Rushmore et al. 2014). In these models, only the R_0, the initial number of infected agents and their interaction rates, mattered. R_0 is the basic reproduction number used to quantify the transmission potential of a parasite, defined as the number of secondary infections caused by a single infected individual introduced into a population made up entirely of susceptible individuals. However, following the acknowledgment that individuals do not move or interact randomly, social networks have been integrated into these models during the last decade or so (Griffin and Nunn 2012; Huang and Li 2007; McCabe and Nunn 2018; Nunn 2009).

Expanding beyond homogenous populations that contain only susceptible individuals that interact randomly, recent studies have integrated network approaches with a classic set of individual-based models used in epidemiology that classify individuals or "agents" into moving between susceptible infected and resistant (or SIR) classes or compartments (Bansal et al. 2007; Brauer 2008; Grimm and Railsback 2005; Kohler and Gumerman 2000). Such integrated network-based SIR models now form the basis of many epidemiological assessments in primate systems (Griffin and Nunn 2012; Kohler and Gumerman 2000; McCabe and Nunn, 2018;

Rushmore et al. 2014). For instance, the likelihood of parasite transmission from A to B may be a function of (1) whether A is already infected, (2) the likelihood of a link between A and B in their social network, and (3) the per-contact transmission probability "Beta." Simulations are run until set criteria are reached, e.g., either the extinction or saturation of infection throughout the group, at which time the average outbreak size or $R_{infinity}$ is calculated (Diekmann et al. 1998; Keeling 1999). The implemented social network might be theoretical (Sueur et al. 2012), based on empirical data (Romano et al. 2016; Rushmore et al. 2014), or both (Griffin and Nunn 2012) depending on the aims of the study.

In a comparative evolutionary analysis that used both natural primate datasets and simulated networks, Griffin and Nunn (2012) used an SIR model to reveal how increased community modularity in the social network, despite a larger group size, negatively impacts parasite success. In a more applied example of the implementation of agent-based models, Rushmore et al. (2014) modeled the social network of a wild population of chimpanzees in order to target specific individuals to vaccinate. Based on their social centrality, they revealed that it was sufficient to target fewer, more central individuals for the same result regarding controlling the spread of infection. In this study, social network position was a model parameter, but it might also be the study object if the model includes feedback loops between the social network and parasite acquisition. Corner et al. (2003), for instance, not only showed that social networks of Australian possums (*Trichosurus vulpecula*) have an effect on the transmission rate of tuberculosis (TB) but also that the spread of TB had a feedback effect on the social network: infected possums showed higher proximity degree and betweenness centrality than non-infected possums. To our knowledge, agent-based models have not been applied to assess parasite transmission at human-macaque interfaces. In order to do so, we recommend that future research implement the SIR approach, but focus on using multimodal network models in the place of social networks which may account for an added level of complexity to model the heterogeneity between human-macaque and macaque-macaque interactions.

13.6 Conclusions

Our goals in writing this chapter were threefold. First, we wanted to convey how, despite a recent surge in primate infectious disease ecological research during the last two decades, relatively few efforts have focused on the impact of humans and anthropogenic factors on disease risk in wild primates. Second, we hope to have illustrated how our current knowledge of human-macaque interactions in particular present "starting points" from which long-term, in-depth assessments of the ecology of parasite acquisition and transmission at human-primate interfaces may be conducted. We believe that such efforts would need to integrate ethnoprimatology and infectious disease ecology in order to be successful. Finally, we hope to have convinced readers of how research on infectious disease ecology at human-macaque interfaces are also in need of the implementation of both novel conceptual frameworks (e.g., One Health, Coupled Systems) and cutting-edge analytical approaches (e.g., Network Analyses, mathematical modeling). Such approaches not only bolster the scope of epidemiological research but are also imperative for the conservation and management of both threatened and potentially problematic free-living primate (and indeed other wildlife) populations.

References

Abkallo HM, Liu W, Hokama S, Ferreira PE, Nakazawa S, Maeno Y, Quang NT, Kobayashi N, Kaneko O, Huffman MA, Kawai S, Marchand RP, Carter R, Hahn BH, Culleton R (2014) DNA from pre-erythrocytic stage malaria parasites is detectable by PCR in the faeces and blood of hosts. Int J Parasitol 14:467–473

Amoroso CR, Frink AG, Nunn CL (2017) Water choice as a counterstrategy to faecally transmitted disease: an experimental study in captive lemurs. Behaviour 154:1239–1258

An L, Lopez-Carr D (2012) Understanding human decisions in coupled natural and human systems. Ecol Model 229:1–4

Balasubramaniam KN, Dittmar K, Berman CM, Butovskaya M, Cooper MA, Majolo B, Ogawa H, Schino G, Thierry B, de Waal FBM (2012) Hierarchical steepness, counter-aggression, and macaque social style scale. Am J Primatol 74(10):915–925

Balasubramaniam KN, Beisner BA, Vandeleest J, Atwill ER, McCowan B (2016) Social buffering and contact transmission: network connections have beneficial and detrimental effects on Shigella infection risk among captive rhesus macaques. PeerJ 4:e2630

Balasubramaniam KN, Beisner BA, Guan J, Vandeleest J, Fushing H, Atwill ER, McCowan B (2018) Social network community structure is associated with the sharing of commensal *E. coli* among captive rhesus macaques (*Macaca mulatta*). PeerJ 6:e4271

Bansal S, Grenfell BT, Meyers LA (2007) When individual behaviour matters: homogeneous and network models in epidemiology. J R Soc Interface 4:879–891

Barua M, Bhagwat SA, Jadhav S (2013) The hidden dimensions of human-wildlife conflict: health impacts, opportunity and transaction costs. Biol Conserv 157:309–316

Bascompte J, Jordano P (2014) Mutualistic networks. Princeton University Press, Princeton, NJ

Beisner BA, Heagarty A, Seil S, Balasubramaniam KN, Atwill ER, Gupta BK, Tyagi PC, Chauhan NPS, Bonal BS, Sinha PR, McCowan B (2014) Human-wildlife conflict: proximate predictors of aggression between humans and rhesus macaques in India. Primates 57:459–469

Beisner BA, Balasubramaniam KN, Fernandez K, Heagerty A, Seil SK, Atwill ER, Gupta BK, Tyagi PC, Chauhan PS, Bonal BS, Sinha PR, McCowan B (2016) Prevalence of enteric bacterial parasites with respect to anthropogenic factors among commensal rhesus macaques in Dehradun, India. Primates 57:459–469

Bente D, Gren J, Strong JE, Feldmann H (2009) Disease modeling for ebola and marburg viruses. Dis Model Mech 2:12–17

Berman CM, Li J, Ogawa H, Ionica C, Yin H (2007) Primate tourism, range restriction, and infant risk among *Macaca thibetana* at Mt. Huangshan, China. Int J Primatol 28:1123–1141

Brauer F (2008) Compartmental models in epidemiology, chapter 2. In: Brauer F, van den Driessche P, Wu J (eds) Mathematical epidemiology. Springer, Berlin

Brent LJN, Lehmann J, Ramos-Fernandez G (2011) Social network analysis in the study of nonhuman primates: a historical perspective. Am J Primatol 73(8):720–730

Bublitz DC, Wright PC, Rasambainarivo FT, Arrigo-nelson SJ, Bodager JR, Gillespie TR (2015) Pathogenic enterobacteria in lemurs associated with anthropogenic disturbance. Am J Primatol 77:330–337

Cagnolo L, Salvo A, Valladares G (2011) Network topology: patterns and mechanisms in plant-herbivore and host-parasitoid food webs. J Anim Ecol 80:342–351

Carne C, Semple S, MacLarnon A, Majolo B, Marechal L (2017) Implications of tourist–macaque interactions for disease transmission. EcoHealth 14:704–717

Caugant DA, Levin BR, Selander RK (1981) Genetic diversity and temporal variation in the *E. coli* population of a human host. Genetics 98:467–490

Chapman CA, Gillespie TR, Goldberg T (2005) Primates and the ecology of their infectious diseases: how will anthropogenic change affect host-parasite interactions? Evol Anthropol 14:134–144

Chapman CA, Speirs ML, Gillespie TR, Holland T, Austad KM (2006a) Life on the edge: gastrointestinal parasites from the forest edge and interior primate groups. Am J Primatol 68:397–409

Chapman CA, Wasserman MD, Gillespie TR, Speirs ML, Lawes MJ, Saj TL et al (2006b) Do food availability, parasitism, and stress have synergistic effects on red colobus populations living in forest fragments? Am J Phys Anthropol 131:525–534

Chapman CA, Rothman JM, Hodder SAM (2009) Can parasite infection be a selective force influencing primate group size? A test with red colobus. In: Huffman MA, Chapman CA (eds) Primate parasite ecology. Cambridge University Press, Cambridge, pp 423–440

Chiyo PI, Grieneisen LE, Wittemyer G, Moss CJ, Lee PC, Douglas-hamilton I, Archie EA (2014) The influence of social structure, habitat, and host traits on the transmission of *Escherichia coli* in wild elephants. PLoS One 9:e93408

Cohen S, Janicki-Deverts D, Turner RB, Doyle WJ (2015) Does hugging provide stress-buffering social support? a study of susceptibility to upper respiratory infection and illness. Psychol Sci 26:135–147

Cords M (2013) The behavior, ecology, and social evolution of Cercopithecine monkeys. In: Mitani JC, Call J, Kappeler PM, Palombit RA, Silk JB (eds) The evolution of primate societies. University of Chicago Press, Chicago, pp 91–112

Corner LAL, Pfeiffer DU, Morris RS (2003) Social-network analysis of *Mycobacterium bovis* transmission among captive brushtail possums (*Trichosurus vulpecula*). Prev Vet Med 59:147–167

Craft ME (2015) Infectious disease transmission and contact networks in wildlife and livestock. Philos Trans R Soc Lond B Biol Sci 370:1–12

Craft ME, Caillaud D (2011) Network models: an underutilized tool in wildlife epidemiology. Interdiscip Perspect Infect Dis 2011:1–12

Croft DP, James R, Krause J (2008) Exploring animal social networks. Princeton University Press, Princeton, NJ

Destoumieux-Garzon D, Mavingui P, Boetsch G, Biossier J, Darriet F, Duboz P, Fritsch C, Giradoux P, Le Roux P, Morand S, Paillard C, Pontier D, Sueur C, Voituron Y (2018) The one health concept: 10 years old and a long road ahead. Front Vet Sci 5:14

Dickman AJ (2010) Complexities of conflict: the importance of considering social factors for effectively resolving human-wildlife conflict. Anim Conserv 13:458–466

Dickman AJ (2012) From cheetahs to chimpanzees: a comparative review of the drivers of human-carnivore conflict and human-primate conflict. Folia Primatol 83:377–387

Diekmann O, De Jong MCM, Metz JJ (1998) A deterministic epidemic model taking account of repeated contacts between the same individuals. J Appl Probab 35:448–462

Dilley M, Chen SR, Deichmann U, Lerner-Lam LA, Arnold M, Agwe J, Buys P, Kjekstad O, Lyon B, Yetman G (2005) Natural disaster hotspots: a global risk analysis. In: Bank W (ed) Disaster risk management. World Bank and Columbia University, Washington, DC

Dore KM, Riley EP, Fuentes A (2017) Ethnoprimatology: a practical guide to research on the human-nonhuman primate interface. Cambridge University Press, Cambridge

Dormann C, Frund J, Schaefer MH (2017) Identifying causes of patterns in ecological networks: opportunities and limitations. Annu Rev Ecol Evol Syst 48:559–584

Drewe JA (2010) Who infects whom? Social networks and tuberculosis transmission in wild meerkats. Proc Biol Sci 277:633–642

Drewe JA, Perkins SE (2015) Disease transmission in animal social networks. In: Krause J, James R, Franks DW, Croft DP (eds) Animal social networks. Oxford University Press, Oxford, pp 95–110

Duboscq J, Romano V, Sueur C, MacIntosh AJJ (2016) Network centrality and seasonality interact to predict lice load in a social primate. Sci Rep 6:22095

Dunbar RIM (1992) Time: a hidden constraint on the behavioural ecology of baboons. Behav Ecol Sociobiol 31:35–49

Ekanayake DK, Arulkanthan A, Horadagoda NU, Sanjeevani M, Kieft R, Gunatilake S, Dittus WP (2006) Prevalence of cryptosporidium and other enteric parasites among wild non-human primates in Polonnaruwa, Sri Lanka. Am J Trop Med Hyg 74:322–329

Engel GA, Jones-Engel L (2011) The role of *Macaca fascicularis* in infectious disease transmission. In: Gumert MD, Fuentes A, Jones-Engel L (eds) Monkeys on the edge: ecology and management of long-tailed macaques and their interface with humans. Cambridge University Press, Cambridge, pp 183–203

Engel GA, Pizarro M, Shaw E, Cortes J, Fuentes A, Barry P, Lerche N, Grant R, Cohn D, Jones-Engel L (2008) Unique pattern of enzootic primate viruses in Gibralter macaques. Emerg Infect Dis 14:1112–1115

Epstein J, Axtell R (1996) Growing artificial societies: social science from the bottom Up. MIT Press, Cambridge, MA

Farine DR, Whitehead H (2015) Constructing, conducting and interpreting animal social network analysis. J Anim Ecol 84:1144–1163

Feeroz MM, Soliven K, Small CT, Engel GA, Pacheco MA, Yee JL, Wang X, Kamrul Hasan M, Oh G, Levine KL, Rabiul Alam SM, Craig KL, Jackson DL, Lee EG, Barry PA, Lerche NW, Escalante AA, Matsen Iv FA, Linial ML, Jones-Engel L (2013) Population dynamics of rhesus macaques and associated foamy virus in Bangladesh. Emerg Microbes Infect 2:e29

Fehlmann G, O'Riain MJ, Kerr-Smith C, King AJ (2016) Adaptive space use by baboons (*Papio ursinus*) in response to management interventions in a human-changed landscape. Anim Conserv 20:101–109

Fiennes R (1967) Zoonoses of primates. Cornell University Press, Ithaca, NY

Finn KR, Silk MJ, Porter MA, Pinter-Wollman N (2019) The use of multilayer network analysis in animal behaviour. Anim Behav 149:7–22

Foley JA, DeFries R, Asner GP, Barford C, Bonan G, Carpenter SR, Chapin FS, Coe MT, Daily GC, Gibbs HK, Helkowski JH, Holloway T, Howard EA, Kucharik CJ, Monfreda C, Patz JA, Prentice IC, Ramankutty N, Snyder PK (2005) Global consequences of land use. Science 309 (5734):570–574

Frias L, MacIntosh AJJ (2018) Threatened hosts, threatened parasites? Parasite diversity and distribution in Red-Listed Primates. In: Behie AM, Teichroeb J, Malone N (eds) Primate research and conservation in the anthropocene. Cambridge University Press, Cambridge

Fuentes A (2006) Patterns and context of human-macaque interactions in Gibraltar. In: Hodges JK, Cortes J (eds) The Barbary macaque: biology, management, and conservation. Nottingham University Press, Nottingham, pp 169–184

Fuentes A (2012) Ethnoprimatology and the anthropology of the human-primate interface. Annu Rev Anthropol 41:101–117

Fuentes A, Hockings KJ (2010) The ethnoprimatological approach in primatology. Am J Primatol 72(10):841–847

Fuentes A, Rompis ALT, Arta Putra IGA, Watiniasih NL, Nyoman Suartha I, Wandia IN et al (2011) Macaque behavior at the human-monkey interface: the activity and demography of semi-free-ranging *Macaca fascicularis* at Padangtegal, Bali, Indonesia. In: Gumert MD, Fuentes A, Jones-Engel L (eds) Monkeys on the edge: ecology and management of long-tailed macaques and their interface with humans. Cambridge University Press, Cambridge, pp 159–179

Fushing H, Wang H, VanderWaal K, McCowan B, Koehl P (2013) Multi-scale clustering by building a robust and self correcting ultrametric topology on data points. PLoS One 8(2):e56259

Gillespie TR, Chapman CA, Greiner EC (2005) Effects of logging on gastrointestinal parasite infections and infection risk in African primates. J Appl Ecol 42:699–707

Godfrey SS (2013) Networks and the ecology of parasite transmission: a framework for wildlife parasitology. Int J Parasitol Parasites Wildl 2:235–245

Godfrey SS, Bull CM, James R, Murray K (2009) Network structure and parasite transmission in a group-living lizard, the gidgee skink, *Egernia stokesii*. Behav Ecol Sociobiol 63:1045–1056

Goldberg T, Gillespie TR, Rwego IB, Wheeler E, Estoff EL, Chapman CA (2007) Patterns of gastrointestinal bacterial exchange between chimpanzees and humans involved in research and tourism in western Uganda. Biol Conserv 135:511–517

Gomez JM, Nunn CL, Verdu M (2013) Centrality in primate–parasite networks reveals the potential for the transmission of emerging infectious diseases to humans. PNAS 110:7738–7741

Gorski L, Parker CT, Liang A, Cooley MB, Jay-Russell MT, Gordus AG et al (2011) Prevalence, distribution, and diversity of *Salmonella enterica* in a major produce region of California. Appl Environ Microbiol 77(8):2734–2748

Grenfell BT, Dobson AP (1995) Ecology of infectious diseases in natural populations. Cambridge University Press, Cambridge

Griffin RH, Nunn CL (2012) Community structure and the spread of infectious disease in primate social networks. Evol Ecol 26(4):779–800

Grimm V, Railsback SF (2005) Individual based modeling in ecology. Princeton University Press, Princeton, NJ

Gumert MD, Fuentes A, Jones-Engel L (eds) (2011) Monkeys on the edge: ecology and management of long-tailed macaques and their interface with humans. Cambridge University Press, Cambridge

Hannibal DL, Bliss-Moreau E, Vandeleest J, McCowan B, Capitanio J (2017) Laboratory rhesus macaque social housing and social changes: implications for research. Am J Primatol 79:1–14

Hennessy MB, Kaiser S, Sachser N (2009) Social buffering of the stress response: diversity, mechanisms, and functions. Front Neuroendocrinol 30:470–482

Hinde RA (1976) Interactions, relationships and social structure. Man 11:1–17

Hockings KJ, Anderson JR, Matsuzawa T (2012) Socioecological adaptations by chimpanzees (*Pan troglodytes*) verus inhabiting an anthropogenically impacted habitat. Anim Behav 83:801–810

Hoffman TS, O'Riain MJ (2012) Monkey management: using spatial ecology to understand the extent and severity of human-baboon conflict in the Cape Peninsula, South Africa. Ecol Soc 17:13

Hopkins ME, Nunn CL (2007) A global gap analysis of infectious agents in wild primates. Divers Distrib 13:561–572

Huang W, Li C (2007) Epidemic spreading in scale-free networks with community structure. J Stat Mech 2007:P01014

Hudson PJ, Rizzoli A, Grenfell BT, Heesterbeek H, Dobson AP (2002) The ecology of wildlife diseases. Oxford University Press, New York

Huffman MA (2016) Primate self-medication, passive prevention and active treatment—a brief review. Int J Multidiscip Stud 3:1–10

Huffman MA, Chapman CA (2009) Primate parasite ecology: the dynamics and study of host-parasite relationships. Cambridge University Press, Cambridge

Huffman MA, Nahallage CAD, Hasegawa H, Ekanayake S, De Silva LDGG, Athauda IRK (2013a) Preliminary survey of the distribution of four potentially zoonotic parasite species among primates in Sri Lanka. J Natl Sci Found 41:319–326

Huffman MA, Satou M, Kawai S, Maeno Y (2013b) New perspectives on the transmission of malaria between macaques and humans: the case of Vietnam. Folia Primatol 84:288–289

Hussain S, Ram MS, Kumar A, Shivaji S, Umapathy G (2013) Human presence increases parasitic load in endangered lion-tailed macaques (*Macaca silenus*) in its fragmented rainforest habitats in southern India. PLoS One 8:e63685

IUCN (2019) The IUCN Red list of threatened species. Version 2019-2. http://www.iucnredlist.org. Accessed 18 July 2019

Jaman MF, Huffman MA (2013) The effect of urban and rural habitats and resource type on activity budgets of commensal rhesus macaques (*Macaca mulatta*) in Bangladesh. Primates 54:49–59

Jones-Engel L, Engel GA, Schillaci MA, Kyes K, Froehlich J, Paputungan U, Kyes RC (2004) Prevalence of enteric parasites in pet macaques in Sulawesi, Indonesia. Am J Primatol 62:71–82

Jones-Engel L, Engel GA, Schillaci MA, Rompis ALT, Putra A, Suaryana KG, Fuentes A, Beer B, Hicks S, White R, Wilson B, Allan JS (2005) Primate to human retroviral transmission in Asia. Emerg Infect Dis 7:1028–1035

Jones-Engel L, Engel GA, Heidrich J, Chalise M, Poudel N, Viscidi R, Barry PA, Allan JS, Grant R, Kyes R (2006) Temple monkeys and health implications of commensalism, Kathmandu, Nepal. Emerg Infect Dis 12:900–906

Kaburu SSK, Marty PR, Beisner B, Balasubramaniam KN, Bliss-Moreau E, Kaur K, Mohan L, McCowan B (2019) Rates of human-macaque interactions affect grooming behavior among urban-dwelling rhesus macaques (*Macaca mulatta*). Am J Phys Anthropol 168:92–103

Kane GC, Alavi M (2008) Casting the net: a multimodal network perspective on user-system interactions. Inf Syst Res 19:253–272

Kannan K, Yun SH, Rudd RJ, Behr M (2010) High concentrations of persistent organic pollutants including PCBs, DDT, PBDEs and PFOS in little brown bats with white-nose syndrome in New York, USA. Chemosphere 80:613–618

Kaplan BS, O'Riain MJ, van Eeden R, King AJ (2011) A low-cost manipulation of food resources reduces spatial overlap between baboons (*Papio ursinus*) and humans in conflict. Int J Primatol 32:1397–1412

Kappeler PM, Van Schaik CP (2002) Evolution of primate social systems. Int J Primatol 23:707–740

Kappeler PM, Cremer S, Nunn CL (2015) Sociality and health: impacts of sociality on disease susceptibility and transmission in animal and human societies. Philos Trans R Soc Lond Ser B Biol Sci 370:20140116

Kasper C, Voelkl B (2009) A social network analysis of primate groups. Primates 50:343–256

Kaur T, Singh J (2009) Primate-parasitic zoonoses and anthropozoonoses: a literature review. In: Huffman MA, Chapman CA (eds) Primate parasite ecology: the dynamics and study of host-parasite relationships. Cambridge University Press, Cambridge, pp 199–230

Kaur T, Singh J, Humphrey C, Tong S, Clevenger D, Tan W, Szekely B, Wang Y, Li Y, Alex Muse E, Kiyono M, Hanamura S, Inoue E, Nakamura M, Huffman MA, Jiang B, Nishida T (2008) Descriptive epidemiology of fatal respiratory outbreaks and detection of a human-related metapneumovirus in wild chimpanzees (*Pan troglodytes*) at Mahale Mountains National Park, western Tanzania. Am J Primatol 70:755–765

Kaur T, Singh J, Huffman MA, Petrzelkova KJ, Taylor NS, Xu S, Dewhirst FE, Paster BJ, Debruyne L, Vandamme P, Fox JG (2011) *Campylobacter troglodytes* sp. nov., isolated from feces of human-habituated wild chimpanzees (*Pan troglodytes schweinfurthii*) in Tanzania. Appl Environ Microbiol 77:2366–2373

Keeling MJ (1999) The effects of local spatial structure on epidemiological invasions. Proc Biol Sci 266:859–867

Keeling MJ (2005) The implications of network structure for epidemic dynamics. Theor Popul Biol 67:1–8

Kilonzo C, Atwill ER, Mandrell R, Garrick M, Villanueva V (2011) Prevalence and molecular characterization of *Escherichia coli* O157:H7 by multiple locus variable number tandem repeat analysis and pulsed field gel electrophoresis in three sheep farming operations in California. J Food Prot 74:1413–1421

Klovdahl AS (1985) Social networks and the spread of infectious diseases: the AIDS example. Soc Sci Med 21:1203–1216

Kohler TA, Gumerman GJ (2000) Dynamics of human and primate societies: agent-based modeling of social and spatial processes. Oxford University Press, Oxford

Kowalewski MM, Salzer JS, Deutsch JC, Raño M, Kuhlenschmidt MS, Gillespie TR (2011) Black and gold howler monkeys (*Alouatta caraya*) as sentinels of ecosystem health: patterns of zoonotic protozoa infection relative to degree of human–primate contact. Am J Primatol 73:75–83

Krause J, Croft DP, James R (2007) Social network theory in the behavioural sciences: potential applications. Behav Ecol Sociobiol 62:15–27

Kumara HN, Singh M, Sharma AK, Santhosh K, Pal A (2014) Impact of forest fragment size on between-group encounters in lion-tailed macaques. Primates 55:543–548

Kyes R, Iskandar E, Onibala J, Paputungan U, Laatung S, Huettmann F (2012) Long-term population survey of the Sulawesi black macaques (*Macaca nigra*) at Tangkoko Nature Reserve, North Sulawesi, Indonesia. Am J Primatol 75:88–94

Laliberte E, Tylianakis JM (2010) Deforestation homogenizes tropical parasitoid–host networks. Ecology 91:1740–1747

Lane KE, Holley C, Hollocher H, Fuentes A (2011) The anthropogenic environment lessens the intensity and prevalence of gastrointestinal parasites in Balinese long-tailed macaques (*Macaca fascicularis*). Primates 52:117–128

Latapy M, Magnien C, Del Vecchio N (2008) Basic notions for the analysis of large two-mode networks. Soc Networks 30:31–48

Lee K, Divis PCS, Zakaria SK, Matusop A, Julin RA, Conway DJ, Cox-Singh J, Singh B (2011) *Plasmodium knowlesi*: reservoir hosts and tracking the emergence in humans and macaques. PLoS Pathog 7:1–11

Leendertz SAJ, Wich SA, Ancrenaz M, Bergl RA, Gonder MK, Humle T, Leendertz FH (2017) Ebola in great apes—current knowledge, possibilities for vaccination, and implications for conservation and human health. Mammal Rev 47:98–111

Li J (1999) The Tibetan macaque society: a field study. Anhui University Press, Hefei

Liu J, Dietz T, Carpenter SR, Alberti M, Folke C, Moran E et al (2007) Complexity of coupled human and natural systems. Science 317(5844):1513–1516

Loudon JE, Patel ER, Faulkner C, Schopler R, Kramer RA, Williams CV, Herrera JP (2017) An ethnoprimatological assessment of human impact on the parasite ecology of silky sifaka (*Propithecus candidus*). In: Fuentes A, Riley EP, Dore KM (eds) Ethnoprimatology: a practical guide to research at the human-nonhuman primate interface. Cambridge University Press, Cambridge, pp 89–110

Lowe SJ, Browne M, Boudjelas S (2000) Published by the IUCN/SSC Invasive Species Specialist Group (ISSG), Auckland

Lusseau D, Newman MEJ (2004) Identifying the role that individual animals play in their social network. Proc Biol Sci 271:S477–S481

MacIntosh AJJ (2016) Pathogen. In: Fuentes A, Bezanson M, Campbell CJ (eds) The international encyclopedia of primatology. Wiley, Hoboken, NJ

MacIntosh AJJ, Frias L (2016) Coevolution of hosts and parasites. In: Fuentes A, Bezanson M, Campbell CJ (eds) The international encyclopedia of primatology. Wiley, Hoboken, NJ

MacIntosh AJJ, Hernandez AD, Huffman MA (2010) Host age, sex and reproduction affect nematode parasitism among wild Japanese macaques. Primates 51:353–364

MacIntosh AJJ, Jacobs A, Garcia C, Shimizu K, Mouri K, Huffman MA, Hernandez AD (2012) Monkeys in the middle: parasite transmission through the social network of a wild primate. PLoS One 7:e51144

Marechal L, Semple S, Majolo B, Qarro M, Heistermann M, MacLarnon A (2011) Impacts of tourism on anxiety and physiological stress levels in wild male Barbary macaques. Biol Conserv 144:2188–2193

Marechal L, Semple S, Majolo B, MacLarnon A (2016) Assessing the effects of tourist provisioning on the health of wild Barbary macaques in Morocco. PLoS One 11:e0155920

Marty PR, Beisner B, Kaburu SSK, Balasubramaniam KN, Bliss-Moreau E, Ruppert N, Sah S, Ahmad I, Arlet ME, Atwill RA, McCowan B (2019) Time constraints imposed by anthropogenic environments alter social behaviour in long-tailed macaques. Anim Behav 150:157–165

Matheson MD, Sheeran LK, Li J, Wagner S (2013) Tourist impact on Tibetan macaques. Anthrozoos 26:158–168

McCabe CM, Nunn CL (2018) Effective network size predicted from simulations of pathogen outbreaks through social networks provides a novel measure of structure-standardized group size. Front Vet Sci 5:71

McLennan MR, Mori H, Mahittikorn A, Prasertbun R, Hagiwara K, Huffman MA (2017) Zoonotic enterobacterial pathogens detected in wild chimpanzees. EcoHealth 15:143–147

Meyers LA (2007) Contact network epidemiology: bond percolation applied to infectious disease prediction and control. Bull Am Math Soc 44:63–86

Minsky M (1965) Matter, mind, and models. In: Minsky M (ed) Semantic information processing. MIT Press, Cambridge, pp 425–432

Moore C, Newman MEJ (2000) Epidemics and percolation in small-world networks. Phys Rev E 61:5678–5682

Moyes CL, Henry AJ, Golding N, Huang Z, Singh B, Baird JK, Newton PA, Huffman MA, Duda KA, Drakeley CJ, Elyazar IRF, Anstey NM, Chen Q, Zommers Z, Bhatt S, Gething PW, Hay SI (2014) Defining the geographical range of the *Plasmodium knowlesi* reservoir. PLoS Negl Trop Dis 8:e2780

Muehlenbein MP, Ancrenaz M (2010) Minimizing pathogen transmission at primate ecotourism destinations: the need for input from travel medicine. J Travel Med 16:229–232

Nahallage CAD, Huffman MA (2013) Macaque-human interactions in past and present-day Sri Lanka. In: Radhakrishna S, Huffman MA, Sinha A (eds) The macaque connection: cooperation and conflict between humans and macaques. Springer, New York, pp 135–148

Nahallage CAD, Huffman MA, Kuruppu N, Weerasingha T (2008) Diurnal primates of Sri Lanka and people's perception of them. Primate Conserv 23:1–7

Nesse RM, Bergstrom CT, Ellison PT, Flier JS, Gluckman P, Govindaraju DR, Niethammer D, Omenn GS, Perlman RL, Schwartz MD, Thomas MG, Stearns SC, Valle D (2010) Making evolutionary biology a basic science for medicine. PNAS 107:1800–1807

Newman MEJ (2002) Spread of epidemic disease on networks. Phys Rev E66:016128

Newman MEJ (2004) Analysis of weighted networks. Phys Rev E 70:056131

Newman MEJ (2006) Finding community structure in networks using the eigenvectors of matrices. Phys Rev E 74:036104

Nunn CL (2009) Using agent-based models to investigate primate disease ecology. In: Huffman M, Chapman CA (eds) Primate parasite ecology: the dynamics and study of host-parasite relationships. Cambridge University Press, Cambridge, pp 83–110

Nunn CL (2011) The comparative approach in evolutionary anthropology and biology. University of Chicago Press, Chicago

Nunn CL (2012) Primate disease ecology in comparative and theoretical perspective. Am J Primatol 74:497–509

Nunn CL, Altizer SM (2005) The global mammal parasite database: an online resource for infectious disease records in wild primates. Evol Anthropol 14:1–2

Nunn CL, Altizer SM (2006) Infectious diseases in primates: behavior, ecology and evolution. Oxford University Press, Oxford

Nunn CL, Thrall PH, Stewart K, Harcourt AH (2008) Emerging infectious diseases and animal social systems. Evol Ecol 22:519–543

Nunn CL, Thrall PH, Leendertz FH, Boesch C (2011) The spread of fecally transmitted parasites in socially structured populations. PLoS One 6:e21677

Nunn CL, Jordan F, McCabe CM, Verdolin JL, Fewell JH (2015) Infectious disease and group size: more than just a numbers game. Philos Trans R Soc Lond Ser B Biol Sci 370:20140111

Palacios JFG, Engelhardt A, Agil M, Hodges K, Bogia R, Waltert M (2012) Status of, and conservation recommendations for, the critically endangered crested black macaque *Macaca nigra* in Tangkoko, Indonesia. Oryx 46:290–297

Pasquaretta C, Levé M, Claidière N, van de Waal E, Whiten A, MacIntosh AJJ, Pele M, Bergstrom ML, Borgeaud C, Brosnan S, Crofoot MC, Fedigan LM, Fichtel C, Hopper LM, Mareno MC,

Petit O, Schnoell AV, di Sorrentino EP, Thierry B, Tiddi B, Sueur C (2014) Social networks in primates: smart and tolerant species have more efficient networks. Sci Rep 4:7600
Paterson JD, Wallis J (2005) Commensalism and conflict: the human-primate interface. American Society of Primatologists, Norman, OK
Patterson JEH, Ruckstuhl KE (2013) Parasite infection and host group size: a meta-analytical review. Parasitology 140:803–813
Petrášová J, Petrželková KJ, Huffman MA, Mapua MI, Bobáková L, Mazoch V, Singh J, Kaur T, Petrášová KJ (2010) Gastrointestinal parasites of indigenous and introduced primate species of Rubondo Island National Park, Tanzania. Int J Primatol 31:920–936
Petrášová J, Uzliková M, Kostka M, Petrželková KJ, Huffman MA, Modrý D (2011) Diversity and host specificity of Blastocystis in syntopic primates on Rubondo Island, Tanzania. Int J Primatol 11:1113–1120
Phillips KA, Bales KL, Capitanio JP, Conley A, Czoty PW, Hart BA, Hopkins WD, Hu SL, Miller LA, Nader MA, Nathanielsz PW, Rogers J, Shively CA, Voytko ML (2014) Why primate models matter. Am J Primatol 76:801–827
Poirotte C, Massol F, Herbert A, Willaume E, Bomo PM, Kappeler PM, Charpentier MJE (2017) Mandrills use olfaction to socially avoid parasitized conspecifics. Sci Adv 3:e1601721
Poirotte C, Sarabian C, Ngoubangoye B, MacIntosh AJJ, Charpentier M (2019) Faecal avoidance differs between the sexes but not with nematode infection risk in mandrills. Anim Behav 149:97–106
Postel SL, Daily GC, Ehrlich PR (1996) Human appropriation of renewable fresh water. Science 271:785–788
Poulin R, Krasnov BR, Pilosof S, Thieltges DW (2013) Phylogeny determines the role of helminth parasites in intertidal food webs. J Anim Ecol 82:1265–1275
Priston NEC, McLennan MR (2013) Managing humans, managing macaques: human–macaque conflict in Asia and Africa. In: Radhakrishna S, Huffman MA, Sinha A (eds) The macaque connection, developments in primatology: progress and prospects. Springer, New York, pp 225–250
Priston NEC, Wyper RM, Lee PC (2012) Buton macaques (*Macaca ochreata brunnescens*): crops, conflict, and behavior on farms. Am J Primatol 74:29–36
Radhakrishna S, Sinha A (2011) Less than wild? Commensal primates and wildlife conservation. J Biosci 36:749–753
Radhakrishna S, Huffman MA, Sinha A (2013) The macaque connection. Springer, New York
Ram S, Venkatachalam S, Sinha A (2003) Changing social strategies of wild female bonnet macaques during natural foraging and on provisioning. Curr Sci 84:780–790
Rezende EL, Albert EM, Fortuna MA, Bascompte J (2009) Compartments in a marine food web associated with phylogeny, body mass, and habitat structure. Ecol Lett 12:779–788
Rifkin J, Nunn CL, Garamszegi LZ (2012) Do animals living in larger groups experience greater parasitism? A meta-analysis. Am Nat 180:70–82
Riley EP (2003) "Whose woods are these?" Ethnoprimatology and conservation in Sulawesi, Indonesia. Am J Phys Anthropol:179–179
Riley EP (2007) The human-macaque interface: conservation implications of current and future overlap and conflict in lore Lindu National Park, Sulawesi, Indonesia. Am Anthropol 109:473–484
Riley EP, Fuentes A (2011) Conserving social–ecological systems in indonesia: human–nonhuman primate interconnections in Bali and Sulawesi. Am J Primatol 73:62–74
Rimbach R, Bisanzio D, Galvis N, Link A, Di Fiore A, Gillespie TR (2015) Brown spider monkeys (*Ateles hybridus*): a model for differentiating the role of social networks and physical contact on parasite transmission dynamics. Philos Trans R Soc Lond Ser B Biol Sci 370:20140110
Romano V, Duboscq J, Sarabian C, Thomas E, Sueur C, MacIntosh AJJ (2016) Modeling infection transmission in primate networks to predict centrality-based risk. Am J Primatol 78:767–779
Romano V, Shen M, Pansanel J, MacIntosh AJJ, Sueur C (2018) Social transmission in networks: global efficiency peaks with intermediate levels of modularity. Behav Ecol Sociobiol 72:154

Rushmore J, Caillaud D, Hall RJ, Stumpf RM, Meyers LA, Altizer S (2014) Network based vaccination improves prospects for disease control in wild chimpanzees. J R Soc Interface 11:0349

Rushmore J, Bisanzio D, Gillespie TR (2017) Making new connections: insights from primate-parasite networks. Trends Parasitol 33:547–560

Rwego IB, Isabirye-basuta G, Gillespie TR, Goldberg T (2007) Gastrointestinal bacterial transmission among humans, mountain gorillas, and livestock in Bwindi Impenetrable National Park, Uganda. Conserv Biol 22:1600–1607

Rwego IB, Gillespie TR, Isabirye-basuta G, Goldberg TL (2008) High rates of *Escherichia coli* transmission between livestock and humans in rural Uganda. J Clin Microbiol 46:3187–3191

Salathe M, Jones JH (2010) Dynamics and control of diseases in networks with community structure. PLoS Comput Biol 6:e1000736

Sapolsky RM, Romero LM, Munck AU (2000) How do glucocorticoids influence stress responses? Integrating permissive, suppressive, stimulatory, and preparative actions. Endocr Rev 21:55–89

Sarabian C, MacIntosh AJJ (2015) Hygienic tendencies correlate with low geohelminth infection in free-ranging macaques. Biol Lett 11:20150757

Sarabian C, Ngoubangoye B, MacIntosh AJJ (2017) Avoidance of biological contaminants through sight, smell and touch in chimpanzees. R Soc Open Sci 4:170968

Schwabe C (1984) Medicine and human health. William & Wilkins, Baltimore

Sears HJ, Brownlee I, Uchiyama JK (1950) Persistence of individual strains of *Escherichia coli* in the intestinal tract of man. J Bacteriol 59:293–301

Sears HJ, Janes H, Saloum R, Brownlee I, Lamoreaux LF (1956) Persistence of individual strains of *Escherichia coli* in man and dog under varying conditions. J Bacteriol 71:370–372

Sha JCM, Gumert MD, Lee B, Jones-Engel L, Chan S, Fuentes A (2009) Macaque-human interactions and the societal perceptions of macaques in Singapore. Am J Primatol 71:825–839

Shah KV, Southwick CH (1965) Prevalence of antibodies to certain viruses in sera of free-living rhesus and of captive monkeys. Indian J Med Res 53:488–500

Siex KS (2005) Habitat destruction, population compression, and overbrowsing by the Zanzibar red colobus monkey (*Procolobus kirkii*). In: Paterson JD, Wallis J (eds) Commensalism and conflict: the human-primate interface. American Society of Primatologists, Norman, OK, pp 294–337

Silk MJ, Croft DP, Delahay RJ, Hodgson DJ, Boots M, Weber N, McDonald RA (2017) Using social network measures in wildlife disease ecology, epidemiology, and management. Bioscience 67:245–257

Singh M, Kumara HN, Kumar MA, Sharma AK (2001) Behavioural responses of lion-tailed macaques (*Macaca silenus*) to a changing habitat in a tropical rain forest fragment in the Western Ghats, India. Folia Primatol 72:278–291

Singh B, Kim Sung L, Matusop A, Radhakrishnan A, Shamsul SS, Cox-Singh J, Thomas A, Conway DJ (2004) A large focus of naturally acquired *Plasmodium knowlesi* infections in human beings. Lancet 363:1017–1024

Sinha A, Mukhopadhyay K, Datta-Roy A, Ram S (2005) Ecology proposes, behaviour disposes: ecological variability in social organization and male behavioural strategies among wild bonnet macaques. Curr Sci 89:1166–1179

Sinton L, Hall C, Braithwaite R (2007) Sunlight inactivation of *Campylobacter jejuni* and *Salmonella enterica*, compared with *Escherichia coli*, in seawater and river water. J Water Health 5:357–365

Southwick CH, Siddiqi MF (1994) Primate commensalism: the rhesus monkey in India. Rev Ecol 49:223–231

Southwick CH, Siddiqi MF (1998) The rhesus monkey's fall from grace. In: Ciochon RL, Nisbett RA (eds) The primate anthology. Prentice Hall, Upper Saddle River, pp 211–218

Southwick CH, Siddiqi F (2011) India's rhesus population: protection versus conservation management. In: Gumert MD, Fuentes A, Jones-Engel L (eds) Monkeys on the edge: ecology and management of long-tailed macaques and their interface with humans. Cambridge University Press, Cambridge, pp 275–292

Southwick CH, Siddiqi MF, Oppenheimer JR (1983) Twenty-year changes in rhesus macaque populations in agricultural areas of Northern India. Ecology 64:434–439

Spiesman BJ, Gratton C (2016) Flexible foraging shapes the topology of plant-pollinator interaction networks. Ecology 97:1431–1441

Springer A, Mellmann A, Fichtel C, Kappeler PM (2016) Social structure and Escherichia coli sharing in a group-living wild primate, Verreaux's sifaka. BMC Ecol 16:6

Stang M, Klinkhamer PGL, Waser NM, Stang I, van der Meijden E (2009) Size-specific interaction patterns and size matching in a plant–pollinator interaction web. Ann Bot 103:1459–1469

Stephens PR, Pappalardo P, Huang S, Byers JE, Farrell MJ, Gehman A, Ghai RR, Haas SE, Han B, Park AW, Schmidt JP, Altizer S, Ezenewa VO, Nunn CL (2017) Global mammal parasite database version 2.0. Ecology 98(5):1476

Sterck EHM (1998) Female dispersal, social organization, and infanticide in langurs: are they linked to human disturbance? Am J Primatol 44(4):235–254

Sueur C, Petit O, De Marco A, Jacobs AT, Watanabe K, Thierry B (2011a) A comparative network analysis of social style in macaques. Anim Behav 82(4):845–852

Sueur C, Jacobs A, Amblard F, Petit O, King AJ (2011b) How can social network analysis improve the study of primate behavior? Am J Primatol 73:703–719

Sueur C, Deneubourg JL, Petit O (2012) From social network (centralized vs. decentralized) to collective decision-making (unshared vs. shared consensus). PLoS One 7:e32566

Sueur C, Romano V, Sosa S, Puga-Gonzalez I (2018) Mechanisms of network evolution: a focus on socioecological factors, intermediary mechanisms, and selection pressures. Primates 60:1–15

Suleyman G, Tibbetts R, Perri MB, Vager D, Xin Y, Reyes K, Samuel L, Chami E, Starr P, Pietsch J, Zervos MJ, Alangaden G (2016) Nosocomial outbreak of a novel extended-spectrum beta-lactamase *Salmonella enterica* serotype isangi among surgical patients. Infect Control Hosp Epidemiol 37:954–961

Tenaillon O, Skurnik D, Picard B, Denamur E (2010) The population genetics of commensal *Escherichia coli*. Nat Rev Microbiol 8:207–217

Thierry B (2007a) The macaques: a double-layered social organization. In: Campbell CJ, Fuentes A, MacKinnon KC, Panger M, Bearder SK, Stumpf RM (eds) Primates in perspective. Oxford University Press, New York, pp 224–239

Thierry B (2007b) Unity in diversity: lessons from macaque societies. Evol Anthropol 16:224–238

Thierry B (2013) The macaques: a double-layered social organization. In: Campbell CJ, Fuentes A, MacKinnon KC, Bearder SK, Stumpf RM (eds) Primates in perspective. Oxford University Press, Oxford, pp 229–240

Thierry B, Aureli F, Nunn CL, Petit O, Abegg C, de Waal FBM (2008) A comparative study of conflict resolution in macaques: insights into the nature of covariation. Anim Behav 75:847–860

Tutin CEG (2000) Ecologie et organisation des primates de la foret tropicale africaine: aide a la comprehension de la transmission des retrovirus. For Trop Emerg Virales 93:157–161

Vallo P, Petrželková KJ, Profousová I, Petrášov J, Pomajbíková K, Leendertz F, Hashimoto C, Simmons N, Babweteera F, Piel A, Robbins MM, Boesch C, Sanz C, Morgan D, Sommer V, Furuichi T, Fujita S, Matsuzawa T, Kaur T, Huffman MA, Modry D (2012) Molecular diversity of entodiniomorphid ciliate Troglodytella abrassarti and its coevolution with chimpanzees. Am J Phys Anthropol 48:525–533

VanderWaal KL, Ezenwa VO (2016) Heterogeneity in pathogen transmission: mechanisms and methodology. Funct Ecol 30:1606–1622

VanderWaal KL, Atwill ER, Isbell LA, McCowan B (2013) Linking social and pathogen transmission networks using microbial genetics in giraffe (*Giraffa camelopardalis*). J Anim Ecol 83:406–414

VanderWaal KL, Atwill ER, Isbell LA, McCowan B (2014) Quantifying microbe transmission networks for wild and domestic ungulates in Kenya. Biol Conserv 169:136–146

Vazquez DP, Chacoff N, Cagnolo L (2009) Evaluating multiple determinants of the structure of plant–animal mutualistic networks. Ecology 90:2039–2046

Waters CN, Zalasiewicz J, Summerhayes C, Barnosky AD, Poirier C, Gauszka A, Cearreta A, Edgeworth M, Ellis EC, Ellis M, Jeandel C, Leinfelder R, McNeill JR, Richter DB, Steffen W, Syvitski J, Vidas D, Wagreich M, Williams M, Zhisheng A, Grinevald J, Odada E, Oreskes N, Wolfe AP (2016) The Anthropocene is functionally and stratigraphically distinct from the Holocene. Science 351:aad2622

Wenz-Mücke A, Sithithaworn P, Petney TN, Taraschewski H (2013) Human contact influences the foraging behaviour and parasite community in long-tailed macaques. Parasitology 140:709–718

Wey T, Blumstein DT, Shen W, Jordan F (2008) Social network analysis of animal behaviour: a promising tool for the study of sociality. Anim Behav 75:333–344

Whitehead H, Dufault S (1999) Techniques for analyzing vertebrate social structure using identified individuals: review and recommendations. Adv Study Behav 28:33–74

Wilensky U, Stroup W (1999) NetLogo HubNet Disease model [computer software]. Center for Connected Learning and Computer-Based Modeling, Northwestern University, Evanston, IL

Wolfe ND, Dunavan CP, Diamond J (2007) Origins of major human infectious diseases. Nature 447:279–283

Young C, Majolo B, Heistermann M, Schülke O, Ostner J (2014) Responses to social and environmental stress are attenuated by strong male bonds in wild macaques. PNAS 111:18195–18200

Zhang X, Kadir KA, Quintanilla-Zarinan LF, Villano J, Houghton P, Du H, Singh B, Glenn Smith D (2016) Distribution and prevalence of malaria parasites among long-tailed macaques (*Macaca fascicularis*) in regional populations across Southeast Asia. Malar J 15:450

Zhao QK (1996) Etho-ecology of Tibetan macaques at Mount Emei, China. In: Fa J, Lindburg DG (eds) Evolution and ecology of macaque societies. Cambridge University Press, Cambridge

Zhao QK (2005) Tibetan macaques, visitors, and local people at Mt. Emei: problems and countermeasures. In: Paterson JD, Wallis J (eds) Commensalism and conflict: the human–primate interface. American Society of Primatologists, Norman, OK, pp 376–399

Zinsstag J, Schelling E, Waltner-Toews D, Tanner M (2011) From "one medicine" to "one health" and systemic approaches to health and well-being. Prev Vet Med 101:148–156

Zinsstag J, Schelling E, Waltner-Toews D, Tanner M (2015) One health, the theory and practice of integrated one health approaches. CAB International, Oxfordshire

Part V
Emerging Technologies in Primatology

Chapter 14
MRI Technology for Behavioral and Cognitive Studies in Macaques In Vivo

Yong Zhu and Paul A. Garber

14.1 Introduction

Primate behavior, especially social behavior, might seem a strange place to begin in trying to advance our understanding of the brain or human mind (Opstal 1996; Critchley and Harrion 2013). Primate brains differ in structural details, proportion among functional areas, as well as in overall size, implying specific adaptations to the challenges posed by particular ecological and social environments across that species' evolutionary history. Given that brains require a disproportionate amount of nutrients and energy relative to other body organs (Aiello and Wheeler 1995), an understanding of the functional implications of evolutionary changes in brain organization is critical for evaluating relationships between cognition, decision-making, and behavior (Rilling 2006) (Fig. 14.1).

The recent decade has seen an explosion of interest and information about brain connectivity and functions over a wide range of spatial scales, including macroscopic, microscopic, and mesoscopic levels (Essen et al. 2016). Noninvasive imaging studies with magnetic resonance imaging (MRI) technologies have been used for over 30 years (Vanduffel 2018) and have made fundamental contributions to our understanding of animal behavior and cognition, especially in our understanding of the relationships between brain structures and brain function. For example, the brain's default mode network (DMN) consists of discrete, bilateral, and symmetrical cortical areas in the medial and lateral parietal, medial prefrontal, and medial and lateral temporal cortices of the human, nonhuman primate, cat, and rodent brains

Y. Zhu (✉)
High Magnetic Field Laboratory, Chinese Academy of Sciences, Hefei, China

School of Life Sciences, Hefei Normal University, Hefei, Anhui, China

P. A. Garber
Department of Anthropology, Program in Ecology, Evolution, and Conservation Biology, University of Illinois, Urbana, IL, USA
e-mail: p-garber@illinois.edu

J.-H. Li et al. (eds.), *The Behavioral Ecology of the Tibetan Macaque*, Fascinating Life Sciences, https://doi.org/10.1007/978-3-030-27920-2_14

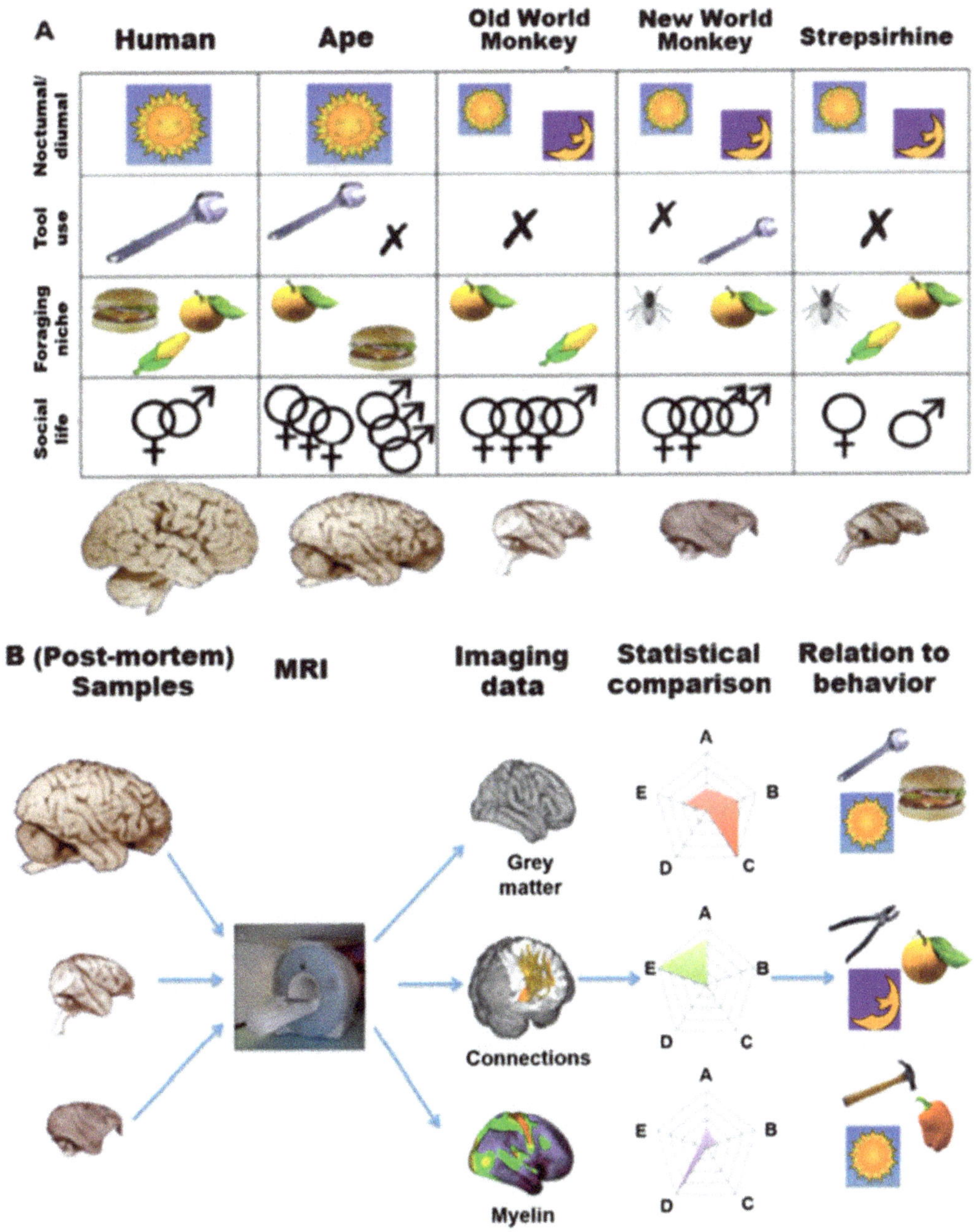

Fig. 14.1 Analyzing the relationship between brains and behavior (Mars et al. 2014). (**a**) In a multivariate comparative approach, each brain is viewed as a unique combination of variables, including whether the animal is active during night or day; whether it uses tools regularly, occasionally, or not at all; what its diet is; or how complex its social life is. By using a whole-brain and multivariate approach, it is possible for us to investigate how differences in specific aspects of brain organization are related to different ecological and social variables. (**b**) MRI of whole-brain (postmortem) samples allows a number of measures to be collected, for which comparative analysis techniques have now been developed and validated

(Raichle 2015). The DMN consistently decreases its activity when compared with the activity of other brain areas during relaxed states. Using fMRI in macaques, researchers have detected that a cortical network, activated during shifts in cognitive activity, largely overlapped with the DMN and therefore have proposed that cognitive shifting in primates generally recruits activity in DMN regions (Arsenault et al. 2018). Prior to 1991, it was virtually impossible to map brain activation rapidly and noninvasively with full brain coverage and relatively high spatial and temporal resolution (Bandettini 2009). The use of fMRI along with positron emission tomography (PET), has revolutionized cognitive neuroscience (Logothetis 2008). Using these noninvasive techniques, we can ask research questions such as the relationship between brain and behavior in primates, which could not have been studied otherwise.

Nonhuman primates (NHPs), especially macaques, have been used traditionally as a model for studying many aspects of human behavior, health, and biology. This includes social structure, social behavior, and social cognition. In this chapter, we describe recent advances in MRI technology (including the state-of-the-art high field MRI), review several fMRI and PET studies on macaques, and indicate how these studies have contributed new insights into an understanding of nonhuman and human primate cognition and behavior.

14.2 Magnetic Resonance Imaging (MRI)

14.2.1 Background of MRI

MRI is a painless, noninvasive tool commonly used to diagnose disease progression, injury, or other ailments. Since its discovery in 1945, nuclear magnetic resonance (NMRI) has been used extensively as an analytic tool in chemistry and physics. In the early 1970s, interest in localized tissue measurements and the realization that highly accurate internal images could be obtained expanded the potential applications of NMRI in medical research. By the 1980s, the quality of NMRI in humans had been improved to the point that the radiological community considered MRI as the next high technology imaging modality.

Today, we refer to "NMRI" as MRI. The use of word "nuclear" in the acronym was dropped to avoid a negative association with the potential exposure to nuclear radiation. Its working principle is based on the fact that certain atomic nuclei are able to absorb and emit radio-frequency energy when placed in an external magnetic field. In both its clinical application and in MRI research, hydrogen atoms are most often used to generate a detectable radio-frequency signal that is received by antennas in close proximity to the anatomical structure being examined. Hydrogen atoms are naturally abundant, particularly in water and fat. For this reason, most MRI scans essentially map the location of water and fat in the body. Pulses of radio waves excite the nuclear spin energy transition, and magnetic field gradients localize the signal in space. By varying the parameters of the pulse sequence, different contrasts may be generated between tissues based on the relaxation properties of the hydrogen atoms.

14.2.2 The Advantages of MRI

Compared with other medical imaging techniques such as X-ray imaging, ultrasonic imaging, and computed tomography (CT), MRI is a painless and noninvasive method to view human or animal tissue and obtain anatomical and functional diagnostic information. While the hazards of X-rays are now well-controlled in most medical contexts, MRI may still be seen as a better choice than CT. MRI scanners are designed to visualize non-bony parts or "soft tissue" areas such as muscles, ligaments, and tendons. In particular, the brain, spinal cord, and nerves are seen much more clearly with MRI than with regular X-rays and CT scans.

What are the advantages of an MRI scan? To summarize, there are several as follows:

1. High accuracy

   In clinical diagnostics and studies, MRI is used as the preferred medical examination tool owing to its high accuracy in the detection of serious ailments such as tumors and cancers. In research, it can be used in structural and functional connectivity studies. In addition, the MRI scan parameters can be adjusted for a more comprehensive view of the area of interest.
2. Less confining

   An MRI scanner has a design that makes it less confining for the patient or research subject. It is possible to assign tasks to the subject in the scanner, for example, evaluate light or sound stimulation, and this facilitates directly linking behavior and functional anatomy.
3. Noninvasive

   MRI does not involve X-rays or the use of ionizing radiation, distinguishing it from CT or CAT scans. This removes all ethical concerns regarding harm that could be done to patients or study subject. This permits the ethical use of fMRI research studies because normal subjects face no risk of harm or injury.

   With these four main advantages, MRI represents a powerful tool well suited for studying in vivo behavioral responses, especially functional studies that link neuroanatomy and cognitive behavior in primates.

14.2.3 State of the Art at High Field MRI

As far as MRI is concerned, the need to achieve the highest possible magnetic field in clinical settings has been a major motivation for scientists and engineers to improve the imaging technology. As a result, MRI scanners have evolved from the first generation (0.5 Tesla or 0.5 T) to the conventional (1.0–1.5 T) scanner and then to the high field (3 T) scanners in clinical applications. Most recently, ultra-high field MRI scanners ranging from 7 to 11.7 T have been designed. Generally speaking, the higher the magnetic field, the better the resolution of the images.

Using MRI, a higher magnetic field intensity leads to a higher signal-to-noise ratio (SNR), resulting in a higher image resolution. However, higher magnetic field intensity can lead to a smaller caliber machine (the diameter of the machine that allows the animal to be scanned) meaning that animals larger than macaques (e.g., gorilla) could not be examined. For example, the 3 T MRI scanner in clinical applications usually has a 70 cm caliber, suited for human or gorilla body scans. The 9.4 T MRI scanner with a caliber ranging from 30 to 40 cm is suitable for NHPs ranging in size from a mouse lemur to a macaque. The 9.4 T MRI scanner obtains enhanced image resolution compared to the 3 T MRI scanner. Therefore, considering the image resolution, a 9.4 T MRI scanner is the most suitable for studying mammals ranging from rodents to most species of NHPs.

There are many 9.4 T MRI scanners currently in use. For example, a 9.4 T MRI scanner with a 40 cm diameter bore (Agilent, US) is operated at the High Magnetic Field Laboratory (HMFL), Chinese Academy of Sciences, Hefei. This scanner is dedicated to the study of larger mammals (e.g., dog, sheep, and macaque). Another 9.4 T/30 cm MRI scanner (Bruker, Germany) is operated at the Shanghai Institute of Biological Sciences of the Chinese Academy of Sciences. This scanner focuses on scientific research requiring a high magnetic field. At the University of Chicago, Chicago, Illinois, a 9.4 T/30 cm MRI facility is used for basic science research within the Department of Radiology. The goal is to advance state-of-the-art animal imaging technology at the anatomical and functional level (https://mris.uchicago.edu/). Many experimental studies of NHPs that focus on research questions including brain development, brain structural changes, and functional studies of different brain regions have been conducted using other types of 9.4 T MRI scanners such as the 9.4 T/35 cm scanner in Germany (Goebel et al. 2009) and France (Même et al. 2015) and the 9.4 T/39 cm scanner at University College London Centre for Advanced Biomedical Imaging, London (Ramasawmy et al. 2016).

Some primate studies have been carried out on 1.5 T or 3 T scanners (Nelissen and Vanduffel 2017), with a sub-optimal resolution in the range of 2–3 mm. In vivo studies of structural brain imaging at 7 T have been reported in the rhesus macaque (*Macaca mulatta*), with resolution from 0.3 to 0.5 mm (Zitella et al. 2015). A much finer image resolution (under 100 μm) is needed to pick up minor changes, such as in brain aging that would be required to examine questions of ontogenetic changes in social behavior related to aging. However, only a limited number of primate studies have used 9.4 T MRI scanners, which would generate in vivo data with a resolution on the order of 0.1 mm. This level of resolution is needed to examine questions related to the substructure of the hippocampus. Because dynamic brain alterations can be detected on a much finer scale, high field MRI greatly facilitates studies integrating primate brain structure and behavioral change.

14.3 In Vivo MRI Study in Macaques

Investigations into the structure and function of the NHP brain have significantly contributed to our overall understanding of cognition.

14.3.1 Structural MRI in Brain Imaging Study

To understand the anatomical localization and functional activity in different cortical areas, it is necessary to obtain an accurate map of neural architectonic areas. In most MRI studies, primates are anesthetized when scanned. The latest research, published in the journal *NeuroImage*, has created an anatomical MRI brain template derived from 31 rhesus macaques (the macaques were juveniles and adults between 3.2 and 13.2 years old). The template also includes tissue maps, surfaces, and transformation scripts to assist in data analysis (Seidlitz et al. 2018). These data can be used to determine the variance of cortical topographies in the same individual over time and also compare cortical differences between infants, juveniles, and adults. This can be used as a framework for examining correlational relationships between changes in behavior and behavioral and brain development.

Unlike morphological studies using postmortem tissue samples, MRI allows the noninvasive, in vivo assessment of many different brain parameters including the topography and volume of gray and white matter in brain structures. At the level of the cerebral hemisphere, gray matter is mainly distributed in the periphery (cortex) while the white matter is located deep within the cortex.

Phillips and Sherwood (2008) described growth patterns in total brain volume, cortical gray and white matter volume, frontal lobe gray and white matter volume, and corpus callosum area in 29 brown capuchin monkeys (*Sapajus nigritus*, formerly *Cebus apella*) ranging in age from 4 days to 20 years. Of the total subjects, 12 were adults (≥5 years) and 17 were juveniles (between 4 days to 5 years). The results revealed that nonlinear age-related changes in total brain volume, cortical white matter volume, and frontal white matter volume occur from birth to 5 years of age (subadult period of development). The implications of these results is the rapid increase of frontal lobe white matter during the first few years of life corresponds with opportunities for social learning and acquiring technical skills related to object manipulation, prey search, possibly tool use, and other complex foraging behaviors (Phillips and Sherwood 2008). Similarly, Wisco et al. (2008) studied the age-related white and gray matter volume changes in eight young adult (5–12 years), six middle-aged adult (16–19 years) and eight old adult (24–30 years) rhesus macaques. The results found an overall decrease in the total forebrain (5.01%), forebrain parenchyma (5.24%), forebrain white matter (11.53%), forebrain gray matter (2.08%), caudate nucleus (11.79%), and *globus pallidus* (18.26%) with increasing age. Corresponding behavioral data for five of the younger, five of the middle-aged, and seven of the old adults on the delayed non-matching to sample (DNMS) task, the delayed recognition span task (DRST), and the cognitive impairment index (CII) found no correlation between these cognitive measures and ROI volume changes.

In a recent study, Scott et al. (2016) longitudinally assessed normative brain growth patterns using MRI in rhesus macaques. Cohort A consisted of 24 individuals (12 males, 12 females) and cohort B of 21 individuals (11 male, 10 female). They scanned the macaques at 1, 4, 8, 13, 26, 39, and 52 weeks of age. Cohort A had additional scans at 156 weeks (3 years) and 260 weeks (5 years) (Fig. 14.2). The

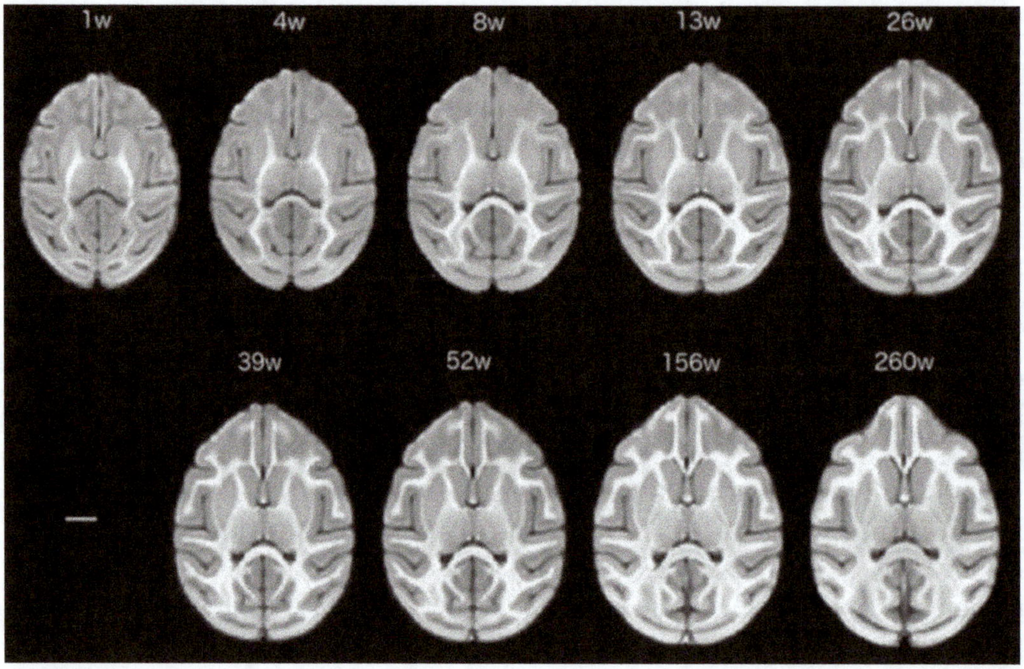

Fig. 14.2 Brain MRI at each study time point (Scott et al. 2016). Age-specific, horizontally oriented templates are shown for each study time point. Each template represents an average brain image constructed from individual subject scans. Images display enhanced signal-to-noise and optimal shape characteristics relative to individual scans (scale bar represents 1 cm)

results showed that total brain volume at 1 week was approximately 64% of that of the adult. Brain volume was larger in male rhesus macaques compared to females. While brain volume generally increased between any two imaging time points, there was a transient plateau of brain growth between 26 and 39 weeks in both males and females (Scott et al. 2016). This study serves as a starting point for more extensive analyses into the relationship between structural development of the brain and behavioral development of the rhesus macaques. The image database is available to behavioral scientists for addressing additional questions examining the relationship between changes in behavior and changes in neural development, including the emergence of sex and species typical behavior.

To evaluate hippocampal development in rhesus macaques, Hunsaker et al. (2014) obtained longitudinal structural MRI scans at 9 time points between 1 week and 260 weeks (5 years) of age in 24 rhesus macaques (12 males, 12 females) (Fig. 14.3). The results showed that the hippocampus reached 50% of its adult volume by 13 weeks of age and full adult volume by 52 weeks in both males and females. The hippocampus appears to be slightly larger at 3 years than at 5 years of age, and damage to the hippocampus deficit can result in permanent changes in behavior such as learning and memory, as well as neurocognitive function. Male rhesus macaques have a 5% larger hippocampi than females from 8 weeks of age onward. Neuroimaging studies in rhesus macaques can provide critical information about the relationships between MRI volumetric changes and behavior during individual development.

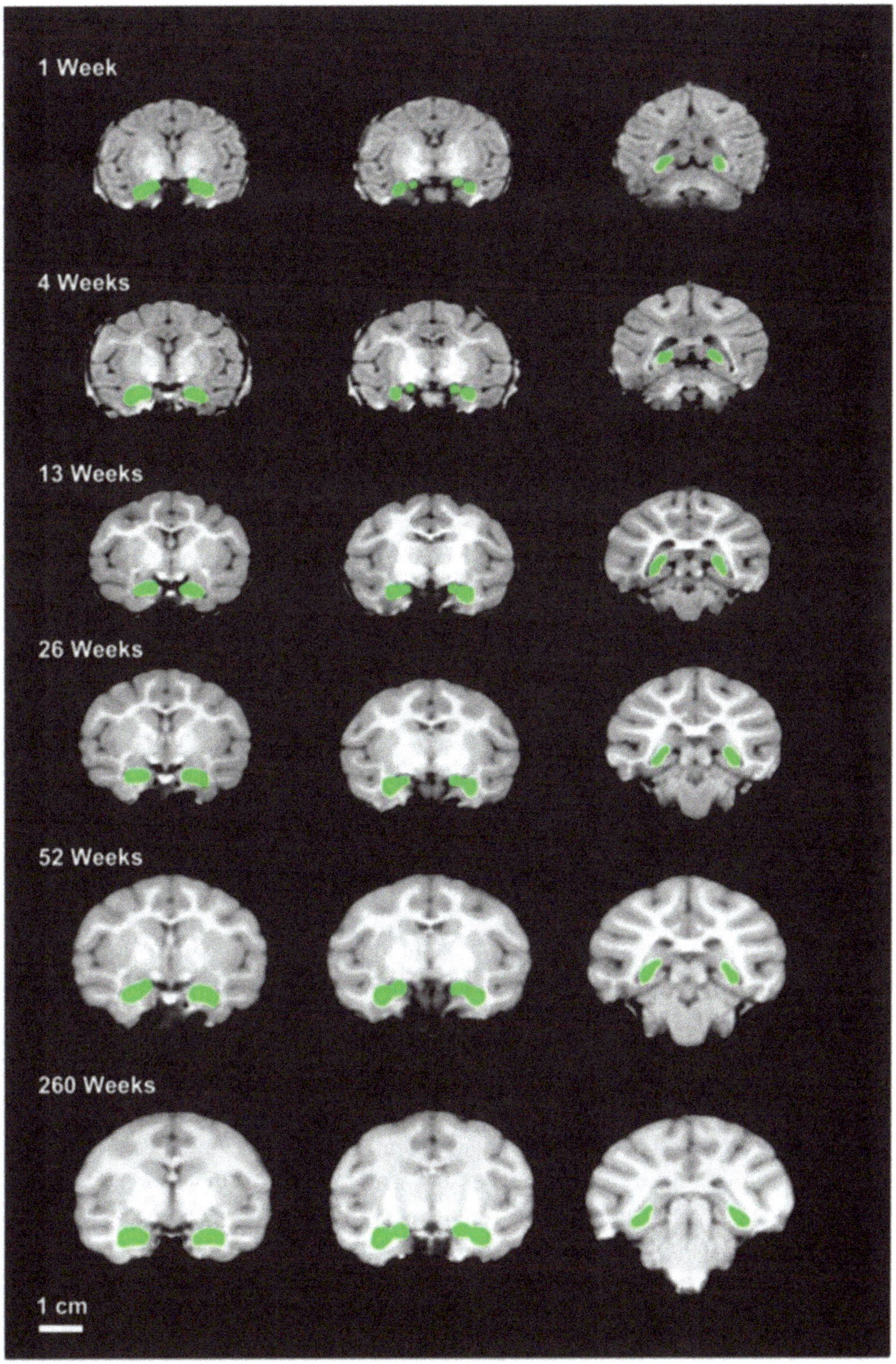

Fig. 14.3 MRI for hippocampal tracings at different ages (Hunsaker et al. 2014). Sample hippocampal tracings from a single male rhesus macaque. Shown are scans at 1 week, 4 week, 26 weeks, 39 weeks, and 260 weeks of age. Note the difference in white matter at the different ages. All scans are shown at the same scale for direct visual comparison. Note the general shape of the traced hippocampus and size relative to the rest of brain at different ages (scale bar = 1 cm)

Liu et al. (2015) examined brain development in 14 male rhesus macaques from 6 to 16 months of age. The results showed rapid growth (6.21%) in brain volume during early development between 6 and 10 months of age compared to a 2.81% growth rate between 10 and 16 months of age. Early expansion is mainly the result of a significant increase in white matter volume, while the later decline can be partly explained by a significant decrease of gray matter volume after 10 months of age. Compared with macaque brain development, human brain volume increases by 100% during the first year of life and then maintains cubic growth into adolescence. The pattern of human brain growth differs from the rhesus macaque in that humans have a less mature brain at birth (approximately 25% of adult size) and a longer period of early rapid brain growth during the first year of life (Phillips and Sherwood 2008).

14.3.2 Functional MRI in Brain Imaging Study

Functional magnetic resonance imaging (fMRI) has been used extensively in comparative sensory and cognitive experiments associated with brain activation mapping in NHPs and humans (Logothetis et al. 1999; Nakahara et al. 2002; Mantini et al. 2012). Moreover given that the subject is alert, one can study relationships between behavior, decision-making, and neural activity in real time. What exactly does fMRI tell us? We know that its signals arise from changes in local brain hemodynamics that, in turn, result in alterations in neuronal activity. However, exactly how neuronal activity, hemodynamics, fMRI signals, and decision-making are related is still unclear. It has been assumed that the fMRI signal is proportional to the local average neuronal activity, but many factors can influence this relationship (Heeger and Ress 2002).

For comparative studies of human and nonhuman primates, subjects are exposed to the same set of images as their brains are scanned using fMRI. The time course of fMRI activity during viewing is extracted from "seed" regions in human participants and correlated with the fMRI signal in NHP, or vice versa, with an adjustment made for interspecific differences in vascular hemodynamics. After statistical thresholding, which is a method of image processing, the resulting maps for each species show areas that are stimulated and potentially homologous with the targeted seed region in the other species (Wager and Yarkoni 2012) (Fig. 14.4). fMRI has been used to study the primate visual system (Logothetis et al. 1999; Nakahara et al. 2002; Russ and Leopold 2015), the auditory system (Mantini et al. 2012; Ortiz-Rios et al. 2015), and the motor system (Bauman et al. 2013), again principally in macaques. We present the results of some of these studies below.

Ortiz-Rios et al. (2015) investigated how species-specific vocalizations are represented in auditory and auditory-related regions of the brain using fMRI in rhesus macaques. The results indicated that these vocalizations preferentially activated the auditory ventral stream and in particular areas of the anterolateral belt and parabelt.

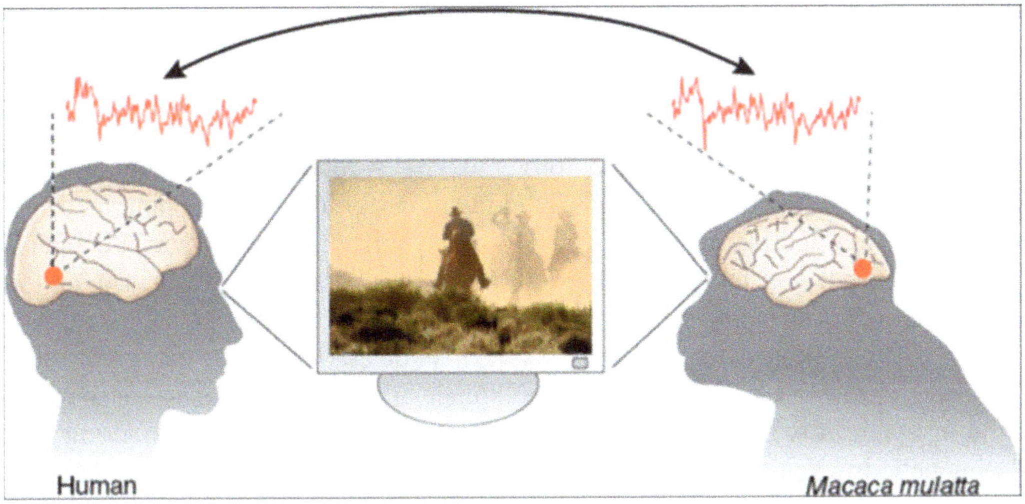

Fig. 14.4 Humans and monkeys watch the same film as their brains are scanned with fMRI (Wager and Yarjoni 2012). The time course of fMRI activity during viewing is extracted from "seed" regions in human participants and correlated with the fMRI signal in monkeys, and vice versa, with an adjustment made for interspecific differences in vascular hemodynamics

A major limitation of fMRI studies of nonhuman primates is the difficulty of training monkeys and apes to remain calm when placed in the restraining devices required to limit movement during scanning. Srihasam et al. (2010) developed a technique for holding subjects' heads motionless during scanning using a custom-fitted plastic helmet, chin strap, and mild suction supplied by a vacuum blower. This vacuum helmet method appeared to have few adverse effects on subject physical health (although the monkeys were stressed) even after repeated use for several months.

14.4 Conclusion

Studies of the brain of nonhuman primates are vital for understanding aggressive and cooperative interactions (Rilling 2014). Noninvasive imaging technologies, such as fMRI, are an important research tool that provide data that field primatologists can use to better understand how age, sex, and species differences in neural development, social experience, and neurohormones affect behavior and cognitive processes. For example, differences in patterns and rates of brain growth may help explain taxonomic differences in maternal pre- and postnatal offspring investment; differences in the manner in which species encode, store, and integrate temporal, spatial, and quantitative information in deciding where to feed; and species-specific differences in cooperative behavior. For example, oxytocin, a neurohormone synthesized in the hypothalamus, can increase cooperative behavior and reduce fear of cheaters in humans by stimulating brain regions that associate cooperative interactions with feelings of pleasure or reward (Rilling 2014). In the case of wild chimpanzees, Wittig et al. (2014) found that during periods of food sharing, levels of urinary oxytocin concentration increased significantly compared to periods when chimpanzees fed alone. Moreover this effect was evident in chimpanzees who shared food with close partners and in chimpanzees who shared food with nonpartners (Wittig et al. 2014). Similarly, Crockford et al. (2013) found that oxytocin levels in wild chimpanzees increased equally when individuals groomed kin or nonkin. Thus, there exists an important neurohormonal mechanism for promoting social cooperation between both related and nonrelated group members. In the case of common marmosets (*Callithrix jacchus*), a small-bodied New World monkey, Finkenwirth et al. (2016) examined relationships between nonmaternal infant caregiving and urinary oxytocin levels. In this species, both male and female helpers were found to exhibit significantly higher levels of urinary oxytocin when caring for offspring than during periods when not providing infant care. Finally, different species of macaques vary in dominance style from highly aggressive to highly affiliative. Although early experience may contribute to the ways in which individuals interact with others, a study by Rosenblum et al. (2002) reported that bonnet macaques (*M. radiata*), which are highly affiliative, were characterized by higher levels of oxytocin than pig-tailed macaques (*M. nemistrina*) which are highly aggressive. Combined, these studies highlight the integrated role of neurohormones, neuroreceptors, and social experience in understanding primate social interactions and behavior.

Neurodevelopmental research is expected to expand in the future, offering new insights and understanding into links between neural anatomy and behavior, both within and between species. For example, neurodevelopmental studies may allow us to understand how the brain neuronal functions organize and mature during different stages of development. This will allow researchers to address questions linking behavior, ontogeny, and social cognition in primates.

Acknowledgments We would like to thank Dr. K. Zhong, H. Y. Yang, J. Zhang, Q. J. Zhu, and H. Y. Tong at the High Magnetic Field Laboratory, the Chinese Academy of Sciences. We also thank Dr. L. X. Sun at Central Washington University for his many helpful comments and suggestions on this chapter. PAG wishes to thank Chrissie, Sara, and Jenni for their love and support. This study was supported by the National Natural Science Foundation of China (31501866) and the Science and Technology Innovation Program of the Ministry of Science and Technology of the People's Republic of China (2014BAL03B00).

References

Aiello LC, Wheeler P (1995) The expensive-tissue hypothesis: the brain and the digestive system in human and primate evolution. Curr Anthropol 36:199–221

Arsenault JT, Caspari N, Vandenberghe R, Vanduffel W (2018) Attention shifts recruit the monkey default mode network. J Neurosci 38:1202–1217

Bandettini PA (2009) Functional MRI limitations and aspirations. In: Kraft E, Gulyás B, Pöppel E (eds) Neural correlates of thinking. Springer-Verlag, Berlin, pp 15–38

Bauman MD, Losif AM, Ashwood P, Braunschweig D, Lee A, Schumann CM, Water JV, Amaral DG (2013) Maternal antibodies from mothers of children with autism alter brain growth and social behavior development in the rhesus monkey. Transl Psychiatry 3(7):e278

Critchley HD, Harrison NA (2013) Visceral influences on brain and behavior. Neuron 77:624–638
Crockford C, Wittig RM, Langergraber K, Ziegler TE, Zuberbuhler K, Deschner T (2013) Urinary oxytocin and social bonding in related and unrelated wild chimpanzees. Proc Biol Sci 280:20122765
Essen DCV, Donahue C, Dierker DL, Glasser MF (2016) Parcellations and connectivity patterns in human and macaque cerebral cortex. In: Kennedy H, Essen DCV, Christen Y (eds) Micro-, meso- and macro-connectomics of the brain. Springer International, Cham, pp 89–106
Finkenwirth C, Martins E, Deschner T, Burkart JM (2016) Oxytocin is associated with infant-care behavior and motivation in cooperatively breeding marmoset monkeys. Horm Behav 80:10–18
Goebel H, Spehl T, Paul D, Markl M, Harloff A (2009) Histopathological correlation and feasibility of atherosclerotic carotid lesion classification using T_2* weighted imaging at 9.4t MRI. Virchows Archiv 455:50–51
Heeger DJ, Ress D (2002) What does fMRI tell us about neuronal activity? Nat Rev Neurosci 3:142–151
Hunsaker MR, Scott JA, Bauman MD, Amaral DG (2014) Postnatal development of the hippocampus in the Rhesus macaque (*Macaca mulatta*): a longitudinal magnetic resonance imaging study. Hippocampus 24:794–807
Liu CR, Tian XG, Liu HL, Mo Y, Bai Y, Zhao XD, Ma YY, Wang JH (2015) Rhesus monkey brain development during late infancy and the effect of phencyclidine: a longitudinal MRI and DTI study. NeuroImage 107:65–75
Logothetis NK (2008) What we can do and what we cannot do with fMRI. Nature 453:869–878
Logothetis NK, Guggenberger HG, Peled S, Pauls J (1999) Functional imaging of the monkey brain. Nat Neurosci 2:555–562
Mantini D, Hasson U, Betti V, Perrucci MG, Romani GL, Corbetta M, Orban GA, Vanduffel W (2012) Inter-species activity correlations reveal functional correspondences between monkey and human brain areas. Nat Methods 9:277–282
Mars RB, Neubert FX, Verhagen L, Sallet J, Miller KL, Dunbar RIM, Barton RA (2014) Primate comparative neuroscience using magnetic resonance imaging: promises and challenges. Front Neurosci 8:298
Même S, Joudiou N, Yousfi N, Szeremeta F, Lopes-Pereira P, Beloeil JC, Herault Y, Meme W (2015) *In vivo* 9.4T MRI and ^{1}H MRS for evaluation of brain structural and metabolic changes in the Ts65Dn mouse model for Down syndrome. World J Neurosci 4:152–163
Nakahara K, Hayashi T, Konishi S, Miyashita Y (2002) Functional MRI of macaque monkeys performing a cognitive set-shifting task. Science 295:1532–1536
Nelissen K, Vanduffel W (2017) Action categorization in rhesus monkeys: discrimination of grasping from non-grasping manual motor acts. Sci Rep 7:15094
Opstal VA (1996) Dynamic patterns: the self-organization of brain and behavior. Complexity 2:253–254
Ortiz-Rios M, Kuśmierek P, Dewitt I, Archakov D, Azevedo FAC, Sams M, Jääskeläinen IP, Keliris GA, Rauschecker JP (2015) Functional MRI of the vocalization-processing network in the macaque brain. Front Neurosci 9:113
Phillips KA, Sherwood CC (2008) Cortical development in brown capuchin monkeys: a structural MRI study. NeuroImage 43:657–664
Raichle ME (2015) The brain's default mode network. Annu Rev Neurosci 38:433–447
Ramasawmy R, Johnson SP, Roberts TA, Stuckey DJ, David AL, Pedley RB, Lythgoe MF, Siow B, Walker-Samuel S (2016) Monitoring the growth of an orthotopic tumour xenograft model: multi-modal imaging assessment with benchtop MRI (1T), high-field MRI (9.4T), ultrasound and bioluminescence. PLoS One 11(5):e0156162
Rilling JK (2006) Human and nonhuman primate brains: are they allometrically scaled versions of the same design? Evol Anthropol 15:65–77
Rilling JK (2014) Comparative primate neuroimaging: insights into human brain evolution. Trends Cogn Sci 18:46–55

Rosenblum LA, Smith ELP, Altemus M, Scharf BA, Owens MJ, Nemeroff CB, Gorman JM, Coplan JD (2002) Differing concentrations of corticotropin-releasing factor and oxytocin in the cerebrospinal fluid of bonnet and pigtail macaques. Psychoneuroendocrinology 27:651–660

Russ BE, Leopold DA (2015) Functional MRI mapping of dynamic visual features during natural viewing in the macaque. NeuroImage 109:84–94

Scott JA, Grayson D, Fletcher E, Lee A, Bauman MD, Schumann CM, Buonocore MH, Amaral DG (2016) Longitudinal analysis of the developing rhesus monkey brain using magnetic resonance imaging: birth to adulthood. Brain Struct Funct 221:2847–2871

Seidlitz J, Sponheim C, Glen D, Ye FQ, Saleem KS, Leopold DA, Ungerleider L, Messinger A (2018) A population MRI brain template and analysis tools for the macaque. NeuroImage 170:121–131

Srihasam K, Sullivan K, Savage T, Livingstone MS (2010) Noninvasive functional MRI in alert monkeys. NeuroImage 51:267–273

Vanduffel W (2018) Long-term value memory in primates. PNAS 115:1956–1958

Wager TD, Yarkoni T (2012) Establishing homology between monkey and human brains. Nat Methods 9:237–239

Wisco JJ, Killiany RJ, Guttmann CR, Warfield SK, Moss MB, Rosene DL (2008) An MRI study of age-related white and gray matter volume changes in the rhesus monkey. Neurobiol Aging 29:1563–1575

Wittig RM, Crockford C, Deschner T, Langergraber KE, Ziegler TE, Zuberbuhler K (2014) Food sharing is linked to urinary oxytocin levels and bonding in related and unrelated wild chimpanzees. Proc Biol Sci 281(1778):20133096

Zitella LM, Xiao YZ, Teplitzky BA, Kastl DJ, Duchin Y, Baker KB, Vitek JL, Adriany G, Yacoub E, Harel N, Johnson MD (2015) *In vivo* 7T MRI of the non-human primate brainstem. PLoS One 10(5):e0127049

Correction to: The Behavioral Ecology of the Tibetan Macaque

Jin-Hua Li, Lixing Sun, and Peter M. Kappeler

Correction to: J.-H. Li et al. (eds.), *The Behavioral Ecology of the Tibetan Macaque*, Fascinating Life Sciences, https://doi.org/10.1007/978-3-030-27920-2

In the original version of the book,

The following sentence **"The Chinese National Natural Science Foundation sponsored the meeting and also provided funding to support open access publication of this volume."** in Acknowledgements section has been updated as follows **"The National Natural Science Foundation of China sponsored the meeting and also provided funding to support open access publication of this volume, as well as the publishing fund of Hefei Normal University. Hefei Normal University also provided a fund to defray the cost of publication including book purchase."**

The following sentence **"This book is mainly based on research papers presented in a spirited international primatology symposium held in the scenic area of Mt. Huangshan, China, in the summer of 2017. The chapters were grouped into four logical parts."** in Preface has been updated as follows **"This book is mainly based on research papers presented in a spirited international primatology symposium held in the scenic area of Mt. Huangshan, China, in the summer of 2017. The chapters were grouped into five logical parts."**

The following sentence **"Many recent discoveries in primatology involve technological advancements in research, which is the content of Part IV. In a single chapter (Chap. 14), Yong Zhu and Paul A. Garber explore the great potential of the high field MRI technology in the study of primate behavior and**

The updated online version of this book can be found at https://doi.org/10.1007/978-3-030-27920-2

J.-H. Li et al. (eds.), *The Behavioral Ecology of the Tibetan Macaque*, Fascinating Life Sciences, https://doi.org/10.1007/978-3-030-27920-2_15

cognition." in Preface has been updated as follows **"Many recent discoveries in primatology involve technological advancements in research, which is the content of Part V. In a single chapter (Chap. 14), Yong Zhu and Paul A. Garber explore the great potential of the high field MRI technology in the study of primate behavior and cognition."**

www.ingramcontent.com/pod-product-compliance
Lightning Source LLC
LaVergne TN
LVHW082118200926
848484LV00008B/1974